Smart Energy Systems

The book series Smart Energy Systems comprehensively explores the application of advanced theories and cutting-edge technologies in the evolving energy landscape. The series emphasizes the strategic practical application in smart grid management, renewable energy integration, and multi-source energy coordination. The series explores topics such as data analysis, system optimization, machine learning, and AI-driven control methods, highlighting their crucial role in ensuring efficient and sustainable energy system operations. Additionally, the series broadens its focus to encompass hybrid energy solutions and distributed energy management, offering a comprehensive perspective on how smart, integrated energy systems can transform future power networks.

The book series Smart Energy Systems seeks to provide researchers, engineers, and professionals with in-depth insights into the implementation of integrated energy frameworks and renewable energy technologies. Each volume focuses on distinct aspects such as advancements in smart grid technology, real-time energy optimization, and the integration of renewable into traditional energy systems. Collectively, the series acts as an indispensable resource for academia, energy industry professionals, and policymakers, providing technical depth to bridge the gap between research and practical implementation in pursuing a sustainable energy future.

Tianguang Lu · Ming Yang · Yu Guo · Qian Ai ·
Ran Hao

Operation Modeling and Collaborative Regulation of Distributed Energy Grid Clusters

Springer

Tianguang Lu
School of Electrical Engineering
Shandong University
Jinan, Shandong, China

School of Engineering and Applied Sciences
Harvard University
Cambridge, USA

Ming Yang
School of Electrical Engineering
Shandong University
Jinan, Shandong, China

Qian Ai
Shanghai Jiao Tong University
Shanghai, China

Yu Guo
School of Electrical Engineering
Shandong University
Jinan, Shandong, China

Ran Hao
State Grid Corporation of China
Beijing, China

ISSN 3059-4359 ISSN 3059-4367 (electronic)
Smart Energy Systems
ISBN 978-981-96-9941-4 ISBN 978-981-96-9942-1 (eBook)
https://doi.org/10.1007/978-981-96-9942-1

This Springer imprint is published by the registered company Springer Nature Singapore Pte Ltd.
The registered company address is: 152 Beach Road, #21-01/04 Gateway East, Singapore 189721, Singapore

Preface

Driven by decarbonization and renewables, energy systems are evolving into distributed energy grid clusters. These clusters combine various energy sources, including electricity, heat, and gas, into a localized and adaptable grid capable of balancing supply and demand autonomously. However, as energy systems become increasingly decentralized, critical challenges arise in managing and optimizing the interactions between the various energy sources within distributed energy grid clusters. The interaction mechanisms among heterogeneous energy flows are complex, involving efficiency and cost optimization across the production, conversion, storage, and consumption stages of each energy type. At the same time, it is necessary to deal with the temporal and spatial distribution characteristics and the uncertainty of supply and demand. Moreover, because each distributed energy grid represents a different stakeholder, the inter-grid interaction within the clusters becomes even more intricate.

This book presents a series of modeling and coordinated control methods designed to address these challenges, with a focus on achieving efficient, reliable, and economical system operation. The content is organized into eight chapters.

Chapter 1: "Overview of Distributed Energy Grid Clusters". This chapter begins by outlining the macro background of the global low-carbon transition and the rapid growth of renewable energy, clarifying the core value of distributed energy grid clusters in multi-energy coordination and autonomous balancing. It then systematically reviews domestic and international research progress in the fields of modeling and regulation for clusters. Finally, it presents several demonstration project cases.

Chapter 2: "Modeling Techniques for the Operation of Distributed Energy Grid Clusters". Chapter 2 centers on real-time sensing and coordinated control of distributed energy grid clusters. It first constructs a physical–information integrated CPES architecture. Then proposes an electricity-heat-gas co-modeling method based on an energy-hub coupling matrix to uniformly describe the coupling and conversion of multiple energy streams. Further analyzes the cluster's energy-use behaviors. Finally introduces cloud-edge computing–based data processing techniques. This chapter lays a solid data foundation for subsequent operational optimization.

Chapter 3: "Electric-Thermal-Gas Coordinated Control Technology of Distributed Energy Grid Clusters". This chapter first establishes a hierarchical system framework based on an energy router to support the coupling and scheduling of electricity, heat, and gas energy flows. It then designs a coordinated controller for distributed energy grids, detailing its structural and functional aspects. Finally, on this basis, it proposes a Proxy Module-based coordinated controller for distributed energy grid clusters, discusses the development and application of the agent modules using real engineering data.

Chapter 4: "Day-Ahead Hierarchical Regulation and Control Technology for Distributed Energy Grid Clusters". Chapter 4 first proposes a coordinated scheduling framework for distributed energy grid clusters, leveraging a dynamic cooperative-game model and energy hubs to uniformly describe the coupling and conversion of electricity, heat, and gas flows. It then designs the day-ahead hierarchical operation model, finally conducts simulation experiments to verify the algorithm's convergence and economic performance, laying a solid foundation for real-time optimization and coordinated control of cluster-level energy systems.

Chapter 5: "Adaptive Adjustment Technology for Distributed Energy Grid Clusters". Chapter 5 develops a hierarchical real-time optimization control model, introduces a proactive power-tracking control strategy based on a multi-agent consensus protocol, and integrates it with the overall cost-minimization objective to enable autonomous distributed decision-making. Building on this, it designs a distributed, multi-area real-time optimization dispatch strategy and validates—through numerical experiments using real engineering data—the algorithm's effectiveness in reducing inter-unit power deviations, achieving cost convergence, and maintaining robustness against communication delays and disturbances.

Chapter 6: "Distributed Coordinated Scheduling of Distributed Energy Grid Clusters". Chapter 6 first introduces a dynamic cooperative-game model and a bi-level game model, and on this basis constructs a bi-level evolutionary game decision-making model to unify multi-energy flow coupling, economic optimization, and scheduling decisions. It then designs a multi-agent hierarchical control framework and a multi-agent distributed scheduling model, validating the algorithm's convergence and economic performance through typical cases. Finally, it develops a distributed demand response framework, an online parameter identification method, a hierarchical demand response optimization approach, and distributed communication algorithms, demonstrating their effectiveness through case studies.

Chapter 7: "Market Trading Strategies for Distributed Energy Grid Clusters". Chapter 7 develops a multi-region joint interaction model that unifies each distributed energy grid's electricity and gas trades into coalition trading variables to minimize the alliance's overall operating cost. It then constructs competitive bidding models for each grid operator covering costs such as upstream grid transactions and natural gas procurement. Next, it explains the five key properties of cooperative game theory and establishes a cooperative-game-based trading model. Finally, it validates the model's effectiveness in reducing total costs and enhancing alliance social welfare.

Chapter 8: "Outlook on Modeling and Collaborative Control Technology for Future Distributed Energy Grid Cluster Operations". Chapter 8 looks ahead to future technological evolution. It first maps out key development trends for distributed energy grid clusters from three angles—digitalization, intelligence, and decarbonization—highlighting edge–cloud collaborative sensing, AI-enabled adaptive scheduling, and negative-carbon technologies. It then summarizes the core challenges that must be overcome, including cross-domain data sharing, the interpretability and scalability of real-time optimization algorithms, and market-policy coordination among multiple stakeholders, thereby pointing the way for subsequent research.

Overall, by emphasizing the importance of energy complementarity, adaptive adjustment, and intelligent decision-making, the book presents a clear path forward for creating more sustainable, resilient, and efficient energy grids. As the world continues to transition toward cleaner, more decentralized energy solutions, distributed energy grid clusters will play an increasingly vital role in shaping the future of global energy systems. This book serves as an essential resource for researchers, practitioners, and policymakers seeking to understand and harness the potential of distributed energy grid clusters to meet the challenges of a rapidly changing energy landscape.

The completion of this book would not have been possible without the support and assistance of many teachers, collaborators, friends, and students. I wish to express my heartfelt gratitude to: Prof. Michael B. McElroy and Mr. Chris P. Nielsen of Harvard University, for their patient guidance on the preparations for writing this book; Prof. Ming Yang and Dr. Yu Guo of Shandong University, for their invaluable suggestions on theoretical development and experimental design; Prof. Qian Ai of Shanghai Jiao Tong University, for his crucial support in data collection, literature review, and case analysis; Engineer Ran Hao of the State Grid Corporation, for providing first-hand materials and insights for the engineering practice sections of this book; and Mr. Shaocong Wu, Ms. Xinning Yi, Mr. Hao Liu, Mr. Yuhao Zhang, Mr. Haibin Zeng, and Mr. Qinzheng Wu of the School of Electrical Engineering, Shandong University, whose contributions to testing, technical discussion, and manuscript proofreading have greatly enhanced the quality of this work. In addition, the authors gratefully acknowledge the support of the National Natural Science Foundation of China (Nos. U22B20102, 52177095 and 52107110) and the Harvard China Project received during the preparation of this book.

Jinan, China Tianguang Lu

Contents

Chapter 1
Overview of Distributed Energy Grid Clusters

Energy supply is often planned, designed, and operated independently, lacking coordination, which leads to low energy utilization efficiency, weak overall safety and self-healing capabilities of the energy supply system, and other issues. Under the "dual carbon" goal, the new generation of power systems will face large-scale and high-proportion renewable energy grid connection, wide-area diversified distributed generation (DG), and collaborative interaction of the main body of multi-energy consumption. Distributed energy grid clusters, as an integrated means of DG, loads, energy storage (ES), and energy conversion devices, can fully tap the potential of distributed resources, improve the capacity to absorb renewable energy, and reduce user costs through local autonomous control methods (Lu et al. 2016). In the distributed energy grid clusters, the energy supply shows diversity, and different forms of energy are coupled to provide users with various forms of energy, such as cold, heat, electricity, gas, and other forms of energy. Using the characteristic differences of different energy sources and mutual conversion can promote the consumption of renewable energy, such as electricity to hydrogen, wind power (WP) heating, etc. At the same time, the optimal complementarity of multiple energy sources can also improve the reliability of energy use by users, and provide more flexible resources for the operation of the power grid (Ai and Hao 2018). Distributed energy grid clusters can realize the coordination and complementarity of multiple energy sources, integrating different energy grids such as natural gas grids, heat grids, etc., which is the connection between distributed energy sources and the power system.

This chapter provides an overview of distributed energy grid clusters, first introducing the definition, functions, and composition elements of distributed energy grids (DEGs) and grid clusters, then discussing the current status of modeling and control research on distributed energy grid clusters, and finally reviewing the technical application cases. This chapter serves as a general overview of the subject matter of this book and is the foundation for subsequent chapters.

T. Lu et al., *Operation Modeling and Collaborative Regulation of Distributed Energy Grid Clusters*, Smart Energy Systems, https://doi.org/10.1007/978-981-96-9942-1_1

1.1 Concept and Composition of Distributed Energy Grid Clusters

As traditional fossil energy resources are increasingly depleted, improving energy utilization efficiency, developing new energy sources, strengthening the comprehensive utilization of renewable energy, and promoting the coordinated development of electricity, heat, cold, and gas are inevitable choices to solve the contradictions between the growing energy demand and energy shortage in the process of socio-economic development. At the same time, with the multi-level and diversified development of society, the traditional power system, mainly based on fossil fuels, is gradually unable to meet the increasing energy requirements of users for cold, heat, electricity, and gas in terms of energy supply quality, economy, reliability, and flexibility, and it is also difficult to provide economical and reliable energy supply to remote areas. In 2015, the National Development and Reform Commission of China, the Chinese Energy Bureau in promoting the development of smart grid guidance put forward "Strengthen energy interconnection and promote the optimal complementarity of multiple energy sources". Meanwhile, building a clean, low-carbon, safe, and efficient modern energy system is the goal and requirement of the energy revolution and the "13th Five-Year Plan for Energy Development". Distributed energy generation is one of the main ways of new energy utilization, with advantages such as green and clean power generation, reducing end-user costs, and improving power supply reliability. However, the distribution of renewable energy is usually more dispersed and subject to a variety of factors, resulting in renewable energy is usually stochastic and intermittent and other characteristics, so that it cannot be accessed by the power grid on a large-scale power generation schedule. The coordination and complementarity of various energies can help to provide flexibility for the energy system and achieve economic and energy efficiency goals. Therefore, the distributed energy grid, which integrates electricity, heat, cold, and gas, is widely regarded as an effective way to promote the use of distributed power and the collaborative development of multi-energy. Each distributed energy grid with certain autonomous characteristics within a certain area can interact with adjacent DEGs to form larger distributed energy grid clusters. As an advanced stage in the development of DEGs, distributed energy grid clusters can carry out inter-grid multi-energy mutual benefit and collaborative regulation of diversified subjects, and through the flexible interaction with the higher-level energy supply grid, it can fundamentally improve the operational reliability, reduce the operation and maintenance costs, and achieve the maximum revenue.

1.1.1 Concept of Distributed Energy Grids

In recent years, with the rapid development of centralized energy grids, the disadvantages of centralized energy grids, such as huge investment, low operation efficiency, large losses, and poor flexibility, have become increasingly prominent. Therefore, DEGs have emerged. Compared with centralized energy grids, DEGs are closer to the user side and have the advantages of small scale and strong flexibility.

A distributed energy grid is a kind of distributed power supply, multiple loads, ES system, and energy conversion device organically integrated and brought together the production, distribution, and storage of multiple energy forms of electricity, heat, cold, and gas, which is a small-scale energy supply system, and it is an autonomous system capable of realizing self-control, protection and management (Ban and Tang 2023). A distributed energy grid as a complete energy system, relying on its independent control and management of energy supply to achieve power balance control, system operation optimization, fault detection and protection, energy supply quality management, and other aspects of the function, can be operated in parallel with the external energy grids, but also can be operated in isolation (Lu et al. 2007). In addition, the distributed energy grid can also make use of internal electricity, heat, cold, gas all kinds of energy in time and space coupling complementary relationships, through the energy conversion devices for all kinds of energy joint optimization and complementary cooperation, thus achieve the efficient and flexible use of multiple energy sources. Therefore, DEGs are mainly used to solve the access and consumption problems of distributed renewable energy with high penetration rates, to provide safe and reliable electricity, heat, and natural gas for remote areas with weak connections to large energy grids, and to meet the special requirements of some users for the quality and reliability of energy supply (Fig. 1.1).

Distributed energy grid transforms the problem of grid-connecting a large number of DG sources into a controllable energy grid-connecting problem, weakening the adverse effects of random and fluctuating intermittent power sources on the safe and stable operation of the energy grid and the quality of the energy supply, etc. (Carrasco et al. 2006). For a large energy grid, the distributed energy grid can be regarded as a controllable unit in the energy grid, which can act within seconds to meet the demands of the external energy grid. For the users, the distributed energy grid can

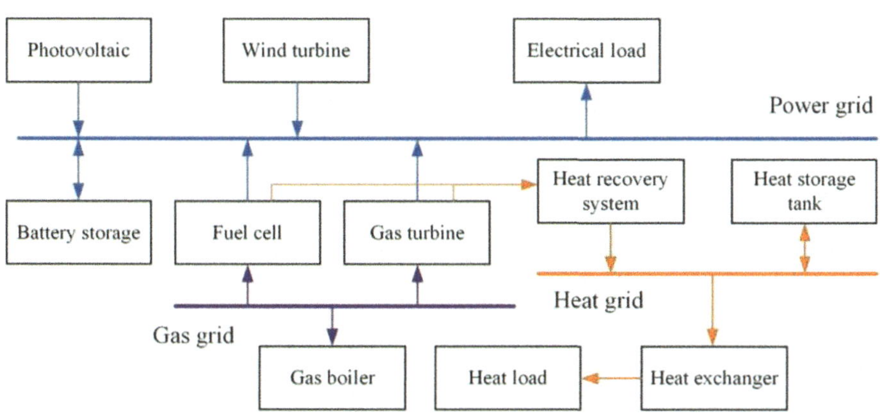

Fig. 1.1 Schematic diagram of the distributed energy grid

satisfy their specific needs such as increasing local reliability, reducing feeder losses, maintaining local voltage stability, improving the efficiency of energy utilization by using of waste heat, and providing an uninterruptible power supply. DEGs and large energy grids exchange energy through interconnecting lines, with both sides acting as reserves for each other, thus increasing the reliability of the energy supply.

There are various forms of DEGs, and combined cooling heating and power (CCHP) for regional or building clusters or independent large and medium-sized buildings is one of the very important ways (Jiang et al. 2017). The gas-fired CCHP system is a cogeneration and combined-supply system based on the concept of gradient utilization of energy, using natural gas as a primary energy source to generate heat, electricity, and cooling. It uses natural gas as fuel, utilizing small gas turbine (GT), gas engines, micro-gas engines, and other equipment to generate electricity from the high-temperature flue gas obtained after the combustion of natural gas, then uses the waste heat for heating in winter; in summer, it drives absorption air conditioning machines for cooling; it can also provide domestic hot water, making full use of exhaust heat. The primary energy utilization rate can be increased to about 80%, saving a large amount of primary energy.

There are various ways of gas CCHP system with a combustion engine as the core, and there are two basic ways. One is the gas engine (including internal combustion engines, GTs, etc.) + waste heat absorption air conditioning machine (waste heat direct firing machine), where natural gas is used as fuel to generate electricity after being sent into the GT for combustion, and the high-temperature exhaust is then entered into the waste heat absorption air conditioning machine (waste heat direct firing machine), providing cooling in summer and heating in winter, and supplementary combustion of natural gas can be used according to the needs of cold and heat loads. The other is the gas engine (including internal combustion engines, and GTs) + waste heat boiler (WB) + steam absorption air conditioning + electric air conditioning machine + gas boilers (GBs), where natural gas is sent into the GT for combustion and electricity generation, and the high-temperature exhaust is sent into the WB to produce steam, which is then sent to the steam absorption air conditioning machine through the steam distribution cylinder, and the steam is used to provide hot water for heating through the heat exchanger in winter. According to the cold load of the building complex in summer, insufficient cold is provided by the electric compression chiller; in winter, insufficient heat is provided by the heat pump (HP) and GBs.

1.1.2 Concept of Distributed Energy Grid Clusters

With the rapid development and widespread application of DG and supply technology, energy system monitoring, control, and management technology, as well as new energy transaction methods, energy coupling is becoming increasingly tight, and mutual aid and mutual benefit are being realized. Within a certain area, there can be multiple DEGs, and these DEGs can achieve multi-energy interconnection between the grids and between each distributed energy grid and the large power grid, natural gas grid, and cold and heat pipe grids by establishing electricity, heat and gas energy interconnection channels, forming a multi-energy complementary distributed energy grid clusters (Wang et al. 2016). Each distributed energy grid is independent and interconnected, coordinating and complementing energy through energy interconnection channels, achieving energy balance of electricity, heat, and gas loads in the distributed energy grid clusters, and can also reduce the economic operation costs of the distributed energy grid to a certain extent, improving overall energy utilization efficiency and the absorption rate of renewable energy.

Distributed energy grid clusters as a multi-energy complementary in the regional energy supply system complete functional realization unit, its multiple energy sources, grids, loads in-depth integration, close interaction on the system analysis, design, and operation puts forward new requirements (Ai and Hao 2018). Adjacent DEGs operate collaboratively in the form of clusters, featuring multi-energy coupling, multiple stakeholders, and a large geographical area. Relying on means such as marketing e-commerce, transaction liberalization, and market-oriented investment, multi-energy interaction between distributed energy grid clusters, between distributed energy grid clusters and superior energy grids can reduce system operation risks and user energy costs. On the other hand, distributed energy grid clusters will achieve the transformation from centralized energy production to distributed energy production and consumption, and through the interaction of intelligent entities at various levels and regions, achieve the best state for energy units to be plug-and-play, peer-to-peer interconnected, and collaboratively autonomous, achieving maximum benefits while meeting multi-energy load demands (Fig. 1.2).

1.1.3 Introduction to Composition Elements

1.1.3.1 DG

DG refers to small-scale, environmentally compatible independent power sources with a power range of several kilowatts to 50 MW. These power sources are owned by electric power departments, electricity users, or third parties to meet the specific requirements of the power system and users. They are the new type of energy production unit with low pollution emissions, flexibility, high reliability, and high efficiency.

As renewable energy usually has the characteristics of randomness and intermittency, resulting in its inability to scale access to the grid power generation schedules, it is usually necessary to use the distributed energy grid structure. The operation of DG in the form of a distributed energy grid can improve the acceptance capacity of the energy system for DG, and at the same time improve the utilization efficiency of intermittent renewable energy sources. When meeting the demand of multiple loads such as cold/heat/electricity, by key technologies such as operation control and energy management of the distributed energy grid, it can reduce the adverse

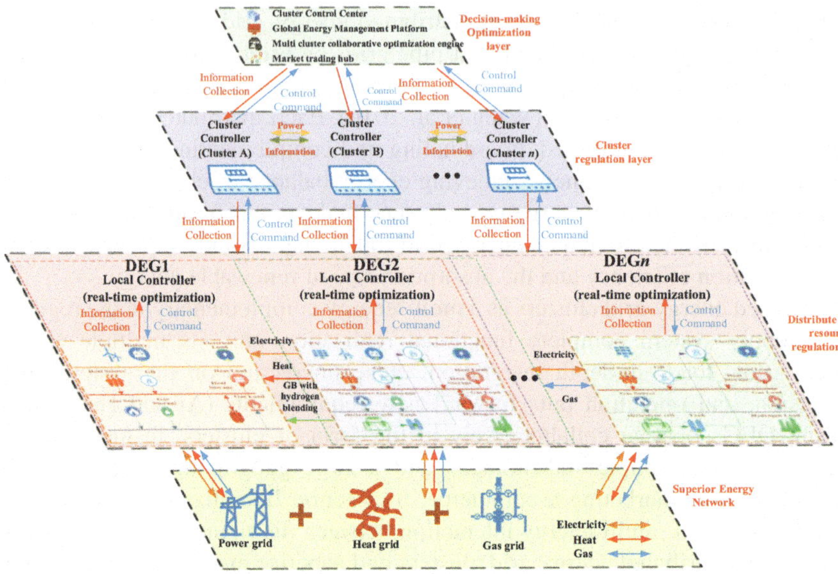

Fig. 1.2 Schematic diagram of the distributed energy grid clusters

effects of randomness, volatility, and other unfavorable elements of the intermittent distributed power supply on the safe and stable operation of the energy grid and the quality of the energy supply, etc. (Lu et al. 2007). Therefore connecting DG to the distribution grid in the form of distributed energy grid is generally regarded as one of the effective ways to utilize DG. Compared with traditional power plants, DG has the following main characteristics:

(1) Economic: Since DG is located on the user side, close to the load center, it greatly reduces the construction costs and losses of the transmission and distribution grid; at the same time, the planning and construction period of DG is short, and the investment takes effect quickly, with relatively small investment risks.
(2) Environmental: DG can widely use clean renewable energy, reducing the consumption of fossil energy and the emission of harmful gases.
(3) Flexibility: DG systems are mostly equipped with advanced medium and small modular equipment, with fast start and stop, convenient maintenance management, flexible regulation, and each power source is relatively independent, which can meet different needs such as peak load reduction, power supply to important users, etc.
(4) Safety: DG has various forms and can reduce the dependence on a single energy source to a certain extent, alleviating the expansion of energy crises; at the same time, the DG is dispersed in location and is not easily affected by unexpected disasters or emergencies, with the potential to resist large-scale power outages.

In recent years, new types of power electronic devices have been continuously applied, and new materials are also being developed. DG will inevitably be widely used across the country, and a large number of DG will be added to the distributed energy grid, which will change the current operation mode of the power grid. Therefore, new and convenient advanced control strategies must be applied to ensure the stability of the power grid and better play the role of DG.

At the same time, according to national policy requirements and future trends, there are four operational modes for DG in China (Wang et al. 2005): full grid-connected mode, priority self-use with surplus electricity grid-connected mode, direct power sales mode, and regional power sales mode. With the advancement of Chinese power market reform, distribution, and sales separation, and the opening up of incremental power grid business, multiple capitals are involved in the construction and operation of distributed new energy generation projects. The participation of distributed new energy generation in market operations will be an inevitable trend in future market development.

1.1.3.2 ES Systems

As the energy in a DG is very temporal and spatial, to rationally use renewable energy and improve energy utilization, it is necessary to use a device that collects and stores the excess energy that is temporarily not used for a certain period in a certain way and then extracts it for use at the time of peak use, which is the ES system. Reducing the power fluctuation in the distributed energy grid caused by the access of the DG system, thus solving the problem of new energy consumption, plays a very important role in the distributed energy grid clusters.

The basic task of an ES system is to overcome temporal or localized differences between energy supply and demand. Two situations cause such differences: one is caused by sudden changes in energy demand, that is, the peak load problem. Using ES methods can play a role in regulation or buffering when the rate of change of load increases. Since the investment cost of an ES system is relatively lower than that of building a peak load plant, although there will be storage losses, the stored energy comes from the surplus energy of the plant or new energy, so it can still reduce fuel costs. The other is caused by reasons such as primary energy and energy conversion devices, and the task of the ES system is to balance the energy output, that is, not only to reduce the peak of energy output but also to fill the valley of output (i.e., fill the valley) (You et al. 2014).

For different application purposes, there are respective ES requirements, but summarized, the common characteristics of a good ES system are as follows: (1) High ES per unit volume (volumetric ES density), that is, the system can store as much energy as possible. For example, high-energy batteries, due to their higher energy density than ordinary batteries and longer service life, are popular with consumers. (2) Good load regulation performance. When using an ES system, it is necessary to regulate the size of the released energy according to the requirements of the energy-using party. The quality of load regulation performance determines the performance of the system. (3) High ES efficiency. ES cannot be separated from energy transfer and conversion technology, so the ES system should be able to receive and release energy at the maximum rate without excessive driving force. At the same time, as much as possible, reduce ES process losses such as leakage, evaporation, friction, etc., and maintain high ES efficiency. (4) Low system cost and reliable long-term operation. If the ES device is not economically reasonable, it will not be widely applied.

ES mainly includes the storage of thermal energy, kinetic energy, electrical energy, electromagnetic energy, chemical energy, etc. The research, development, and application of ES technology are mainly focused on the storage of thermal energy and electrical energy and are widely used in solar energy utilization, power "peak shaving and valley filling", waste heat and residual heat recovery, and energy saving in industrial and civil buildings and air conditioning.

(1) Thermal ES Technology

Thermal ES is a method of storing excess heat that is not needed for some time through some means, and then extracting and using it when needed. It includes sensible heat storage technology, latent heat storage technology, and chemical reaction heat storage technology.

Sensible heat storage technology stores thermal energy by heating the storage medium to increase its temperature. Common sensible heat storage materials include water, soil, and rocks. Under the same temperature change conditions, if heat loss is not considered, then the specific heat storage of water is the largest, followed by soil, and rocks are the smallest. Many countries in the world have conducted tests and applications on these heat storage materials. At present, this is a relatively mature technology with high efficiency and low cost.

Latent heat storage technology stores thermal energy by utilizing the heat of fusion when the storage medium changes from liquid to solid phase. The actual application of latent heat storage media includes sodium sulfate decahydrate (the chemical formula is $Na_2SO_4 \cdot 10H_2O$), sodium thiosulfate pentahydrate (the chemical formula is $Na_2SO_4 \cdot 5H_2O$), and calcium chloride hexahydrate (the chemical formula is $CaCl_2 \cdot 6H_2O$), etc. The characteristic of this technology is that it stores energy at low temperatures, has a high ES density, and can extract heat at a certain phase change temperature, but the storage medium is expensive, prone to corrosion, some media may undergo decomposition reactions, and the storage device is more complex than the sensible heat type, with greater technical difficulty.

Chemical reaction heat storage technology stores energy by decomposing chemical substances and then storing energy separately, and the decomposed substances can release the stored heat when they recombine. It can be achieved by three technologies: reversible decomposition reaction, organic reversible reaction, and hydride chemical reaction, among which hydride chemical reaction technology has the most development potential. Many researchers are conducting in-depth studies. If a breakthrough success can be achieved, it will provide a good way to solve the energy shortage problem.

(2) Electrical ES Technology

There are three main types of electrical ES technology applied in industry, which are pumped hydro storage technology, compressed air ES technology, and flywheel ES technology. Pumped hydro storage technology is the oldest, most mature, and largest capacity commercial technology in the world, with about 500 pumped hydro storage power stations, including 35 with a capacity exceeding 1000 MW. A pumped hydro storage system generally has two large storage reservoirs, one at a lower position and the other at a higher elevation. During off-peak electricity periods, water is sent from the lower reservoir to the higher storage reservoir for storage. When electricity is needed, the potential energy of the water flow from the high reservoir can be used to drive the water turbine to generate electricity.

Compressed air ES is to compress air and transport it to underground salt mines, abandoned stone mines, underground water storage layers, etc., during off-peak electricity periods. When the electricity load is large, compressed air can be burned with fuel to produce high-temperature, high-pressure gas, driving GTs to generate electricity. The capacity of the applied unit equipment has reached hundreds of megawatts. For example, the Fendorf power station in Germany with an installed capacity of 290 MW was put into use in 1980.

Flywheel ES technology is a new technology that connects to the power grid to achieve the conversion of electrical energy. The system mainly consists of a motor, a flywheel, and a power electronic converter. The basic principle of flywheel ES is to convert the electrical energy in the power system into the kinetic energy of the flywheel under conditions of electricity abundance. When the power system is short of electricity, the kinetic energy of the flywheel is converted back into electrical energy for power users. Compared with other ES technologies, flywheel ES technology has the advantages of high efficiency (80–90%), low cost, no pollution, fast ES, and reliable technology, and has attracted the attention of researchers in Japan, the United States, and Germany.

Battery ES system (BESS) is a chemical-based ES system that uses a parallel voltage source converter connected to an AC system (Jin et al. 2009). Its converter can be quickly adjusted to supply power to the AC system or absorb electrical energy from the system. The series–parallel structure of BESS for DC distributed energy grid is shown in Fig. 1.3, where the BESS consists of a bidirectional DC/DC converter and an ES battery. The use of multiple BESS in series can increase the output side voltage to meet the requirements for accessing the DC distributed energy grid. This structure uses a bi-directional DC/DC converter to control low-voltage battery cells, which can reduce the inconsistency and "short-board effect" of series–parallel connection in large-capacity batteries. As shown in Fig. 1.3, there are j parallel groups, each group consists of k BESSs connected in series. All the BESSs in the system are controlled in a decentralized and autonomous manner, which makes the DC distributed energy grid operation highly reliable.

With the development of smart distribution grids, the demand for distributed ES technology will also become more diversified. No one ES technology can fully take into account safety and reliability, high energy density, high efficiency, long life, and low cost. Therefore, using hybrid ES forms and combining ES systems with different performance characteristics can give full play to the advantages of different

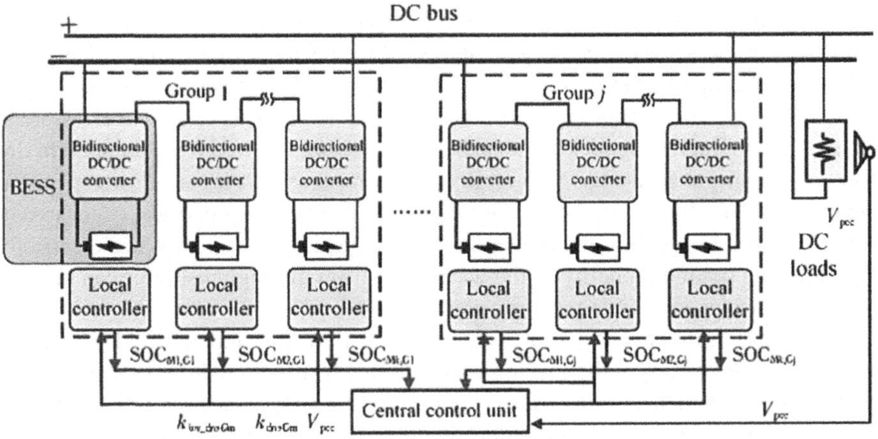

Fig. 1.3 Series–parallel structure

ES technologies to meet the needs of energy in many aspects. This will be the development and application trend of distributed ES technology in the future.

1.1.3.3 Power-to-Gas Systems

Power-to-gas (P2G) technology uses surplus electricity to carry out chemical reactions to produce renewable hydrogen or renewable natural gas, which is then stored in natural gas/hydrogen ES equipment and pipelines for use when needed. The waste heat in the chemical reaction process can also be recovered for heating. Therefore, P2G transforms the unidirectional coupling of traditional electricity/heating/natural gas systems into bidirectional coupling.

At present, there are two mainstream P2G technologies: Power-to-hydrogen and Power-to-methane. The basic principle of Power-to-hydrogen is to use electrical energy to ionize water to produce pure hydrogen, and the commonly used equipment is the electrolyzer, with three mainstream technologies: alkaline electrolyzer (AEC), proton exchange membrane electrolyzer (PEMEC), and solid oxide electrolyzer (SOEC). The principle of Power-to-Methane is based on the principle of Power-to-hydrogen, using carbon dioxide and hydrogen to react under high temperatures and pressure to produce methane, and the commonly used equipment is the methanation equipment reactor. P2G mainly uses surplus renewable energy generation, converting electrical energy into high-energy-density combustible gas and storing it to achieve large-scale and long-term ES and large-capacity and long-distance transportation of energy. The methanation process can also consume carbon dioxide. Therefore, the application of P2G technology can effectively improve the absorption capacity of renewable energy in the energy system and reduce carbon emissions.

With the rapid development of P2G technology, the coupling relationship between the power grid, natural gas grid, and heat grid is deepening. The traditional analysis method that studies different energy grids separately can no longer meet the reliability and economy of the collaborative operation of multi-energy grids. Therefore, using multi-energy complementary distributed energy grid clusters for the collaborative optimization of electricity, heat, gas, and other energies has become an important development trend in future energy systems.

1.2 Current Research on Modeling and Regulation

1.2.1 Current Research on Operation and Modeling

Distributed energy grid clusters operate in two modes: Islanding and grid-connected. Islanding operation refers to a mode in which the distributed energy grid is disconnected from the superior energy grid and relies solely on its internal distributed energy sources, ES, P2G systems, and other devices to provide stable and reliable electricity, heat, and gas to meet the load demand. Grid-connected operation refers to the mode where the distributed energy grid connects to the superior energy grid through a static switch at the point of common coupling (PCC) to operate in parallel.

In both operation modes, the stability of the energy system depends not only on the regulation of electricity but also on the control of the heat and gas grids, which are crucial. The selection of an appropriate operating mode requires a comprehensive consideration of the structure, economy, safety, stability, and quality requirements for electricity, heat, and gas in the distributed energy grid clusters. For example, some DEGs may require minimal fluctuations in the electricity, heat, and gas loads, especially in Islanding operation mode, where the system must have high self-sufficiency and balance supply and demand through precise regulation of the electricity, heat, and gas sources. On the other hand, other distributed energy grid clusters may require higher stability in the supply of electricity, heat, and gas, relying on the external grid to ensure the reliability of the energy supply. In this case, a grid-connected operation mode would be more suitable. In complex distributed energy grid clusters, coordinating electricity, heat, and gas grid scheduling is essential, and it may be necessary to flexibly choose or switch between operating modes based on system demand and external environmental factors.

1.2.1.1 Islanding Operation of Distributed Energy Grid Clusters

Islanding operation refers to a mode in which distributed energy grid clusters operate independently from the superior grid, temporarily disconnecting from it (Xue et al. 2013). In this mode, the power grid, heat grid, and gas grid must work together to meet the system's load demands. To achieve this goal, the operation of the electricity, heat, and gas grids in the distributed energy grid clusters requires high levels of coordination and flexibility. In Islanding operation, the main challenge for the distributed energy grid is how to balance the supply of electricity, heat, and gas to ensure stable operation without superior energy grid support. At this time, the operation control of the electricity, heat, and gas grids is not independent; rather, it requires high integration and collaboration, ensuring that the supply from each grid can be adjusted at any time to meet the constantly changing load demands.

Currently, based on several basic control methods for DG, the Islanding operation control strategies for distributed energy grid clusters commonly used in the power grid include master–slave control, peer-to-peer control, and hybrid control of distributed energy grid operations. Taking the master–slave control strategy as an example, the master–slave control method with a single V/f power source in the distributed energy grid clusters islanding operation mode selects a distributed energy or storage device as the master power source. This system uses V/f control to provide stable voltage and frequency for the entire system, coordinating with other slave power sources that adopt PQ control to achieve power balance. In the grid-connected mode, all micro-sources adopt PQ control to provide constant power output. While this strategy can guarantee voltage and frequency stability to some extent, it places the responsibility of meeting instantaneous load demands solely on the master power source, which requires a high capacity. Additionally, this strategy is highly dependent on the master power source, and a failure of the master source could lead to the collapse of the entire system.

Unlike the power grid, the regulation of the heat and gas grids is not just a simple mapping of electricity load; they also need to dynamically adjust heat and gas supply based on changes in the power grid load. In the heat grid regulation, the principle of thermal balance is applied to adjust the heat source output in real-time to meet the heating demand of the system. Especially during significant load fluctuations, thermal ES devices can store excess heat and release it during peak demand periods, alleviating the pressure on heat supply. The regulation of the gas grid relies on flow control and pressure management. The gas grid dynamically adjusts the flow and pressure of gas based on changes in the power grid load and the heat grid's demand, ensuring a stable supply of gas to all user endpoints. Particularly in Islanding operations, the gas grid must be capable of responding quickly to fluctuations in electricity load or heating demand, which could impact the supply of gas.

In Islanding operation, the operation of the power grid, heat grid, and gas grid must be highly coordinated. To achieve this, the system uses multi-level feedback mechanisms and adaptive control strategies to predict and schedule the loads of all three grids. For example, a combined heat and power (CHP) system can simultaneously provide electricity, heat, and gas, adjusting its output in real-time to meet the demands of the power grid, heat grid, and gas grid. The coordination between the three grids not only requires each grid to adjust independently but also involves real-time optimization of scheduling strategies based on changes in the loads of the electricity, heat, and gas grids, ensuring the efficiency and stability of the energy supply.

1.2.1.2 Grid-Connected Operation of Distributed Energy Grid Clusters

When distributed energy grid clusters are integrated into the superior energy grid, it not only impacts the stability of the power grid but also involves the coordinated regulation of the heat grid and gas grid. To ensure the stability and efficiency of the entire system under grid-connected operation, the operation of the distributed energy grid must meet the following requirements (Yang et al. 2012):

(1) Active Power: When the distributed energy grid clusters operate in a grid-connected mode, its active power regulation must not only account for the load fluctuations in the power grid but also coordinate with the demands of the heat and gas grids. The capacity of the distributed energy grid clusters should consider the protection of the power grid, the pipeline capacity of the heat and gas grids, and be calculated based on the storage capacity of ES devices. In grid-connected operation, the distributed energy grid must deliver power to the power grid according to its load characteristics while also meeting the demands of the heat and gas grids. During this process, the distributed energy grid clusters must adjust the active output of micro sources to maintain power grid frequency stability and also adjust the supply of heat and gas sources according to load changes to ensure the coordinated operation of all three grids.

(2) Reactive Power: The micro-sources and reactive power compensation devices within the distributed energy grid should have the capability to automatically and quickly regulate reactive power. The regulation of reactive power is not only crucial for the voltage stability of the power grid but also directly affects the normal operation of the heat and gas grids. Changes in the temperature and pressure of the heat grid can have a feedback effect on the power grid's voltage, and fluctuations in the flow and pressure of the gas grid may also be influenced by voltage fluctuations in the power grid. Therefore, when operating in grid-connected mode, the distributed energy grid clusters must comprehensively regulate the reactive power of the electricity, heat, and gas grids to improve the power factor, reduce active power loss in the large energy grid, and enhance the system's ability to resist interference. In cases where the reactive power capacity of some micro sources is insufficient, reactive power compensation devices should be installed to meet the voltage regulation needs of the system, ensuring the stability of the three grids during coordinated operation.

(3) Voltage Regulation: The control system of the distributed energy grid clusters should be able to operate under constant power factor or constant reactive power output, and be equipped with automatic voltage regulators. However, in grid-connected mode, voltage regulation of the power grid is generally managed by the large grid, and the distributed energy grid clusters only actively participate in voltage regulation under conditions permitted by the large grid. The distributed energy grid clusters should ensure that their reactive power can be automatically adjusted within allowed ranges to maintain stable voltage and power factors at the PCC. At the same time, the regulation of the heat and gas grids should be synchronized with voltage fluctuations in the power grid to ensure the stability

of temperature and pressure and prevent imbalances in the heat and gas grids caused by voltage fluctuations.

(4) Energy Quality: When distributed energy grid clusters are connected to the superior energy grid, the energy quality indicators at the grid connection point primarily include: harmonics, waveform distortion, voltage fluctuations, flicker, DC injection, voltage imbalance, voltage deviation, and frequency. Integrating the distributed energy grid clusters introduces significant harmonics, and the type and severity of these harmonics depend on the power converter technology, the composition and characteristics of the micro-sources in the distributed energy grid clusters, and the connected load devices. Heat energy quality refers to whether the supply of heat sources in the heat grid is stable, and whether temperature fluctuations remain within a controllable range. The regulation of CHP systems and thermal storage devices must be synchronized with the power grid load changes to avoid temperature fluctuations due to excess or insufficient heat supply, which could impact the heating stability for users. Similarly, gas energy quality primarily concerns the stability of gas supply in the gas grid, including flow and pressure fluctuations. If the gas grid pressure is unstable, it could lead to gas supply interruptions, affecting the regulation efficiency of electricity and heat. Therefore, during grid-connected operation, it is essential to ensure that the quality of electricity, heat, and gas is simultaneously optimized and managed, preventing fluctuations in any single form of energy from disrupting the overall operation of the system.

(5) Safety and Protection: The protection system must be able to respond to faults on both the superior energy grid and the distributed energy grid cluster sides, and quickly isolate the faulted areas. When a fault occurs on the superior energy grid side, the distributed energy grid should quickly disconnect from the superior energy grid to protect the local loads. Conversely, when a fault occurs within the distributed energy grid clusters, the protection system should quickly isolate the smallest possible fault section within the distributed energy grid clusters to prevent disruption to the normal operation of other parts of the system.

1.2.1.3 Operation Mode Transition of Distributed Energy Grid Clusters

In most cases, the distributed energy grid clusters operate in grid-connected mode, with Islanding operation occurring for shorter durations. The distributed energy grid clusters continuously monitor the operating status of various energy grids, such as the power grid, heat grid, and gas grid. When a fault or energy quality issue arises in the superior energy grid, the distributed energy grid clusters will automatically or, based on scheduling instructions, transition from grid-connected mode to islanding mode. Once the superior energy grid is restored to normal, the distributed energy grid clusters will quickly resume grid-connected operation according to preset control mechanisms or scheduling instructions. To ensure a smooth transition of the electricity, heat, and gas grids during this switch, the monitoring system of the distributed energy grid clusters needs to acquire real-time data on the electricity, heat, and gas at the PCC. When abnormal conditions are detected, such as load fluctuations in the power grid, temperature changes in the heat grid, or pressure variations in the gas grid, the system will adjust the power grid frequency, heat source output, and gas flow pressure accordingly to ensure stable and coordinated operation of all three grids during the mode transition.

Most of the distributed power sources in the distributed energy grid clusters need to connect their generated electricity to the superior energy grid through power electronic devices (inverters). These inverters are mainly classified into three types: grid-feeding inverters, grid-forming inverters, and grid-supporting inverters. Grid-feeding inverters require external voltage and frequency support and cannot operate independently; grid-forming inverters can only operate in islanding mode and independently; grid-supporting inverters use a droop control method similar to traditional synchronous generators and can operate both in grid-connected mode and islanding mode. Currently, most of the inverters used in research are grid-supporting inverters.

The coordination of the electricity, heat, and gas grids is crucial during grid connection. In particular, when grid-connecting, issues may arise from load mismatches between the power grid, heat grid, and gas grid. For example, when the power grid load increases, the outputs of the heat and gas grids need to be adjusted accordingly to ensure a stable supply of heat and gas. Similarly, fluctuations in gas grid pressure and flow can affect the power grid voltage and heat grid temperature, so precise load adjustments are required to ensure the coordination of the three grids. Moreover, during grid connection, the operational parameters of the electricity, heat, and gas grids (such as voltage, current, pressure, temperature, etc.) must be kept as close as possible to minimize transient processes and system fluctuations during the transition, preventing equipment damage and system instability caused by load fluctuations.

1.2.2 Current Research on Collaborative Regulation

With the development of distributed energy grid clusters that incorporate various types of distributed energy sources and power-heat-gas energy conversion devices, small-scale distributed energy grid systems are increasingly characterized by wide coverage, clustered distribution, and individual autonomy. Each distributed energy grid cluster, with a certain degree of autonomy within a specific region, can interact with neighboring clusters to form larger distributed energy grid clusters. As a foundational platform for clean energy consumption and energy substitution, distributed energy grid clusters represent highly integrated grids of multiple energy forms, including the power grid, heat grid, gas grid, and information network. They also serve as an efficient transmission, utilization, and conversion platform for various energy types, with the energy grid as the main framework. The system involves communication, information, distributed computing, and other fields, where intelligent entities in different layers and regions interact to achieve the optimal state, thereby obtaining relatively reasonable maximum benefits. Therefore, the scheduling, control, and trading between various entities must be organically coordinated, and hierarchical regulation strategies are effective tools for integrating multiple intelligent entities in distributed energy grid clusters, coordinating operational states, and achieving locally optimal strategies.

1.2.2.1 Control Structure of Distributed Energy Grid Clusters

From the perspective of the control structure of the energy management system (EMS) in distributed energy grid clusters, these clusters can be divided into centralized control and decentralized control (Wu et al. 2014).

Centralized control generally consists of a central controller and local controllers. The central controller sends scheduling instructions to the local controllers after performing optimization calculations, and the local controllers execute these instructions to control the output of the DG. The advantages of centralized control include: a clear division of labor, ease of implementation and maintenance, lower equipment costs, and the ability to control the entire system. It is widely used and technologically mature. However, with the increase in distributed energy sources, ES, and P2G devices, the central controller needs to have strong computational capabilities and high communication capacity. In case of a failure of the central unit, the entire system is at risk of paralysis. Additionally, distributed energy sources are not plug-and-play, making it difficult to expand applications.

Decentralized control is another control approach for the EMS in distributed energy grid clusters. Both centralized and decentralized control methods involve a central controller and local controllers, but decentralized control weakens the dominant role of the central controller by enhancing communication with peripheral devices and decentralizing the control power to local controllers. In decentralized control, each component in the distributed energy grid clusters is controlled by a local controller, which monitors the operation status of micro sources and communicates with other local controllers through the communication network. Local controllers do not need to receive control instructions from the central controller and have the autonomy to determine the operation status of the controlled micro-sources. Since local controllers only need to communicate with nearby devices, their information transmission is less than that of centralized control. The computational load is also distributed across each local controller, reducing the workload of the central controller. In decentralized control, the central controller mainly transmits system load and electricity price information to the local controllers and operates the local controllers in emergencies or fault scenarios.

Decentralized control works by weakening the dominant role of the central controller and delegating control authority to local controllers. These local controllers monitor the operation status of distributed energy sources, heat sources, and gas sources, and exchange information with other controllers via local communication networks. Local controllers have greater autonomy and can independently adjust power output based on real-time load changes. The advantages of decentralized control are that the computational pressure on the central controller is relieved, and the system's flexibility and reliability are improved. However, since the coordination and information transmission between local controllers relies on communication networks, there is a risk of communication delays and protocol conflicts. Especially in large-scale systems, decentralized control may require more complex communication and management technologies.

As interconnected energy systems continue to grow in scale and complexity, and are geographically dispersed over large areas, if centralized control is adopted, substantial amounts of information must be exchanged between system components and the central controller. The ability of the central controller to process data may not keep up with the system's development, which is economically and technologically unsustainable. Furthermore, decentralized control is not conducive to the unified regulation of the superior energy grid, and it requires a more complex and higher-level communication system, which current technologies cannot fully support. Therefore, hierarchical control, which integrates the advantages of both centralized and decentralized control, resolves this contradiction. The key feature of hierarchical control is that control instructions become increasingly detailed as they move down through the levels, while feedback information is refined and transmitted upwards. The monitoring institutions at different levels have a hierarchical relationship, with clearly defined roles and responsibilities.

1.2.2.2 Control Technology of Distributed Energy Grid Clusters

The regulation of distributed energy grid clusters faces complex challenges, especially in the collaborative operation of the power grid, heat grid, and gas grid. The mutual influence between different energy grids means that the regulation model is not limited to the switching between grid-connected and island modes but must also consider diverse energy demands and economic requirements. These factors make the operation of distributed energy grid clusters more complicated, and the correctness and proficiency of regulation significantly impact the maintenance cycle of the grid clusters. Due to the differences between these systems and traditional DEGs, applying previous regulation methods may not be suitable for distributed energy grid clusters and may even increase the difficulty of regulation. Therefore, a new regulation method for distributed energy grid clusters is necessary. Currently, the main control structures for distributed energy grid clusters are divided into three types: master–slave control structure, peer-to-peer control structure, and multi-agent control structure (MAS).

(1) Master–Slave Control Structure

In the regulation of distributed energy grid clusters, the master–slave control structure is a commonly used approach. This structure selects a large-capacity distributed energy unit with ES devices as the master control unit, while other distributed energy units function as slave control units, adjusting the output of the power grid, heat grid, and gas grid under grid-connected conditions. This structure can improve the utilization of renewable energy when superior energy grids provide voltage and frequency support. However, when operating in island mode, the master control unit is responsible for providing voltage and frequency stability, while also regulating the heat grid and gas grid outputs.

(2) Peer-to-Peer Control Structure

Another control approach is the peer-to-peer control structure, where each unit in the distributed energy grid clusters is equal in status and does not rely on central control. Each energy unit adjusts its power output according to load changes, with the regulation of the power grid, heat grid, and gas grid achieved through local measurements and droop control to maintain power balance. In both grid-connected and island modes, the distributed energy grid clusters can autonomously adjust electricity, heat, and gas energy based on local information, ensuring stable operation of all three grids.

(3) Multi-agent System (MAS)

MASs represent a cutting-edge area of research in distributed artificial intelligence, consisting of two or more agents, each a typical knowledge entity with the following characteristics: autonomy, social ability, reactivity, goal orientation, and proactivity. Each agent can plan and make appropriate decisions, exchange data with other agents via TCP/IP protocols, and react to the environment, all for modeling and simulation of complex systems.

The multi-agent strategy, adapted from traditional power system applications, employs agents at different levels—superior energy grid level, distributed energy grid clusters level, and component level—to control distributed energy grid clusters and ensure the economic and efficient operation of distributed energy grid clusters. The MAS is intelligent and can promptly detect changes in the surrounding environment to meet operational needs. Its basic unit is the agent, and agents can communicate and cooperate with each other. At the superior energy grid level, agents analyze the needs of the superior energy grid and each distributed energy grid, then pass control decisions to the distributed energy grid cluster agents. The distributed energy grid cluster agents, as the intermediary layer, pass decisions from the superior energy grid agents to the component-level agents and integrate information from component-level agents to make reasonable judgments, which are then transmitted downward. This layer acts as a communication bridge. Component-level agents are responsible for communicating their information to the distributed energy grid cluster agents, and adjusting their operation according to the received control decisions, based on component constraints.

The MAS for distributed energy grid clusters control is divided into three layers: component agent, distributed energy grid agent, and superior energy grid agent, as shown in Fig. 1.4. Each lower-level component in the distributed energy grid clusters (including generators, distributed energy sources, CHP units, P2G devices, electricity-heat-gas loads, etc.) operates as an independent agent. Each component agent should have the ability to forecast output, process data, and communicate with higher-level agents. The distributed energy grid clusters agent is responsible

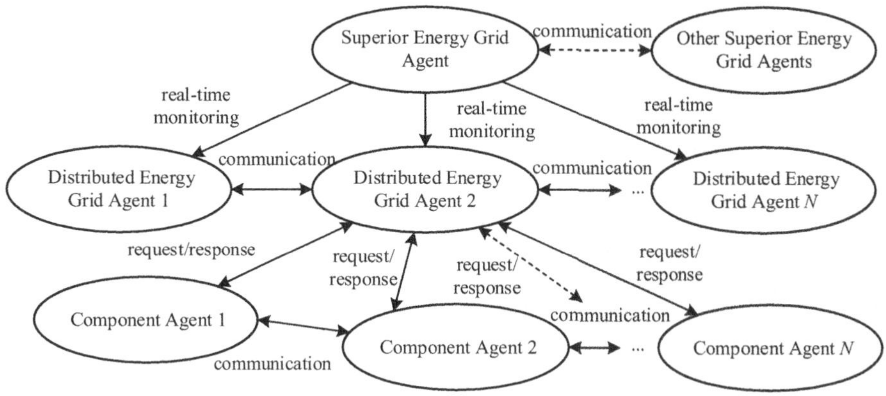

Fig. 1.4 MAS structure diagram

for managing these agents, such as receiving information from component agents and providing corresponding control strategies based on the operation status and adjustment strategies of the distributed energy grid clusters. Distributed energy grid cluster agents can communicate and interact with each other, and coordinate with superior energy grid agents through communication to solve task allocation and resource sharing among agents. The superior energy grid agent is responsible for the energy market and coordinating the scheduling of tasks among all agents, making major decisions based on the information from distributed energy grid agents.

1.2.2.3 Distributed Collaborative Control of Distributed Energy Grid Clusters

In distributed energy grid clusters, the dynamic characteristics of the power grid, such as voltage and frequency, are influenced by changes in distributed energy and thermal and gas loads. Especially in the case of weak connection with or separation from the superior energy grid, how to ensure the coordinated operation of power, heat, and gas grids has become a big challenge. Traditional control methods are often unable to meet the complex needs of modern distributed energy grid clusters. As a result, distributed collaborative control has become a key research focus in distributed energy grid clusters.

Due to the different management entities for the power grid, heat grid, and gas grid, centralized coordination is difficult. As a solution, distributed architectures are widely used for multi-energy collaboration in current research (Wen et al. 2017). Based on whether a coordination center is present, distributed architectures can be divided into two types: "decentralized architecture" and "weakly centralized architecture" (Cui et al. 2015). In these architectures, various distributed algorithms, such as optimality condition decomposition, Benders decomposition, and the alternating direction method of multipliers (ADMM) and its variants, are widely applied to the distributed collaborative control of distributed energy grid clusters (Huang et al. 2017; Yan et al. 2018; Liang et al. 2020).

For instance, Jinbo Huang et al. implemented distributed collaboration between the power system and regional heating grid using the optimality condition decomposition algorithm, effectively tapping into the ES potential of the heating grid, enhancing the flexibility of the power system, and improving WP integration (Huang et al. 2017). Mingyu Yan et al. proposed a two-layer distributed algorithm based on Benders decomposition and column and constraint generation algorithms, achieving fast solving of a robust model considering wind power uncertainty, and enhancing the flexibility of the integrated energy system through power-gas collaborative scheduling (Yan et al. 2018). Xinyu Liang et al. proposed an R-ADMM algorithm with strong anti-packet loss capability for the power-heat collaboration problem, based on the Peaceman-Rachford splitting theory (Liang et al. 2020). Compared with traditional ADMM, this algorithm still converges quickly during packet loss events. Jian Chen applied ADMM variants to the power-gas-heat collaboration, proposing a Gaussian back-substitution-based ADMM algorithm, achieving distributed collaboration among different energy subsystems (Chen et al. 2019). Compared with centralized architectures, both of these distributed architectures and the corresponding distributed algorithms effectively improve solving speed and, to a large extent, protect the privacy of the subsystems.

However, the distributed energy grid clusters involve different operating entities across energy sectors and include complex coupled equipment. It requires consideration of the dynamic characteristics of natural gas and thermal grids, making the design of distributed algorithms with privacy protection significantly more challenging due to the system's intricate internal coupling relationships and mathematical models. Additionally, multi-agent collaboration and the encrypted transmissions required for privacy protection will increase the volume of information exchanged, further burdening communication and potentially causing issues like congestion and packet loss. Distributed algorithms typically rely on continuous information exchange for convergence, and they often exhibit sensitivity to parameters and local convergence. The communication issues mentioned above can slow the convergence rate of the algorithm or, in extreme cases, prevent convergence altogether.

1.3 Technology Application Cases

Countries worldwide have actively explored emerging industries to revitalize their economies in recent years. New energy and technology have developed rapidly. In addition, due to the deepening of environmental crises such as climate change caused by the widespread use of fossil energy, the reserves of fossil energy such as oil are decreasing yearly and the prices are high. A huge change in the energy industry is imminent. Renewable energy such as wind and solar energy, as important new energy carriers, have undertaken an important mission to transform the energy industry.

Distributed energy grid clusters can not only reduce the volatility of distributed power sources through ES and optimized configuration, and achieve a basic balance between production and load, but also further improve the system operation efficiency through the synergy and complementarity of multiple energy flows such as electricity, heat, and gas. For example, waste heat recovery technology can use the waste heat generated during power generation for regional heating or cooling, and the joint dispatch of natural gas and electricity can effectively balance the volatility of a high proportion of renewable energy, providing users with stable and reliable comprehensive energy services. This multi-energy synergy model helps to use resources more efficiently, while further improving the energy experience of end users.

From a global perspective, distributed energy grid clusters are mainly in the experimental and demonstration stages. The technology promotion of distributed energy grid clusters has passed its infancy and the market size has grown steadily. In view of the current global energy and environmental dilemma and the long-term high demand for energy security, the application prospects of distributed energy grid technology are promising. In the next 5–10 years, the market size, regional distribution, and application site distribution of distributed energy grid clusters will significantly change. Among them, the technical application of multi-energy complementarity of electricity, heat, and gas and the coordinated management of cross-energy flows will become important directions for research and development.

1.3.1 Application Cases in the United States

The United States is a leader in the research and practice of DEGs in the world, with the largest number of distributed energy grid demonstration projects in the world, exceeding 200. The Consortium for Electric Reliability Technology Solutions (CERTS) comprises American power groups, Berkeley Lawrence National Laboratory and other research institutions. With funding from the U.S. Department of Energy and the California Energy Commission, it has conducted special research on distributed energy grid technology. The basic concept of a distributed energy grid is defined by CERTS as a collection of loads and micro power sources. The micro power source operates in a system that provides both electricity and heat. Most of these micro power sources must be power electronic and provide the required flexibility to ensure that they can operate as an integrated system. The flexibility of their control enables the distributed energy grid to serve as a controlled unit of the upper energy system to adapt to the reliability and safety requirements of local loads.

The distributed energy grid defined by CERTS proposes a new method for distributed power access systems that is completely different from the previous ones. The traditional method focuses on the impact of distributed power on grid performance when considering the distributed power access system. According to the traditional method, when a problem occurs in the superior energy grid, it is necessary to ensure that the distributed power sources connected to the grid automatically stop operating to avoid adverse effects on the upper energy system. The distributed energy grid defined by CERTS is designed to be seamlessly separated from the upper energy grid or operate as an island when a fault occurs in the upper energy grid, and can be reconnected to the upper energy grid once the fault is removed. The advantage of this distributed energy grid is that it is regarded as a self-controlled entity in the distribution system connected to it, ensuring the uninterrupted energy supply of important users, improving the reliability of the power supply, reducing feeder losses, and supporting and correcting the local voltage. Therefore, the distributed energy grid not only avoids some negative effects of traditional DG on the superior energy grid, but also plays a certain supporting role in the energy grid at the access point of the distributed energy grid.

Figure 1.5 is a schematic diagram of the distributed energy grid structure, which includes three energy flow paths and loads of electricity, heat, and gas, and the grid has a radial structure. The micro power sources in the distributed energy grid can be in the form of micro power sources such as PV power generation, micro GTs, and fuel cell (FC). The micro power sources close to thermal users can also provide heat sources for local users, thereby ensuring full utilization of energy. When the load changes, the local micropower sources of the distributed energy grid adjust the power output automatically. The distributed energy grid is also equipped with an energy manager and a flow controller, which can realize the comprehensive control and optimization of the entire distributed energy grid. The distributed energy grid and the upper energy grid have common connection points for electricity, heat, and gas, and it does not output energy to the upper energy grid. Therefore, from the perspective of the system, the distributed energy grid can also be regarded as a single controllable load. The distributed energy grid in Fig. 1.5 has three types of loads with different requirements for energy supply quality, namely sensitive loads, adjustable loads, and interruptible loads. When the power supply quality of the upper energy grid has problems such as voltage drop, the static switch will trip, and the energy path in the distributed energy grid will form an independently operated system until the upper energy grid returns to normal.

From the perspective of the modernization of the U.S. energy grid, improving the reliability of power supply for important loads, meeting the various power quality requirements customized by users, reducing costs, and realizing intelligence will be the development focus of the U.S. distributed energy grid. The use and control of power electronic devices and many new energy sources in the distributed energy grid provide new ideas for the full development of the potential of renewable energy and the solution of problems such as stability and control. At the same time, the coordinated use of thermal energy and natural gas energy has gradually attracted attention. By organically combining the production, transmission, and use of electricity, heat, and gas energy, the distributed energy grid can more efficiently meet the comprehensive energy needs of users.

The energy system of the future will be an energy supply system that is closely integrated with centralized and distributed energy. Its main framework structure is a unified body that combines centralized energy with long-distance power transmission backbone grids, regional transmission and distribution lines, and micro-energy systems with DEGs as the core. On this basis, cogeneration, gas-electricity coupling,

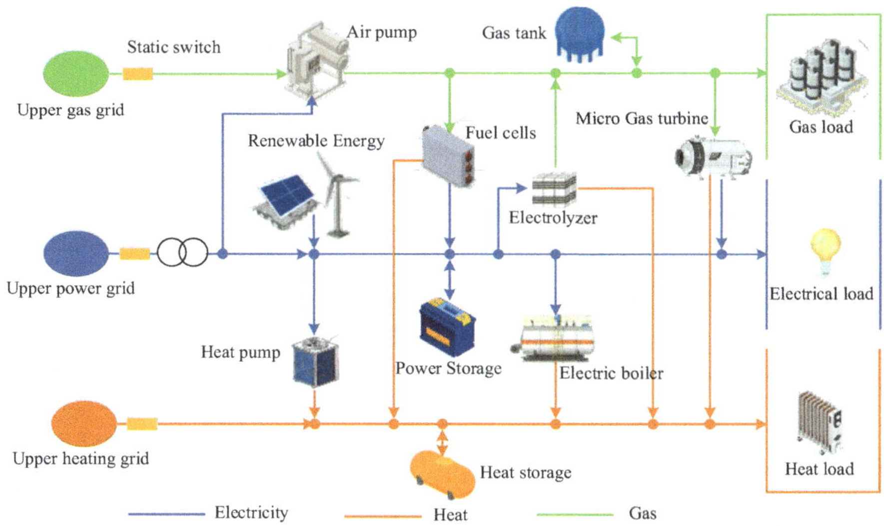

Fig. 1.5 Distributed energy grid structure diagram

and heat and gas storage technologies will be integrated into the energy system to further optimize energy flow and distribution. Through the coordinated optimization of multiple energy flows such as electricity, heat, and gas, it can save investment, reduce energy consumption, and improve energy efficiency, while greatly improving the reliability, flexibility, and energy supply quality of the energy system. Multi-energy synergy and complementarity will become an important development direction of the energy industry in the twenty-first century, providing all-round technical support for achieving green, intelligent, and low-carbon energy goals.

The U.S. Department of Energy also co-funded the second "General Electric Company (GE) Global Research" program with General Electric, with an investment of approximately USD 4 million. GE's goal is to develop an EMS of the distributed energy grid that can provide a unified control, protection and energy management platform for devices in the distributed energy grid. The design of DEM aims to meet various user needs by optimizing the coordinated control of interconnected components in the distributed energy grid, such as the highest operating efficiency and the lowest operating cost. This distributed energy grid program is implemented in two phases. The first phase mainly studies some basic control technologies and energy management technologies and explores the market prospects of the program. The second phase is mainly to simulate the technology of the first phase under a specific model and build a demonstration project for specific implementation. This distributed energy grid program is a good supplement to other distributed energy grid research in this field. CERTS distributed energy grid research mainly focuses on the design and robust control of DERs, while GE distributed energy grid focuses more on the research and development of external monitoring loops and the optimization of energy utilization and operating costs.

The grid-connected distributed energy grid meets the energy needs of Co-Op City, the largest residential complex in the United States, and ensures the energy supply security of the system in extreme weather conditions. The core equipment of the project is the GT, steam turbine and control system produced by Siemens that can achieve CCHP. The total installed capacity of the energy station reaches 40 MW, which can meet the peak power load demand of 24 MW for all 60,000 residents. The remaining 16 MW of electricity is sold to the upper energy grid. In October 2012, when Hurricane Sandy swept across the east coast of the United States and caused large-scale power outages, the distributed energy grid of Co-Op City continued to supply energy, and 60,000 residents were not affected. In addition to Co-Op City, New York University and Princeton University, which are located in the hurricane landing area, are also equipped with DEGs based on natural gas distributed energy stations. The two universities are disconnected from the upper energy grid and switched to "island mode", ensuring the energy supply of the campus during the power outage of the municipal energy grid. This fully demonstrates the stability of the distributed energy grid system.

In addition to the civilian field, the U.S. distributed energy grid demonstration project has also expanded to the military field. The U.S. Department of Defense, in cooperation with the Department of Energy and the Department of Homeland Security, has invested a total of $38.5 million since 2011 to carry out the "Spider" demonstration project—the smart power infrastructure demonstration for energy, reliability, and security (SPIDERS), and built three distributed energy grid demonstration projects at three U.S. military bases (Pearl Harbor-West Ken Joint Base, Fort Carson Base and Smith Base) to support the key loads of the bases. The scale and complexity of the DEGs of these three bases are increasing. The goal is to summarize the distributed energy grid standards and templates suitable for military base applications through demonstration projects.

1.3.2 Application Cases in European

Europe has proposed to make full use of distributed energy, intelligent technology, advanced power electronics technology, etc. to achieve an efficient and close combination of centralized power supply and distributed power generation, and actively encourage all sectors of society to participate in the energy market and jointly promote the development of the energy system. DEGs, with their intelligence and diversified energy utilization, have also become an important part of Europe's future energy system. They can not only optimize power flow through smart energy grid technology, but also combine thermal energy and gas energy (such as natural gas and hydrogen energy) to achieve multi-energy complementarity and efficient coordination of electricity, heat, and gas. At present, Europe has initially formed theories on the operation, control, protection, safety, and communication of DEGs, and verified these theories on laboratory platforms. A number of distributed energy grid demonstration projects have been built one after another, such as the Greek NTUA distributed energy grid, the German Demotec distributed energy grid, the French ARMINES distributed energy grid, the Spanish Labein distributed energy grid and other laboratory energy grids.

The European Energy Matching City project uses multi-agent technology to achieve coordinated interconnection of energy supply and thermal power demand, and coordination of DEGs and demand response (DR) intelligently. The British company FENIX project integrates distributed power sources into virtual large-scale power plants for hierarchical management, optimizing the flexible operation of distributed power sources in the energy system. SIEMENS, Honeywell and other companies have also proposed hierarchical control energy efficiency management systems. Their subsequent tasks will focus on studying more advanced control strategies, formulating corresponding standards, and establishing demonstration projects, etc., to actively prepare for large-scale access to distributed power sources and renewable energy and the initial transition of traditional energy systems to smart energy systems. In response to the problem of coordinated control of multiple energy sources, the Swiss Federal Institute of Technology in Zurich proposed the concept of Energy Hub, which realizes comprehensive optimization control of multiple energy sources through comprehensive management. The EU-funded smart grid integrated research program ELECTRA (Li et al. 2017) proposed the concept of Web of Cell (WoC) system for future (2030+) energy systems with high penetration of renewable energy. ELECTRA conducted in-depth research on its observability and security, and discussed the provision of auxiliary services by DEGs under this system (Li et al. 2017).

The Isle of Eigg in Scotland, UK, is a model for the successful application of off-grid DEGs on islands. The distributed energy grid, which is adapted to local conditions, makes full use of local natural resources. The power generation system is mainly composed of distributed PV, small WP and hydropower facilities, with a total installed capacity of 184 kW. The distributed energy grid also includes two 70-kW diesel generators for emergency use. Although the installed capacity of the entire system is not large, it is enough to meet the energy needs of nearly 100 residents, which can be called a "small and beautiful" distributed energy grid.

In the distributed energy grid, various energy sources operate in coordination in different seasons and at different times, and multi-energy complementarity has become the optimal configuration of the energy system of Isle of Eigg. Thanks to its high latitude, Isle of Eigg can enjoy a long period of sunshine in the summer. In addition, there is less rain in the summer, and the utilization rate of the PV system has also increased. Affected by the weather, the output of WP and hydropower in the summer is not ideal, and the residents' electricity consumption throughout the day comes from PVs and ES batteries. In winter, the island has more rainfall, and three small hydroelectric generators have become the main source of electricity. The island's heating is mainly through a combination of electricity and biomass energy, with a high supply stability. The control system of the Isle of Eigg distributed energy grid can monitor the operation of power generation facilities, optimize the battery charge and discharge cycle, and automatically start the diesel generator when there is an energy shortage. The effective integration of wind, light, water, and storage has freed residents from the limitations of fossil energy. The experience of the Isle of Eigg has also proved that off-grid island DEGs can also meet the energy needs of modern life.

Another relatively successful city is Espoo in Finland, which has achieved regional energy optimization through distributed energy grid clusters efficiently integrated and utilized solar energy, waste heat, geothermal energy, and ES, improving regional energy management efficiency. Rapid electrification and the development of DEGs will accelerate emissions reduction in regional heating and transportation. To achieve Espoo's 2029 carbon–neutral clean heating plan, the city of Espoo and Fortum developed an implementation path in 2014. In the short term, based on traditional energy supply, old thermal power units will be eliminated, large HPs will be developed, and solar panels will be invested; in the medium term, fossil energy thermal power units will be eliminated, and decentralized heat sources such as data centers will be fully utilized; in the long term, various decentralized heat sources will be fully utilized, geothermal technology will be developed, etc., to achieve carbon–neutral regional heating.

1.3.3 Application Cases in China

In recent years, the Chinese DEGs have made great progress. Its characteristics are that it adapts to the needs and direction of China's energy development and has a clear development route. State Grid Corporation of China is the main institution for distributed energy grid technology research.

In March 2016, the National Energy Administration of China issued the "Energy Technology Revolution Innovation Action Plan (2016–2030)", which supports the use and innovation of various new energy sources such as high-efficiency photovoltaic (PV), WP, biomass, geothermal, and hydrogen energy. In November 2017, the National Energy Administration issued the "Notice on the Pilot Program of DG Market Trading", which improved and perfected the public service and management system of distributed energy trading and organized the pilot program of DG market trading. This series of measures provides a strong policy guarantee for the development of DEGs. At the same time, the "dual carbon" goals (carbon peak and carbon neutrality) have become the core driving force for the development of distributed energy grid technology. Through DEGs, efficient consumption of new energy and low-carbon operation in local areas can be achieved. With the advancement of national industry support policies, the advancement of core technologies and the expansion of downstream application fields, China's distributed energy grid industry has huge market potential in the future (Lei et al. 2020).

(1) Distributed Energy Grid in Changdao County, Shandong Province

Changdao County, the only island county in Shandong Province, consists of 32 islands. With its special geographical location, the construction and operation of its energy system face problems such as high cost, difficult construction, and easy damage. Previously, Changdao was mainly connected to the mainland and the main islands of Changdao by submarine cables. However, once the submarine cable fails, emergency repairs are very difficult and it is difficult to complete the repairs and restore power supply in a short time. Changdao has abundant wind resources, but WP generation is intermittent and random with wind changes. When the submarine cable fails and the distributed energy grid cluster is in an isolated grid operation, it is difficult for the wind turbine to control the output and it is impossible to achieve a balance between power generation and power supply. In August 2015, the distributed power generation and distributed energy grid access control project in Changdao County, Yantai City, Shandong Province passed the acceptance of the National Development and Reform Commission and was awarded the title of "National Major Internet of Things (IoT) Application Demonstration Project, officially completed and put into operation." This is the first island energy grid project in northern China. It can achieve isolated grid operation in the event of the collapse of the external energy system, thereby ensuring continuous power supply to important users and greatly improving the energy supply capacity and reliability of the Changdao energy system.

The Changdao County Intelligent Distributed Energy Grid Demonstration Project in Shandong Province started construction in the first half of 2019. Under the unified deployment and leadership of the State Grid Shandong Electric Power Company, all participating units assisted fully in overcoming a series of difficulties brought by the remote island to the allocation of manpower and material resources for the project construction, and completed the civil construction, single equipment installation and grid-connecting at the end of 2019. The single-unit commissioning was completed in May 2020; the system joint commissioning and grid-connecting trial operation were completed in June. Based on the commissioning experience of the first phase of the Changdao Distributed Energy Grid Clusters Project, the project team of the institute minimized the impact of the commissioning process on the normal power supply of the island, carefully formulated the project commissioning plan and implementation plan, evaluated the risk points one by one, and ensured that the risks were fully controllable and re-controllable, and ensured the success rate of the commissioning of the flexible direct current single-end bipolar pairing and single-end on-site networking pairing. At the same time, electric boiler (EB) and HP systems are configured inside the distributed energy grid clusters to use wind and solar surplus power for heating and hot water supply; small biomass treatment equipment is introduced to produce biogas from domestic garbage and agricultural waste to provide clean gas for some residents. The distributed energy grid clusters use an intelligent EMS to efficiently coordinate the dispatch of electricity, heat and gas. The system dynamically allocates different energy flows such as wind and solar power generation, ES, natural gas power generation and waste heat recovery according to real-time demand. By 2024, renewable energy generation in Changdao County will account for more than 70% of the country's energy consumption, and the electricity price will drop from more than 3 ¥ per kilowatt-hour to less than 1 ¥.

Changdao County's distributed energy grid clusters make full use of the abundant renewable energy resources on the island to build a green, efficient, and reliable energy system. Its successful practice not only solves the energy dilemma in the island area, but also provides valuable experience for energy development in other similar areas.

(2) Jiaxing City Energy Internet, Zhejiang Province

In 2019, the comprehensive pilot demonstration project of Jiaxing City Energy Internet in Zhejiang Province passed the acceptance of Zhejiang Provincial Energy Bureau. This is the first city-level energy Internet built in China. The core demonstration area has not only achieved 100% access and absorption of renewable energy, but also achieved extensive open interconnection of clean energy, efficient energy grid, low-carbon buildings, smart energy use, and green transportation, as well as green sharing between the supply side and the consumption side. It is one of the typical application examples of a distributed energy grid cluster.

The project includes sub-projects such as a flexible interconnection converter station project, new energy grid-connected equipment improvement project and a source-grid-load-storage coordinated control system construction project. Among them, the flexible interconnection converter station project uses flexible interconnection technology to transfer the distributed electric energy that cannot be fully absorbed by the 20 kV line to the 10 kV line for absorption; the new energy grid-connected equipment improvement project monitors the power quality of all parties by building a full-scenario power monitoring system. Using the DC power provided by the flexible converter station to build an AC/DC low-voltage hybrid distribution grid, supporting the construction of distributed power sources, ES stations and new power vehicle charging piles, to create a micro-regional energy Internet with interactive coordination of multiple energy sources such as electricity, heat and gas.

The source-load-storage coordinated control system is the "brain" of the entire project. During the peak hours of energy demand every day, the ES system switches from the ES state to the energy supply state, the intelligent building adjusts the flexible load, and the trigeneration facilities increase the power supply and cooling capacity of the building. The flow and distribution of energy provide more favorable guarantees for the city's energy supply.

(3) Distributed Energy Grid Cluster in the Winter Olympics Competition Area of Zhangjiakou City, Hebei Province

Zhangjiakou, Hebei Province is known as the "hometown of wind and ocean of light". It is one of the regions with the richest wind and solar energy resources in North China. The region has more than 40 million kW of exploitable wind energy resources and more than 30 million kW of exploitable solar energy. The unique natural resources have laid a solid foundation for the development and utilization of renewable energy. The distributed energy grid in the Winter Olympics competition area of Zhangjiakou, Hebei Province is one of the important demonstration projects for China's green energy development. Its construction goal is to provide a fully green and zero-carbon energy supply solution for the 2022 Winter Olympics. Relying on Zhangjiakou's rich renewable energy resources, the energy grid mainly uses wind and solar energy to achieve efficient complementarity and comprehensive utilization of multiple energy forms such as electricity, heat, and gas.

The core technologies of the distributed energy grid cluster include distributed energy systems, ES technology, and intelligent energy management platforms. By connecting multiple wind farms, PV power stations, and centralized ES facilities, the distributed energy grid cluster realizes the on-site production, storage, and dispatch of clean energy, ensuring the stable energy supply of the Winter Olympic venues around the clock. The venues are also equipped with a ground source HP system to use the geothermal resources around the venues for heating and cooling. At the same time, hydrogen energy technology has also been widely used. Zhangjiakou has built multiple hydrogen preparation stations and hydrogen refueling stations to provide power for hydrogen fuel vehicles during the Winter Olympics. During the Winter Olympics, more than 600 transportation and support vehicles in the Zhangjiakou competition area all used zero-pollution, zero-emission hydrogen FC vehicles. These hydrogen energy vehicles are equipped with extremely cold-resistant power batteries, which can achieve extremely cold low-temperature startup at -30 °C and extremely cold low-temperature storage and shutdown automatic protection at -40 °C. Zhangjiakou's abundant renewable energy provides sufficient energy guarantee for hydrogen production by electrolysis of water, and is also conducive to creating a scalable and replicable innovative development model of synergistic and complementary development of zero-carbon hydrogen production and hydrogen energy industry.

At the same time, the distributed energy grid cluster achieves precise energy control through an intelligent EMS. The system can dynamically adjust the energy flow between wind energy, solar energy, and ES batteries according to the real-time energy demand of the venue and weather conditions, maximizing energy utilization efficiency. During the event, this system not only ensured uninterrupted energy supply for the Winter Olympics, but also achieved coordinated optimization of electricity, heat, and hydrogen energy, providing safe, reliable, and low-carbon energy services for the event.

In addition, China's typical distributed energy grid cluster projects include the Sino-Singapore Tianjin Eco-City Smart Energy Town. Multi-energy complementarity, multi-dimensional perception, multi-party participation, and diversified services are the core of this distributed energy grid cluster. The distributed energy grid cluster has a complete range of user types and rich application scenarios, covering multiple energy sources such as electricity, cooling, and heating, multiple links of source, grid, load, and storage, and multi-scenario smart energy applications and services in production and life. The regional power supply reliability can reach 99.999%; the Hainan Prefecture Distributed Energy Grid Cluster in Qinghai Province promotes the construction of a comprehensive energy supply system with multi-energy complementarity of "wind, light, water, heat and storage", transforming abundant clean energy resources such as water, wind, light, and heat into "green development power", and successfully built a geothermal clean heating project with a heating area of 220,000 m^2, realizing 24-h uninterrupted heating; there are also many industrial distributed energy grid clusters, civilian DEGs, and campus DEGs.

References

Ai Q, Hao R (2018) Key technologies and challenges for multi-energy complementarity and optimization of integrated energy system (in Chinese). Autom Electr Power Syst 42(4):2–10+46. https://doi.org/10.7500/AEPS20170927008

Ban X, Tang Z (2023) Energy interaction in integrated energy microgrid group and benefit analysison multi-energy market construction (in Chinese). Mod Electr Power 1–13. https://doi.org/10.19725/j.cnki.1007-2322.2022.0462

Carrasco JM, Franquelo LG, Bialasiewicz JT, Bialasiewicz E, PortilloGuisado RC, Prats MAM, Leon JI, Moreno-Alfonso N (2006) Power-electronic systems for the grid integration of renewable energy sources: a survey. IEEE Trans Power Electr 53(4):1002–1016. https://doi.org/10.1109/TIE.2006.878356

Chen J, Zhang W, Lin D, Li Z, Song G, Gu Y (2019) Distributed optimal dispatch of integrated electricity-gas-heating system based on improved alternative direction multiplier method (in Chinese). Autom Electr Power Syst 43(07):50–58. https://doi.org/10.7500/AEPS20180720004

Cui X, Yu X, Liu Y, Lu Z (2015) Distributed stream processing: a survey (in Chinese). J Comput Res Dev 52(02):318–332. https://doi.org/10.7544/issn1000-1239.2015.20140268

Huang J, Li Z, Wu QH (2017) Coordinated dispatch of electric power and district heating networks: a decentralized solution using optimality condition decomposition. Appl Energy 206:1508–1522. https://doi.org/10.1016/j.apenergy.2017.09.112

Jiang Z, Hao R, Ai Q (2017) Interaction mechanism of industrial park based on multi-energy complementation (in Chinese). Electr Power Autom Equip 37(6):260–267. https://doi.org/10.16081/j.issn.1006-6047.2017.06.034

Jin Y, Song Q, Liu W (2009) Nonlinear controller for battery energy storage system (in Chinese). Autom Electr Power Syst 33(7):75–80

Lei L et al (2020) Dynamic energy dispatch based on deep reinforcement learning in IoT-driven smart isolated microgrids. IEEE Internet Things J 8(10):7938–7953

Li J, Guo B, Xue Y, Niu M (2017) Summary of domestic micro-grid demonstration project and operation control and energy management of micro-grid (in Chinese). Electr Energy Manag Technol 18:1–7. https://doi.org/10.16628/i.cnki.2095-8188.2017.18.001

Liang X, Li Z, Huang W et al (2020) Relaxed alternating direction method of multipliers for hedging communication packet loss in integrated electrical and heating system. J Mod Power Syst Clean Energy. 8(5):874–883. https://doi.org/10.35833/MPCE.2020.000163

Lu Z, Wang C, Min Y, Zhou S, Lu J, Wang Y (2007) Overview on microgrid research (in Chinese). Autom Electr Power Syst 31(19):100–107

Lu T, Ai Q, Sun S, Chen Y, Zhao Y (2016) Behavioural analysis and optimal operation of active distribution system with multi-microgrids (in Chinese). Proc CSEE 36(1):122–132. https://doi.org/10.13334/j.0258-8013.pcsee.2016.01.013

Wang J, Li X, Qiu X (2005) Power system research on distributed generation penetration (in Chinese). Autom Electr Power Syst 29(24):90–97

Wang W, Wang D, Jia H, Chen Z, Guo B, Zhou H, Fang M (2016) Review of steady-state analysis of typical regional integrated energy system under the background of energy internet (in Chinese). Proc CSEE 36(12):3292–3305. https://doi.org/10.13334/j.0258-8013.pcsee.152858

Wen Y, Qu X, Li W et al (2017) Synergistic operation of electricity and natural gas networks via ADMM. IEEE Trans Smart Grid 9(5):4555–4565. https://doi.org/10.1109/TSG.2017.2663380

Wu X, Wang X, Liu S, Zhu Z, Liu C, Duan J, Hou F (2014) Summary of research on microgrid (in Chinese). Electr Power Autom Equip 34(10):7–14. https://doi.org/10.3969/j.issn.1006-6047.2014.10.002

Xue G, Zhang Y, Zhu D (2013) Operational control strategy of stand-alone DC microgrid (in Chinese). Electr Power Autom Equip 33(3):112–117. https://doi.org/10.3969/j.issn.1006-6047.2013.03.019

Yan M, He Y, Shahidehpour M et al (2018) Coordinated regional-district operation of integrated energy systems for resilience enhancement in natural disasters. IEEE Trans Smart Grid 10(5):4881–4892. https://doi.org/10.1109/TSG.2018.2870358

Yang Z, Le J, Liu K, Xie X (2012) Study on the standard of the grid-connected microgrids (in Chinese). Power Syst Protect Control 40(02):66–71+76. https://doi.org/10.7667/j.issn.1674-3415.2012.02.013

You Y, Liu D, Zhong Q, Yu N (2014) Multi-objective optimal placement of energy storage systems in an active distribution network (in Chinese). Autom Electr Power Syst 38(18):46–52. https://doi.org/10.7500/AEPS2013072209

Chapter 2
Modeling Techniques for the Operation of Distributed Energy Grid Clusters

The operation control and energy management of distributed energy grid clusters rely on advanced information collection and data processing technologies. Enhancing the perception capability of the distributed energy grid and optimizing the configuration of information flow are particularly critical for the optimized regulation of distributed energy grid clusters. This chapter focuses on the information collection and data processing technologies for distributed energy grid clusters. By leveraging advanced technologies such as artificial intelligence and cloud-edge computing, it provides a detailed discussion on the modeling of physical information and the analysis of energy consumption behavior in distributed energy grid clusters. This lays the foundation for situational awareness (SA) of the operation of distributed energy grid clusters, paving the way for operation control and optimization in subsequent chapters.

2.1 Modeling Method for Fusion of Physical Information in Distributed Energy Grid Clusters

2.1.1 Introduction to Cyber-Physical Systems (CPS) for Physical Information Fusion

The CPS, a new type of system that closely integrates intelligent computing with physical resources, was first proposed by the U.S. National Science Foundation (NSF) in 2006. There is no universally agreed definition of CPS, both domestically and internationally. According to the NSF, CPS is defined as the "tight coupling and deep integration between physical resources and information resources" (CPS Steering Group 2008). Professors Rajkumar and Lee from CMU define CPS as "a system where embedded devices with computing and communication cores enable

T. Lu et al., *Operation Modeling and Collaborative Regulation of Distributed Energy Grid Clusters*, Smart Energy Systems, https://doi.org/10.1007/978-981-96-9942-1_2

coordinated operation and state monitoring of physical and engineering systems" (Rajkumar et al. 2010). Professor Sastry defines CPS as "a system that integrates computational, communication, and storage capabilities to reliably, efficiently, and securely control physical entities".

The proposal of CPS promotes the deep integration of actual physical systems with information and computing systems, utilizing updated information to complete state sensing for better operational control, thus endowing the entire system with new capabilities, achieving the goals of intelligence and automation. The components of CPS include: physical networks, control systems, large-scale sensor networks, large-scale communication networks, and computing systems. Although both the IoT and CPS emphasize the interconnection of physical entities, IoT focuses on the integration of physical components, while CPS aims to break the spatio-temporal limitations of the physical world for overall optimization and real-time control (NSF Cyber-Physical System Program Solicitaion [EB.OB] 2011).

According to NSF Workshop on Cyber-Physical Systems [EB/OB] (2006) and McGrangahan and Goodman (2005) and the definition of CPS, its characteristics can be summarized as follows:

1. Sensors are embedded in each physical entity to collect information and sense the system's state.
2. A network covering the entire system enables the flow and sharing of information within the system.
3. The communication network can transmit useful information according to the characteristics and needs of the physical entities, enabling information filtering and saving communication bandwidth.
4. CPS integrates embedded devices into physical entities and uses large-scale distributed computing, parallel computing techniques, and MASs to process or preprocess large amounts of information.
5. Physical and information systems collaborate in a coordinated manner across the entire network, with the controllers embedded in the physical entities determining the overall system's functions and operational characteristics.
6. CPS has self-organizing and self-adaptive capabilities. The self-organizing ability allows it to automatically detect and search for DERs in a given area, instantly obtain their information, and control them. The self-adaptive ability enables quick responses to failures in the physical and information systems within the area, ensuring safe operation of the system.
7. CPS supports system modularization and dynamic reorganization of system modules, with software capable of online updates.

Traditional analysis methods for distributed energy grid clusters focus on energy flow. However, with the widespread deployment of sensing devices within energy systems, the selection, aggregation, and optimization of control are highly dependent on the information communication network. Therefore, research on information flow becomes particularly important. Moreover, the characteristics of physical systems provide a basis for the extraction, analysis, and transmission capacity of information, while information systems offer diverse data that optimize the scheduling and control of physical systems. Hence, physical information systems, as an integrated whole within a power system, should not be treated in isolation. The fusion of physical and informational systems in multi-energy frameworks organically integrates both domains, offering a new approach toward achieving the intelligence of distributed energy grid clusters.

CPS, as a deep integration of physical and information systems, enables real-time acquisition of operational states of resources and devices at various nodes within a distributed energy system. It facilitates hierarchical control optimization and energy dispatch, achieving globally coordinated optimization and control, thereby providing good controllability and observability for the distributed energy grid clusters. This improves the reliability of the clusters' operation, yielding better performance under the same equipment configuration, reducing energy consumption and emissions, enhancing the efficiency of renewable energy generation, and ultimately realizing the intelligence of the distributed energy grid clusters (Liu et al. 2008).

2.1.2 Modeling Method for the Fusion of Physical Information in Distributed Energy Grid Clusters

The fusion of physical information systems in distributed energy grid clusters, which involves the high integration of primary and secondary functions of the distributed energy grid, aims to organically combine physical models, communication, and computational within the CPS model. To reduce the heterogeneity between the physical system and the information system, the CPS framework should adhere to the following requirements:

1. Due to the scale of the distributed energy grid clusters and the complexity of massive information collection and processing, the CPS should be modeled in a modular approach first, followed by overall system modeling.
2. A mapping relationship must be established to link the physical and information systems.
3. The interfaces between the information and physical components can be defined and implemented, with unified rules for information signal transmission.
4. Since the communication system is discrete and the physical processes are continuous differential equations, the CPS model must be a hybrid system, combining both discrete and continuous elements.

As an important application of CPS, the cyber-physical energy system (CPES) has attracted significant attention both domestically and internationally. This research field includes areas such as automatic generation control (AGC), energy management systems (EMS), supervisory control and data acquisition (SCADA) for distribution grid scheduling, geographic information systems (GIS), temperature measurement systems, gas pressure measurement systems and relay protection information management systems. Data obtained by sensors, connected through communication networks, is aggregated from various distributed data acquisition and control

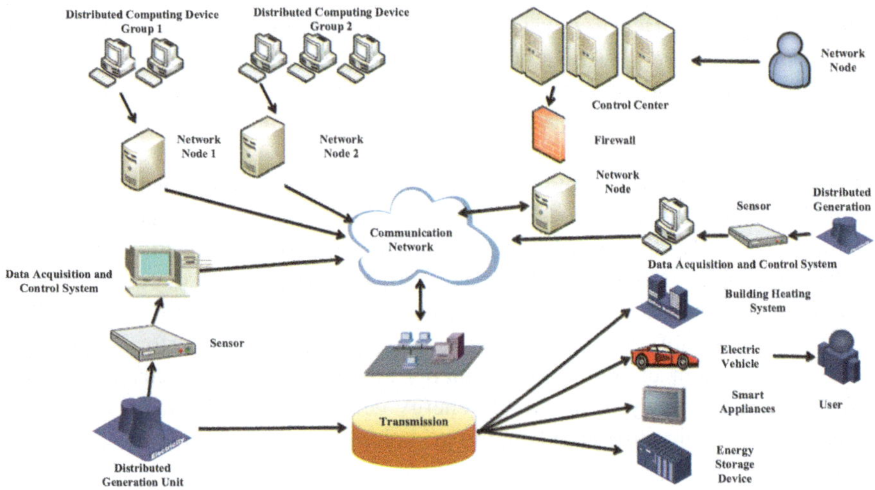

Fig. 2.1 CPES architecture diagram

elements. After deriving global optimization strategies, these strategies are ultimately applied to distributed controllers.

The typical structure of a distributed energy information system is shown in Fig. 2.1 (Revoda et al. 2011).

After the installation of advanced measurement systems and monitoring devices in traditional distributed energy grid clusters, their high reliability, high-speed communication systems, and advanced EMSs significantly enhance the network's ability to perform computational analysis, big data processing, and coordinated operation of distributed intelligence with global optimization. In complex operational scenarios with a large number of decision variables, CPES often achieves better optimization solutions. By utilizing big data analytics, CPES can improve the accuracy of grid state perception and forecasting. As a result, the reliability analysis of distributed energy grid clusters, can more accurately and flexibly monitor abnormal states, conduct safety checks, and compute the security margin for the safe operation of distributed energy grid clusters.

It is evident that, compared to traditional methods, CPES has the following characteristics:

1. A large number of distributed storage and computing devices.
2. A combination of wired and wireless networks.
3. Diverse intelligent user terminals.
4. A large-scale network address requirement.
5. Energy-consuming devices can be directly controlled by the control center when necessary.

2.1.2.1 The Method of Physical Information Fusion

The physical model of a distributed energy grid cluster consists of physical modules and their interactions. The Common information model (CIM) adopts an object-oriented modeling approach, abstracting physical entities to form a set of attributes for entity classes and the relationships between entities, allowing different physical components to map their characteristics onto the CIM platform. On the CIM platform, the attributes of various physical systems and the interrelationships between components are clearly defined. The collection, flow, and acquisition of information from different physical classes are reflected in the model, making information interaction more transparent. Physical quantities of the system no longer reside internally; instead, they are identified through various remote, indirect physical information and are transmitted to the distributed computing center via the information network after being collected by a wide area measurement system (WAMS).

The Electric Power Research Institute (EPRI) in the United States was the first to propose the control center application programming interface (CCAPI) for the CIM model. This interface provides a universal EMS that enables the operation of different hardware and software systems to interact and share data on the same platform.

CIM defines the entities of power control and computer application systems, such as EMS, in terms of class inheritance, association, aggregation, dependency, and the attributes of each class. The objects include distributed energy devices, distributed energy load models, SCADA systems, distributed energy lines, measurement devices, and others. The CIM package describes a specific system, with each package consisting of classes and the relationships between these classes. Each class describes the informational attributes of a particular component in the system.

Based on CIM, the design of a global coordination optimization algorithm involves clearly defining the amount of information required by the algorithm, collecting and transmitting the relevant data in the corresponding physical systems, and then using or filtering the data at distributed computing devices before uploading it to the upper-level scheduling control center. This is a general method for modeling the physical information system in distributed energy grid clusters. It requires the study of physical-cyber fusion modeling methods, where the physical model and the information model are integrated.

The C model is derived based on the analysis of the P model, where the physical processes are segmented to obtain accessible variables and determine their relational structure. The P model abstracts the dynamic processes of a physical system in the time domain, and the physical quantities collected in the P model should partially or entirely correspond to those in the C model. Likewise, the quantities in the C model should have corresponding attributes in the P model. However, some intermediate variables obtained through computational processing in the C model may not have direct physical counterparts in the P model. Additionally, the P model may contain physical quantities that do not need to be considered when developing the information model algorithm.

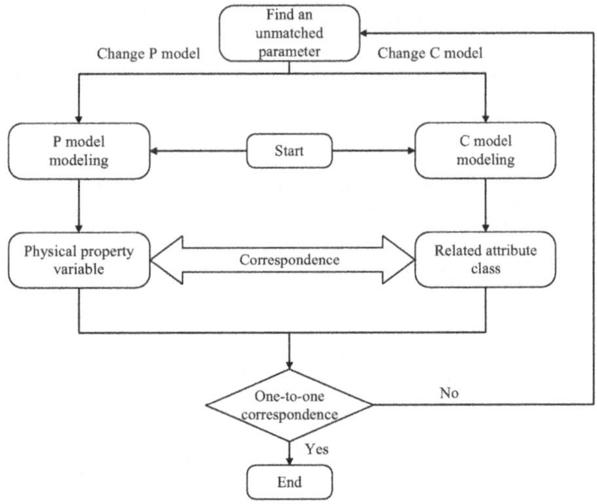

Fig. 2.2 Modelling process of CP fusion

In conclusion, achieving a one-to-one correspondence between the P and C models during the initial modeling phase is challenging, which hinders the global collaborative control of the physical system by the information system. The primary goal of integrating the two models is to establish a one-to-one correspondence between the variables of the P and C models, abstract the information system, and map it to the actual physical system, creating a tight correspondence between them. The CP fusion modeling process is shown in Fig. 2.2.

After completing the modeling of the P model and the C model, if it is found that the physical quantities in the P model and the corresponding attributes in the C model do not correspond one-to-one, the models should be modified accordingly. If there are physical quantities in the P model that do not have corresponding attributes in the C model, the P model should be refined to represent the relevant class attributes, or the C model should be simplified by developing a new algorithm that excludes the attributes not corresponding to the P model. Conversely, if there are attributes in the C model that do not correspond to those in the P model, the C model should be refined or the P model simplified, ultimately achieving a full one-to-one mapping between the C and P models.

2.1.2.2 Fusion Method for a Single Cyber-Physical Module

The CPES is an extremely complex system, and it is unrealistic to achieve system modeling all at once. The concept of CPS module division has already been discussed above. Modeling should first start with individual modules. Based on their dynamic characteristics, each module is divided into multiple components, and a corresponding information layer for the module is established. The output of each process serves as the input for the next process or as the input for the entire module, with

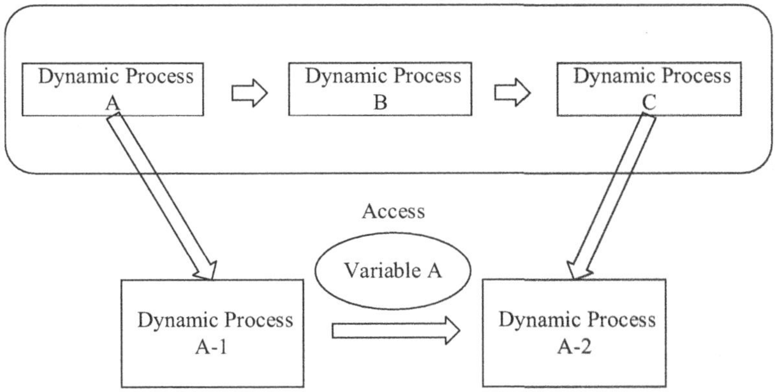

Fig. 2.3 Component segmentation strategy for internal quantity utilization in single CP fusion modelling method

dynamic characteristics connected sequentially. The purpose of this approach is to transform the input and output quantities of the components into extractable information quantities that can be accessed by the information system.

With the development of optimization control algorithms, many internal braking quantities are now reflected in these algorithms and participate in decision-making calculations. This requires the decomposition of the internal dynamic processes of components, transforming internal quantities into input–output variables after the component decomposition, allowing the information system to access and utilize them. Figure 2.3 illustrates the general single CP fusion modeling method and the component segmentation strategy when internal quantities need to be utilized.

2.1.2.3 Fusion Method for Physical Information Systems

To study the fusion method of physical information systems, it is essential to clarify which parts of the physical and information systems are integrated and at what level. In traditional systems, sensors collect information from physical components and transmit it to the control center through optical cables or power line communication. The scheduling center then controls the system centrally.

However, with the widespread use of distributed energy devices and diverse intelligent components, a large amount of information is constrained by transmission and computation limitations. This requires a substantial number of embedded devices to perform tasks such as information collection, communication, and control. These embedded devices, along with their associated wired and wireless communication devices, belong to both the physical system and the information system. Therefore, the fusion of physical and information systems lies in the mapping of physical parameters, information quantities, and communication quantities (Fig. 2.4).

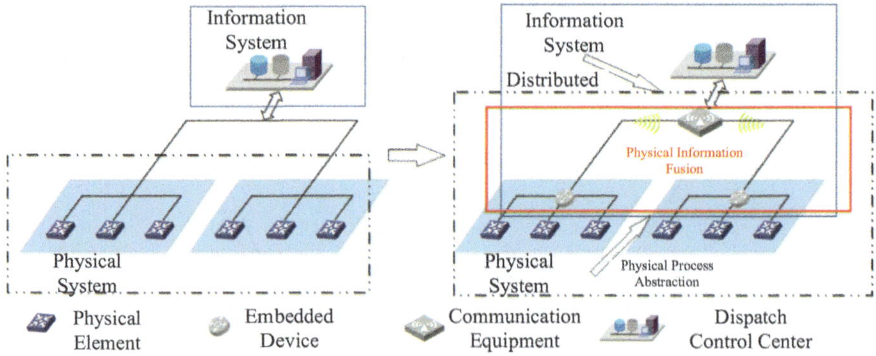

Fig. 2.4 Comparison of traditional system and CPS system architecture

Integrating the already established CP modules involves first abstracting all physical systems as logical nodes and organizing all information systems into functional blocks. Horizontally, each logical node corresponds to a functional block. Vertically, the logical nodes are modeled collectively, followed by the corresponding integration of the functional blocks to achieve the integration of information and physical systems. Finally, the two corresponding integrated modules are merged.

The specific steps are shown in Fig. 2.5.

The reason for integrating C and P models first before fusion, rather than separately integrating them and then performing the final integration, is to optimize overall information transmission and control. By directly accessing specific information from a unified database, the system makes both the information and its corresponding physical structure clearer. Similarly, this approach also supports the fusion of physical information within the modules. The system and modules remain relatively independent, with local control of the modules reducing global computation requirements and preventing conflicts between multiple controllers controlling the same components.

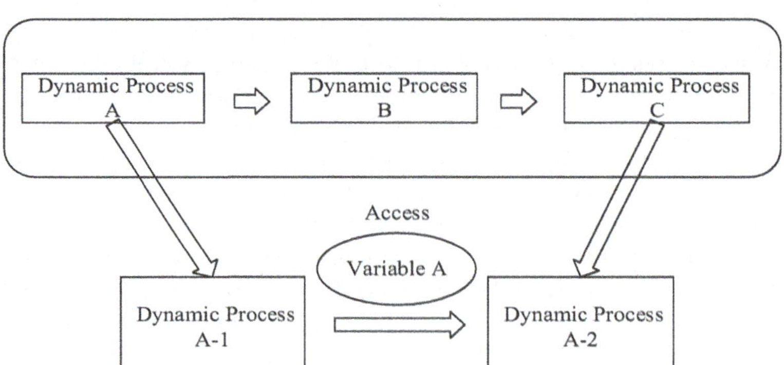

Fig. 2.5 Integration flowchart of multiple CP modules

2.2 Collaborative Modeling Method for Electrical, Thermal, and Gas Systems in Distributed Energy Grid Clusters

2.2.1 *Theory of Electrical, Thermal and Gas Synergy*

The model of distributed energy grid clusters established in this chapter is shown in Fig. 2.6. The distributed energy grid clusters can achieve efficient integration of multiple energy sources through the coupling of electricity, heat, and gas, while also improving the efficiency of these energy forms over the entire operation cycle. Overall, the system can be divided into three parts: energy input, energy conversion and coupling, and energy demand.

From the perspective of energy input, the distributed energy grid clusters include conventional and renewable energy sources, such as WP and PV, along with external power grids and gas grids. In terms of energy conversion and coupling, the system incorporates equipment like electric furnaces (EF), GBs, P2G, and combined CHP units. The energy produced by these devices is used to meet the electricity and heat demand on the user side. Additionally, the electricity and thermal load regulation capabilities of electric energy storage (EES) and thermal energy storage (TES) are also crucial, as they can influence the overall economic operation of the distributed energy grid clusters.

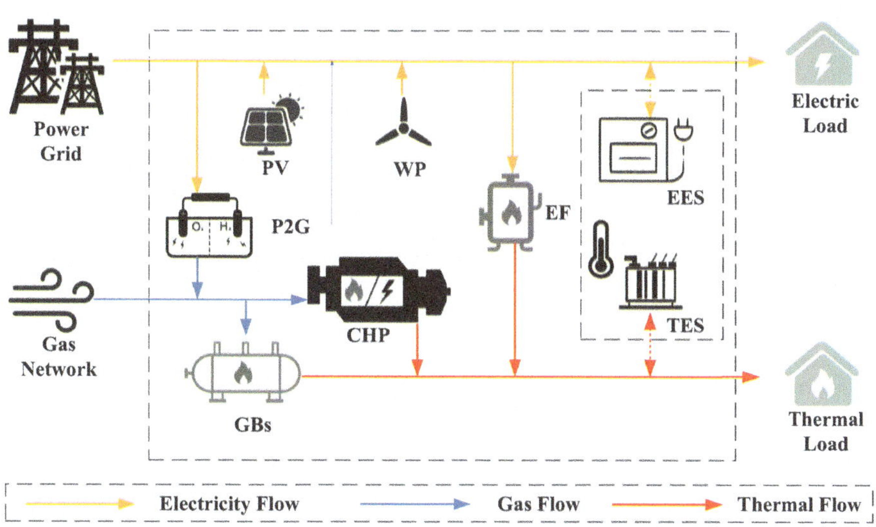

Fig. 2.6 Model of distributed energy grid clusters

2.2.1.1 CHP Unit

The CHP unit is an important component of distributed energy grid clusters. On the one hand, coal-fired power generation can provide electricity to the power grid, and on the other hand, it can utilize the waste heat from steam to provide thermal energy to the district heating network. There are two types of CHP units: back-pressure and extraction-type. Back-pressure CHP units have high energy efficiency, but their controllability is poor, and their capacity is small. The thermal energy output of extraction-type CHP units mainly comes from extracting intermediate steam into the extraction pipe via a steam turbine. Therefore, the output of the CHP unit can be controlled by adjusting the extraction power of the extraction pipe. Extraction-type CHP units offer better controllability within a specified range.

In summary, extraction-type CHP units are more feasible, and their operational characteristics are shown in Fig. 2.7.

From Fig. 2.7, it can be seen that the areas ABCD represent the adjustable ranges of thermal power and electrical power. The maximum and minimum output powers of the CHP unit are represented by AB and DC, respectively. The AD segment indicates the system's operation in pure power generation mode, where only electrical power is output, and no thermal power is produced. Therefore, points A and D represent the upper and lower limits of electrical power output in this mode. When the CHP unit operates in the BC segment, unlike in the AD segment, the unit primarily outputs thermal power. Point B represents the upper limit of thermal power output. When the CHP unit is operating in the middle range of the BC segment, it is capable of adjusting the output of electrical power. From the figure, it is clear that EF represents the adjustable range of electrical power output for the CHP unit.

The electrical and thermal outputs of the extraction-type CHP unit are interdependent, and its characteristic equation is given by (2.1).

$$\left\{P_{CHP}^{\min} - Z_{v1}H_{CHP,t}, X_m\left(H_{CHP,t} - H_{CHP}^{\min}\right)\right\} \leq P_{CHP,t} \leq P_{CHP}^{\max} - Z_{v2}H_{CHP,t} \quad (2.1)$$

where, $P_{CHP}^{\min}$ and $P_{CHP}^{\max}$ represent the minimum and maximum electrical output power of the CHP unit, in MW; Z_{v1} and Z_{v2} represent the corresponding electrical-thermal

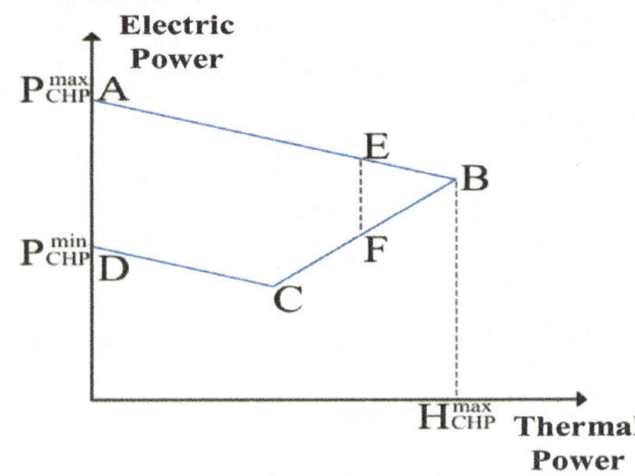

Fig. 2.7 Operating characteristics of extraction-type CHP unit

conversion coefficients at the minimum and maximum output of the CHP unit; $H_{CHP,t}$ represents the thermal power output of the CHP unit during the time period t, in MW; X_m represents the electrical-thermal conversion coefficient of the CHP unit; $H_{CHP}^{\min}$ represents the minimum thermal power output of the CHP unit, in MW; and $P_{CHP,t}$ represents the electrical power output of the CHP unit during the time period t, in MW.

2.2.1.2 GBs

The GBs is a key piece of equipment that connects the gas grid to the thermal grid. The combustion chamber is the core component of the GBs, responsible for burning natural gas or liquefied petroleum gas (LPG) to produce high-temperature flue gases. The heat exchanger facilitates heat transfer between the flue gases and water, heating the water and producing hot water. The control system is responsible for the automatic control of the entire system, including functions such as ignition, shutdown, and temperature regulation. The working principle of the GBs is simple: it uses the high-temperature flue gases generated by combustion to heat water, bringing it to the set temperature. During this process, the heat in the flue gases is absorbed by the water, and the heated water is then transmitted to the thermal load through the heating network.

The GBs model is shown in (2.2).

$$H_{GL,t} = \eta_{GL} E_{GL,t} \tag{2.2}$$

where, $H_{GL,t}$ represents the thermal power generated by the GBs during time period t, in MW; η_{GL} is the conversion coefficient of the GB, in MW/m^3; and $E_{GL,t}$ is the gas consumption of the GBs during time period t, in m^3.

2.2.1.3 EF

The EF is a device that heats water or other thermal media using electrical energy, primarily for heating, hot water supply, and meeting thermal energy demands in certain industrial processes. Unlike GBs and coal-fired boilers, the operating principle of an EF involves using electrical energy to directly heat water or other media through electric heating elements (such as electric heating tubes or heating wires), without the need for combustion of any fossil fuels.

The EF model is shown in (2.3).

$$H_{EF,t} = \eta_{EF} E_{EF,t} \tag{2.3}$$

where, $H_{EF,t}$ represents the electric power generated by the EF during time period t, in MW; η_{EF} is the conversion coefficient of the EF; and $E_{EF,t}$ is the electric consumption of the EF during time period t, in MW.

2.2.1.4 P2G Technology

P2G technology enables the large-scale storage of electrical energy while also facilitating the consumption of surplus renewable energy. It can provide natural gas for the gas grid, further enhancing the coupling between the gas and electricity grids, and expanding the application prospects for energy collaborative planning between the power and natural gas grids.

The specific operation process of P2G technology is shown in Fig. 2.8. In the P2G technology pathway, the first step is the power-to-hydrogen process, typically achieved through water electrolysis. The electrolysis of water converts electricity into hydrogen gas, which is then transported to hydrogen application terminals or integrated into the power grid through FCs. The chemical reaction for water electrolysis is shown in (2.4).

In the second step of the P2G technology pathway, hydrogen undergoes a methanation reaction. The hydrogen produced in the first step reacts with CO_2 to undergo a reduction reaction in a controlled environment, producing synthetic methane and water. The chemical reaction for this process is shown in Eq. (2.5).

The overall energy conversion efficiency of the entire P2G process is 50–65%.

$$2H_2O \xrightarrow{\text{Electrolysis}} 2H_2 \uparrow + O_2 \uparrow \tag{2.4}$$

$$CO_2 + 4H_2 \xrightarrow{\text{Catalyst}} CH_4 + 2H_2O \tag{2.5}$$

The electroconversion of hydrogen to methane requires fewer conversion steps and exhibits higher conversion efficiency compared to electroconversion to hydrogen. Although hydrogen can be injected into natural gas pipelines, it is rarely used in practice in most countries. This is due to the inherent physical and chemical properties of hydrogen. Firstly, hydrogen molecules are smaller compared to other gas molecules, which makes them more likely to permeate the infrastructure of natural gas networks,

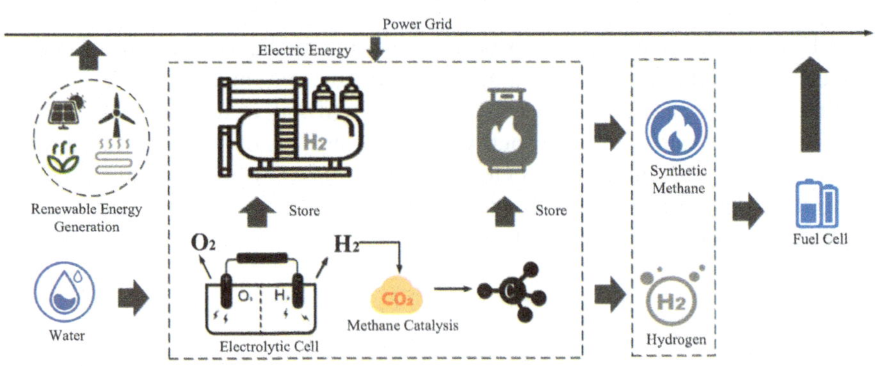

Fig. 2.8 Structure diagram of P2G technology

potentially causing hydrogen embrittlement. This can harm the service life, safety, and strength of pipelines, posing significant risks. Secondly, hydrogen has a lower energy density. To transport and store the same energy as methane, a greater volume of gas storage is needed, increasing the complexity of transportation and storage. Thirdly, the low density of hydrogen makes it prone to leakage, which not only raises costs but, due to its high flammability, could lead to explosions and safety accidents. On the other hand, methane produced by electroconversion (synthetic natural gas) has an energy density four times greater than hydrogen and is easier to store. Additionally, its physical properties are nearly identical to natural gas, enabling it to be directly injected into the natural gas grid. Thus, electroconversion to CH_4 has more practical application value.

In distributed energy grid clusters, the application of P2G technology plays a supportive role in promoting multi-energy complementarity, providing auxiliary services, and reducing carbon emissions. Moreover, the co-production advantages of P2G technology are particularly notable. Specifically, the technology for converting gases into methane or other carbon-based fuels is relatively mature. The source of carbon dioxide can be obtained through collaborative operations with facilities that emit large amounts of carbon, such as biogas power plants, incineration plants, and sewage treatment plants. During the conversion process, the captured CO_2 is fully consumed, which aligns with the current energy strategy goals of the world. The by-products of this process, such as excess hydrogen and methane, can be used in hydrogen FCs, gas-powered vehicles, natural gas power plants, and other industries. This approach not only saves investment but also further enhances the decarbonization potential of industries and optimizes overall benefits. Additionally, another by-product of the methane synthesis process is water, which can be recycled. Therefore, the approach employed in this chapter is the direct injection of electroconverted CH_4 into natural gas pipelines or storage tanks.

2.2.2 Electro-Thermal-Gas Synergistic Modeling Method

The multi-energy synergistic complementarity between different energy grids is crucial for improving the comprehensive energy utilization rate and reducing user energy costs. However, in most existing distributed energy grid clusters, electricity, thermal, and gas are still separated, resulting in low energy efficiency. For example, power generation systems typically convert only 30–40% of fuel energy into electricity, with the remaining energy either being transferred to the heat source or directly discharged as waste, leading to inefficient utilization. In thermal grid, high-temperature steam produced by GBs is not used for power generation but is instead directly used for heating, wasting its potential. Additionally, due to the underdevelopment and high costs of electricity storage technologies, excess electricity in the power grid cannot be fully absorbed and is often wasted. This series of issues leads to significant resource consumption and waste.

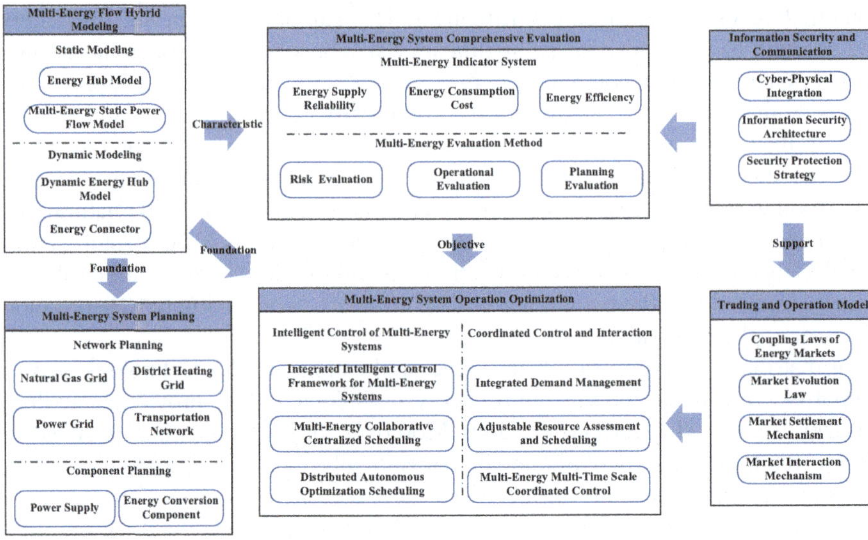

Fig. 2.9 Relationship diagram of multi-energy complementarity and integrated optimization research

In recent years, multi-energy complementarity and integrated optimization of energy grids have become the focus of both project practices and theoretical research. The key issues in this field can be summarized in Fig. 2.9. Multi-energy flow mixed modeling, as a unified description of different energy grids, serves as the research foundation for the planning, scheduling, control, and interaction of multi-energy grids. The comprehensive evaluation of multi-energy flows, based on the characteristics of the multi-energy flow model, provides a set of objectives for planning and operational optimization. Multi-energy flow transactions and business operation models, as the upper-level design rules of the system, offer diverse driving forces for operational optimization. Furthermore, the multi-energy fusion information system supports customized energy trading, and the security of the information system is a crucial factor in evaluating the system's reliability and safety.

The hybrid modeling of multiple energy systems describes the operational and complementary transformation characteristics of each energy grid. As a unified description of multi-energy complementarity systems, hybrid modeling serves as the foundation for integrated optimization and other key technologies.

2.2.2.1 Static EH Model

The EH model reflects the static transformation relationships between energy systems. A significant amount of related research has been conducted to describe the coupling relationships involving cooling, heating, electricity, and gas systems, and it has been widely applied in various integrated energy system studies. For example,

Koeppel and Andersson (2006) and Koeppel (2007) introduces the concept of the EH model to characterize the coupling relationships between different forms of energy (such as electricity, gas, cooling, and heating) in integrated energy systems. This model reflects the static relationships of energy during transmission and conversion processes but cannot describe the complex dynamic behaviors within the energy grid.

The EH can be equivalently considered as a multi-input multi-output conversion structure for energy in a specific region, with C representing the energy conversion matrix, where each coupling coefficient represents the conversion relationship between a certain type of energy and its corresponding load. The input variables are generally represented by the energy vector $\mathbf{P} = [P_a, P_b, \ldots, P_n]^{\mathrm{T}}$, which indicates all the energy entering the region; the output variables are generally represented by the load vector $L = [L_a, L_b, \ldots, L_n]^{\mathrm{T}}$, which represents the loads of all terminals in the region. The multi-input multi-output power conversion formula is as follows:

$$\mathbf{L} = \mathbf{CP} \tag{2.6}$$

$$\mathbf{C} = \begin{bmatrix} C_{aa} & C_{ba} & \cdots & C_{na} \\ C_{ab} & C_{bb} & \cdots & C_{nb} \\ \vdots & \vdots & & \vdots \\ C_{an} & C_{bn} & \cdots & C_{nn} \end{bmatrix} \tag{2.7}$$

$$C_{ij} = \sum^{N}\left(v_{i,n}(t)\prod^{M}\eta^{j}_{n,m}(\Theta, t)\right) \tag{2.8}$$

where, C_{ij} represents the coupling coefficient between the i-th energy type and the j-th load, which is determined by the characteristics of the conversion units and scheduling parameters, and can be unified as Eq. (2.8). $v_{i,n}(t)$ is the scheduling parameter, representing the distribution coefficient of i-th energy across the n-th unit. $\eta^{j}_{n,m}(\Theta, t)$ is the conversion efficiency of the energy conversion unit n in the m-th stage of producing load j, which depends on the unit's operational parameters set Θ and time t. The EH model establishes the corresponding coupling matrix for the power conversion relationships between different energy carriers. From the perspective of coordination theory, redundant energy flow paths provide space for collaborative optimization, and the goal of system optimization is to search for the optimal coupling matrix under the system constraints.

2.2.2.2 Model of Multi-energy Steady-State Hybrid Power Flow

Whether in planning or scheduling operations, energy flow calculation, as a static energy transmission model, has always been a key issue in the static analysis of multi-energy systems. An improved EH model is generally used, considering the coupling units as balancing nodes to account for the impact on power flow in the electricity and natural gas networks, thereby forming the appropriate power flow solution algorithm for the system. The multi-energy complementary static hybrid power flow model for typical scenarios involving electricity, heating, and natural gas systems is generally expressed in the form of (2.9):

$$\begin{cases} 0 = G(x_e, x_g, x_h, x_{eh}) \\ 0 = F(x_e, x_g, x_h, x_{eh}) \\ 0 = H(x_e, x_g, x_h, x_{eh}) \\ 0 = E_H(x_e, x_g, x_h, x_{eh}) \end{cases} \tag{2.9}$$

where, x_e, x_g, x_h, x_{eh} represents the variables for the power system, natural gas system, thermal system, and energy hub; $F(x_e, x_g, x_h, x_{eh})$ is the power flow equation for the electricity system; $G(x_e, x_g, x_h, x_{eh})$ is the algebraic equation for the natural gas system, including compressor equations, branch flow equations, and pressure loop equations; $H(x_e, x_g, x_h, x_{eh})$ is the thermal system, including thermal power calculation equations, node flow equations, pressure loop equations, and the 4-node temperature mixing equation, etc.; $E_H(x_e, x_g, x_h, x_{eh})$ is the input–output conversion equation set for the EH.

The corresponding research can be divided into two categories: unified solution methods and decoupled solution methods. When using the unified solution method, it is necessary to establish a hybrid model of the multi-energy flow grid, create power flow equations for multiple energy grids under a unified framework, and then solve the system's hybrid power flow, which requires higher demands on algorithmic solving.

By combining the static EH model and the steady-state hybrid power flow model, the multi-energy system modeling scheme is shown in Fig. 2.10.

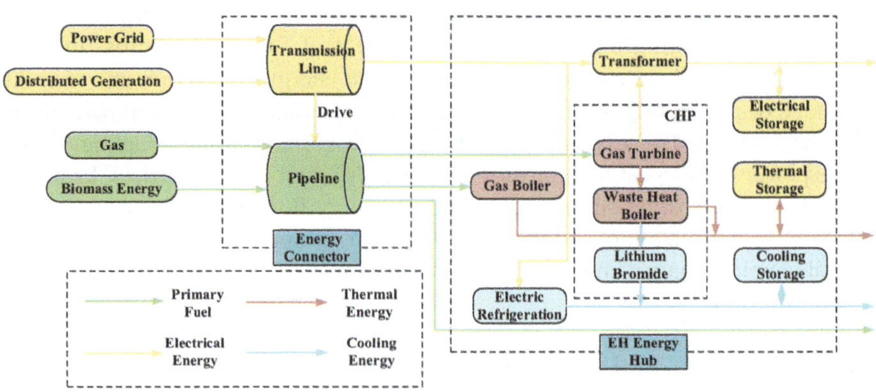

Fig. 2.10 Multi-energy flow steady-state modelling scheme

2.2.2.3 Multi-energy Complementarity Dynamic Modeling

The multi-energy complementarity dynamic model generally includes dynamic EH and dynamic energy connector models, which describe the dynamic characteristics of conversion units and energy flow transmission, respectively. The dynamic EH model, based on the static model, takes into account the dynamic characteristics of the energy conversion units during state switching; the dynamic energy connector describes the static features and dynamic variation laws of electrical energy, liquid working fluids, or gaseous fuel transmission, studying the transfer and coordination feedback mechanisms at both ends, and providing unified and coordinated control for multiple energy transmission links.

In summary, the modeling of multi-energy complementarity static transformation and power flow is relatively mature, with some studies considering multi-energy grid constraints. However, research on the dynamic characteristics of multi-energy flows and multi-time scales is relatively weak, with a lack of dynamic descriptions of the transmission and conversion processes, and mature modeling methods have yet to be developed.

2.3 Energy Consumption Behavior Analysis of Distributed Energy Grid Clusters

The energy consumption behavior analysis of distributed energy grid clusters includes key steps such as data collection and aggregation, energy consumption behavior feature extraction, user group classification, and identification of associated factors. First, data is collected from various terminals such as smart meters, synchronized phasor measurement devices, gas flow meters, gas pressure sensors, and distribution network SCADA systems, as well as big data management systems, to gather user attribute data, electricity consumption data, energy trading data, and environmental data such as weather, socioeconomic factors, and energy pricing policies. Then, effective information is mined from the data using data mining techniques to extract user energy consumption behavior features. The energy consumption demands and behavior characteristics of different users are analyzed, and a user classification model is established to achieve refined user classification. Furthermore, potential associated factors of user energy consumption behavior are analyzed, and correlation is calculated, along with sensitivity analysis of environmental factors. Based on energy consumption behavior analysis, SA technology can be used to evaluate the real-time operational status, detect and locate anomalies, and predict future operational trends of distributed energy grid clusters. The research technical framework for energy consumption behavior analysis of distributed energy grid clusters is shown in Fig. 2.11.

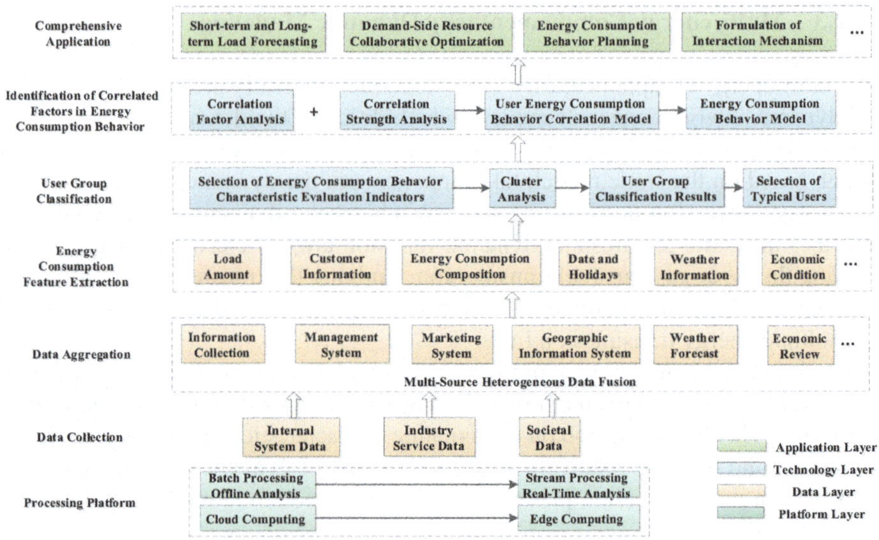

Fig. 2.11 Electricity consumption behavior analysis framework

2.3.1 Energy Consumption Information Collection, Aggregation, and Feature Extraction

With the application of smart sensor devices and systems, the State Grid Corporation of China has implemented information collection for over 155 million households in pilot projects across 26 provinces, establishing a large-scale Advanced Metering Infrastructure (AMI) system. Smart metering terminal devices enable flexible and real-time data access, ensuring interoperability between data, which provides the foundational equipment and technical support for user energy consumption behavior data collection. The diverse sensors correspond to various data structures, and the classification and characteristics of information data from different data sources are shown in Table 2.1.

The energy consumption information collection system provides the main data source and enables data processing, analysis, and application at the network edge layer, making it a crucial component of edge computing. The physical architecture of the energy consumption information collection system is shown in Fig. 2.12, consisting mainly of the master station layer, communication layer, and device collection layer. After the device collection layer gathers energy consumption information, it is transmitted to the master station layer through the communication layer. The server at the master station layer monitors and assesses the real-time operational status and issues corresponding adjustment instructions when anomalies occur. The direct entry point and edge computing node for energy consumption information collection are primarily concentrated on the user-side metering devices.

Table 2.1 Classification and characteristics of user energy consumption information data

Data source		Data information	Data characteristics
Distributed energy grid clusters internal data	User information collection system	Real-time and historical data such as type, power, gas pressure, temperature, etc.	Data volume at the TB and PB levels, large data volume
	Management system	Topology information, operational information, user spatial distribution information	Large data volume, low information density
	Production operation system	Production plans, maintenance and inspection records	Low real-time requirements
	Marketing system	User contract files, remote meter reading data, billing data, financial data, credit evaluation	High data information density, large data volume
	Customer service system	Service acceptance records, service processing records	Data accumulation, rapid growth
Industry service data	Geographic information system	Geographic information data	Slow update
	Weather forecasting system	Temperature, wind speed, humidity, sunlight	High real-time requirements
	Economic audit system	Regional economic conditions, industry development status	Important, slow data update
	Urban management system	Population distribution, urban planning	Important, moderate data volume
Societal data	Date, customs	Workdays, weekends, holidays	Important, small data volume
	Social census	User feedback on interaction plans, user concerns	Direct, important, difficult to obtain

Through data collection, storage, and processing technologies, data that is spatially and temporally dispersed is first distributed for storage, and an update queue is formed to update data in real-time, thus reducing the requirements for storage capacity. Then, data from different dimensions and sources is normalized, and different databases are established. Finally, relevant data from the multi-source big data platform is integrated.

Fig. 2.12 Physical architecture of the energy consumption information collection system

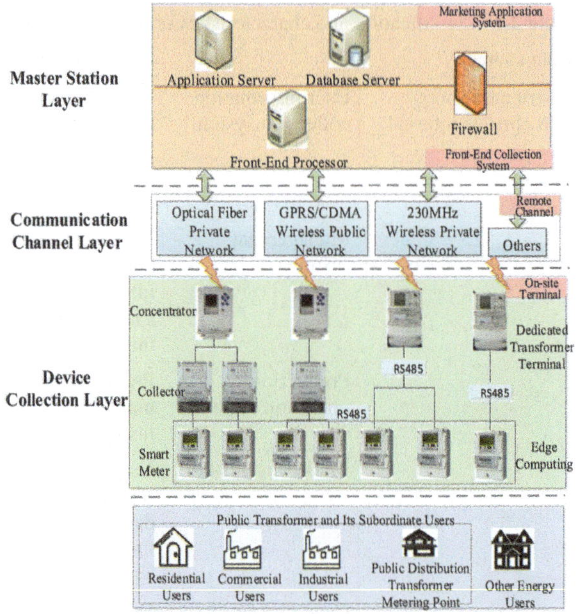

After the collection and aggregation of multi-source data, the feature extraction of energy consumption information relies on the detailed monitoring technology of user energy consumption. Unlike smart meters that only measure total power, load energy consumption detail monitoring can track energy usage at the level of specific appliances within a household. Load energy consumption detail monitoring can be divided into intrusive and non-intrusive types. Intrusive monitoring (ILM) equips each appliance in the total load with sensors capable of digital communication, which collect and send energy consumption information through a local area network. This method is more costly. Non-intrusive monitoring (NILM), on the other hand, involves installing a total sensor at the entry point of the household, which collects and analyzes energy consumption data and decomposes the total load to monitor individual appliance energy usage. The energy consumption information collection and analysis framework based on NILM is shown in Fig. 2.13. NILM technology provides real-time feedback on sub-metered energy consumption data, breaking down simple energy usage curves into real-time sub-metered consumption data, which provides quantifiable references for formulating node strategies and reasonable peak-shaving plans for users. Compared to ILM, NILM uses decomposition algorithms to provide real-time feedback on sub-metered energy data. It offers advantages such as simplicity, cost-effectiveness, reliability, good data integrity, and ease of large-scale deployment. Additionally, the usability and value density of energy information provided by NILM technology are superior to traditional collection methods, providing quantifiable support for advanced intelligent energy services such

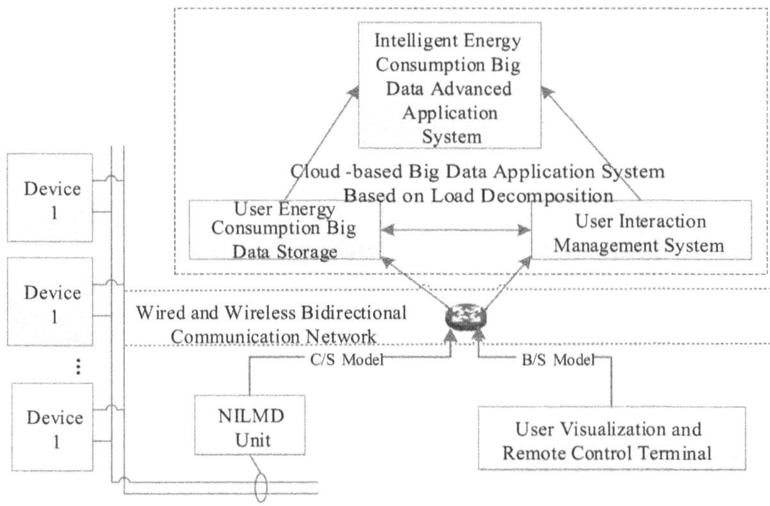

Fig. 2.13 NILM energy consumption information collection and analysis cloud architecture

as energy behavior analysis, energy efficiency analysis, and demand-side management, and has the potential to become the core technology for the next-generation AMI.

2.3.2 Energy Consumption Behavior Pattern Recognition

In distributed energy grid clusters, after collecting and aggregating user energy consumption data, a user group classification model is established to identify energy consumption behavior patterns. This approach aids in the detailed analysis of different users' energy consumption demands and behavioral characteristics.

Traditional user classification methods are typically based on empirical rules. For example, users may be classified according to attributes such as industry, sector, and household size, thereby creating a "user-label" network and constructing a relational weight model for classification. While this traditional approach is simple and fast, it is heavily influenced by empirical rules and may yield imprecise classification results. As a result, recent studies have increasingly focused on utilizing clustering techniques from data mining to achieve more refined energy user classification.

Clustering involves searching for similarities among samples based on selected features and weights, classifying the samples according to their characteristics. This can be refined further by electricity consumption data. User energy behavior patterns are typically recognized by calculating similarity using load curves or load characteristic indicators, and clustering algorithms are used to implement this. Common clustering methods for energy behavior patterns include partition-based methods such as k-means and K-medoids, model-based methods like COBWEB and self-organizing neural networks, density-based methods such as DBSCAN, as well as fuzzy clustering and hierarchical clustering.

However, traditional clustering algorithms tend to exhibit issues such as unstable clustering results, slow processing speeds, poor effectiveness, and excessive memory consumption when applied to high-dimensional, large-scale data with highly varied cluster shapes. Researchers have often combined dimensionality reduction algorithms to improve clustering methods. For instance, Wang and Zhou (2018) proposes a load curve clustering method based on an unsupervised extreme learning machine for high-dimensional input, which shows more reasonable results compared to k-means clustering in high-dimensional data. Chen et al. (2018) introduces a load curve clustering method based on an improved fast-density peak algorithm, which enhances the local density and distance calculation criteria of the original algorithm, improving clustering stability and effectively reducing execution time and memory consumption. Wang et al. (2017) utilizes information entropy segmentation and aggregation approximation to re-express load data with variable time resolution, which employs spectral clustering methods for load classification, offering advantages in terms of data dimensionality reduction, clustering effectiveness, stability, and reduced computational workload. Furthermore, Yi et al. (2017) shifts focus from the load curve shape and proposes a new dynamic clustering method for energy behavior. This method models energy consumption using a Markov model, transforming the load curve dataset into multiple state transition matrices. By applying fast search and density peak methods, it classifies users into multiple clusters.

In practical applications, clustering results are influenced by multiple factors and stages in the extraction process, which must be considered. Depending on the specific data type, an appropriate classification method for energy consumption behavior patterns should be chosen. With proper classification, further analysis of user cluster energy behavior can be conducted.

2.3.3 Analysis of Correlated Factors Based on Understanding of Energy Consumption Behavior

The energy endowment and demand of users in distributed energy grid clusters determine their energy consumption behavior preferences. As a result, user energy demands and behaviors are influenced by various environmental factors, primarily including natural factors such as weather, seasons, and geography, as well as social factors like price, demand-side management policies, economic structure and development level, residential consumption levels, and population. These factors influence energy consumption behavior in both spatial and temporal dimensions, affecting both short- and long-term behaviors. Conducting correlation analysis to identify strong correlations and remove redundant factors is of great significance for understanding user energy consumption behavior patterns and improving the accuracy of load forecasting and DR modeling. The methods for analyzing environmental impact factors are mainly based on two models: causal analysis and data-driven approaches.

Causal analysis-based methods typically involve mechanism analysis based on certain clear physical concepts. From the earliest experimental verification methods to later studies on laws and theorems (e.g., expert experience methods), these research approaches fundamentally aim to analyze the mechanisms of causal relationships, exploring the connections between influencing factors and energy consumption behavior.

In recent years, with the construction of ubiquitous IoT endpoints and advancements in metering technologies, the volume of data in energy internet systems has increased significantly (Bedi et al. 2018; Čolaković and Hadžialić 2018). Data-driven methods for identifying correlated factors are now more effective in uncovering the relationships between influencing factors and energy consumption behavior. On the other hand, causal analysis can complement data-driven approaches by enhancing the interpretability of correlation analysis.

Data-driven correlation analysis methods primarily target statistical relational datasets (Han et al. 1993), but these datasets may not capture the underlying mechanisms perceived by users, thus making strict deduction and precise analysis difficult. For instance, human electricity consumption behavior is not entirely rational, so mathematical methods are used to analyze the correlation of influencing factors. Typical data-driven approaches include association rule mining, principal component analysis, mutual information, grey correlation analysis, neural networks, support vector machines, and chaos theory (Jieyang et al. 2023; Wang and Shi 2013). In the context of big data, due to the high dimensionality and complexity of identifying correlated factors, researchers have proposed new data-driven methods based on random matrix theory (RMT) (Xiong et al. 2020), which analyzes correlations by studying the energy spectra and eigenstates of complex systems, revealing the overall correlational behavior characteristics in the data.

Considering the high dimensionality and complexity of massive energy consumption data and influencing factors, traditional data-driven correlation analysis methods require accurate extraction of feature factors, user numbers, sampling points, and other information from the data matrix. The matrix is then decomposed into multiple factor vectors for repeated analysis, leading to low execution efficiency. To address this issue, this chapter proposes a random matrix method suitable for multivariate big data analysis. This method analyzes correlations based on the statistical properties of random matrices, without needing to decompose the original data matrix into multiple vectors for repeated calculations, offering advantages in high-dimensional big data processing.

A brief introduction to the method of analyzing multivariate data correlation based on RMT is as follows.

First, the basic theory of random matrices introduces the single-ring theorem: Let $\widehat{\mathbf{X}} = \{x_{i,j}\}_{N \times T}$ be a random matrix with non-Hermitian characteristics, where each element is an independent and identically distributed random variable, with expectations and variances satisfying $E(x_{i,j}) = 0$ and $E\left(|x_{i,j}|^2\right) = 1$, respectively. For multiple random matrices $\widehat{\mathbf{X}}_i (i = 1, 2, \ldots, L)$ with non-Hermitian characteristics, the matrix product is defined as:

$$\widehat{\mathbf{Z}} = \prod_{i=1}^{L} \widehat{\mathbf{X}}_{u,i} \tag{2.10}$$

where, $\widehat{\mathbf{X}}_{u,i} \in \mathbb{C}^{N \times N}$ is a singular value equivalent matrix of $\widehat{\mathbf{X}}_i$, and $\mathbb{C}$ represents the complex number set. The matrix product $\hat{\mathbf{z}}$ is transformed to obtain the standard matrix product $\tilde{\mathbf{Z}} = \{\tilde{z}_{i,j}\}_{N \times N}$, where each element satisfies $E(\tilde{z}_{i,j}) = 0, E\left(|\tilde{z}_{i,j}|^2\right) = 1/N$. As N and T approach infinity, while keeping $c = N/T$ constant, the empirical spectral distribution of the eigenvalues of $\tilde{Z}$ almost certainly converges to the single-ring theorem, and its probability density function is:

$$f(\lambda) = \begin{cases} \frac{1}{\pi c L} |\lambda|^{\tau^2 - 2}, & (1-c)^{\frac{2}{\tau}} \le |\lambda| \le 1 \\ 0, & \text{other} \end{cases} \tag{2.11}$$

where, $c \in (0, 1]$. According to the single-ring theorem, on the complex plane, the eigenvalues of $\tilde{Z}$ are approximately distributed between two rings, with the inner ring radius being $(1-c)^{\frac{l}{2}}$ and the outer ring radius being 1.

The Linear Eigenvalue Statistic (LES) reflects the eigenvalue distribution of a random matrix. For a random matrix, the LES is defined as:

$$N_n(\varphi) = \sum_{i=1}^{n} \varphi(\lambda_i) \tag{2.12}$$

where, $\lambda_i (i = 1, 2, \ldots, N)$ represents the eigenvalue of X; $\varphi(\bullet)$ is a test function. By choosing different test functions, different LESs can be obtained.

The mean spectral radius (MSR) is a type of LES. It is defined as the average value of the radii of all the eigenvalues of a matrix distributed on the complex plane. MSR is easy to calculate and serves as a statistical characteristic of the system from a high-dimensional perspective. The formula for the MSR is as follows:

$$K_{\text{MSR}} = \frac{1}{N} \sum_{i=1}^{N} |\lambda_i| \tag{2.13}$$

where, $\lambda_i (i = 1, 2, \ldots, N)$ represents the eigenvalue of the matrix; $|\lambda_i|$ is the radius of the distribution of λ_i on the complex plane. In this study, the MSR is used to describe the eigenvalue distribution of the standard matrix product $\mathbf{Z} \in \mathbb{C}^{N \times N}$, as follows:

$$K_{\text{MSR}} = \frac{1}{N}\sum_{i=1}^{N}\left|\lambda_{z,i}\right| \tag{2.14}$$

where, since each eigenvalue of $\tilde{Z}$ is a complex function of the random matrix $\mathbf{X}_i(i = 1, \ldots, L)$, and the eigenvalues are highly correlated random variables, K_{MSR} itself is a random variable.

The LES describes the trace of a random matrix. The law of large numbers and the central limit theorem (CLT) suggest that, for a random matrix, the trace of the matrix can reflect the statistical properties of its elements. However, a single eigenvalue of the matrix, due to its randomness, cannot capture these properties. Therefore, the MSR is used as an indicator for correlation analysis. The greater the difference between the MSR and the theoretical prediction of the single-ring theorem, the stronger the correlation.

Based on the above theory and combined with the denoising functionality of wavelet decomposition, this chapter proposes a correlation analysis of user electricity consumption behavior based on RMT. The analysis process is as follows.

(1) Wavelet Denoising

User energy consumption behavior exhibits significant randomness. This chapter aims to remove the random components of the electricity consumption curve through filtering methods. The orthogonal wavelet transform is equivalently treated as a series of mirror filtering processes. Specifically, the electricity consumption data is passed through a high-pass filter and a low-pass filter, and the Mallet method is used to remove the high-frequency components of the input data, thereby reducing the energy of the Gaussian random signal. Additionally, since different data types have different sampling frequencies, it is assumed that the data types with lower sampling frequencies have equal values within the sampling intervals.

(2) Normalization

The dimensionalities of the characteristic factors from different data sources are often different. To ensure that heterogeneous data from various data platforms have comparable expressiveness, normalization is required.

(3) Singular Matrix Handling

The data matrix is obtained from the data source, and the matrix is then tiled to form a standard non-Hermitian matrix $\tilde{\mathbf{X}}$. Next, the singular matrix $\tilde{\mathbf{X}}$ is transformed into a singularly equivalent matrix $\tilde{\mathbf{X}}_u$.

$$\tilde{\mathbf{X}}_u = \sqrt{\tilde{\mathbf{X}}\tilde{\mathbf{X}}^T U} \tag{2.15}$$

where, $\mathbf{U} = \tilde{\mathbf{X}}\sqrt{\left(\tilde{\mathbf{X}}^T\,\tilde{\mathbf{X}}\right)^{-1}}$ is a unitary matrix.

(4) Constructing the Augmented Matrix

Assume that the energy consumption behavior state of a system is a variable with N-dimensional parameters and M potential influencing factors. By measuring $t_i(i = 1, \ldots, T)$ over a certain period, the state-related N-dimensional vector can naturally form the basic state matrix $\mathbf{B} \in \mathbb{C}^{N \times T}$, and each influencing factor can also be assigned a value for that time segment—the factor vector $c_j \in \mathbb{C}^{1 \times T}(j = 1, 2, \ldots, M)$.

Two matrices (or vectors) of the same length can be concatenated to form a new matrix. Based on this principle, the basic state matrix B and the factor vector c_j can be concatenated in some manner to form a composite matrix $\mathbf{A}_j$.

To facilitate the analysis of the impact of influencing factors on the basic state, it is necessary to amplify the influence of the factors. The factor vector c_j is selected and replicated K times (denoted as $0.4 \times N$) in a certain manner to form a matrix $\mathbf{D}_j$ that matches the scale of the state matrix, as shown below:

$$\mathbf{D}_j = \left[c_j^H c_j^H \ldots c_j^H\right]^H \in \mathbb{C}^{K \times T} \tag{2.16}$$

The next step is to introduce white noise into $\mathbf{D}_j$ to eliminate the internal correlation, as shown in the following equation:

$$\mathbf{C}_j = \mathbf{D}_j + m_{ej}\mathbf{R} \quad (j = 1, 2, \ldots, m) \tag{2.17}$$

where, $\mathbf{R}$ is a standard Gaussian random matrix, and $m_{e,j}$ is related to the signal-to-noise ratio (SNR) ρ:

$$\rho_j = \frac{\mathrm{Tr}\left(\mathbf{D}_j \mathbf{D}_j^H\right)}{\mathrm{Tr}\left(\mathbf{R}\mathbf{R}^H\right) \times m_{e,j}j^2} \quad (j = 1, 2, \ldots, m) \tag{2.18}$$

In this way, each factor vector c_j can be parallelly processed to form the composite matrix $\mathbf{A}_j$ through matrix concatenation:

$$\mathbf{A}_j = \begin{bmatrix} \mathbf{B} \\ \mathbf{C}_j \end{bmatrix} \quad (j = 1, 2, \ldots, m) \tag{2.19}$$

(5) Calculate the MSR and Perform Correlation Measurement

Calculate the eigenvalues λ of each $\mathbf{A}_j$ and the MSR $K_{\mathrm{MSR}} = \frac{1}{N} \sum_{i=1}^{N} |\lambda_i|$. Integrate the MSR and perform a comparison. The larger the MSR, the greater the regularity of the factor matrix relative to the state matrix. That is, the degree of correlation between the state variables and the factor variables is higher. Factors with higher correlation are selected as the sensitive factors influencing the state.

The above random matrix correlation analysis process can be summarized in the flowchart shown in Fig. 2.14.

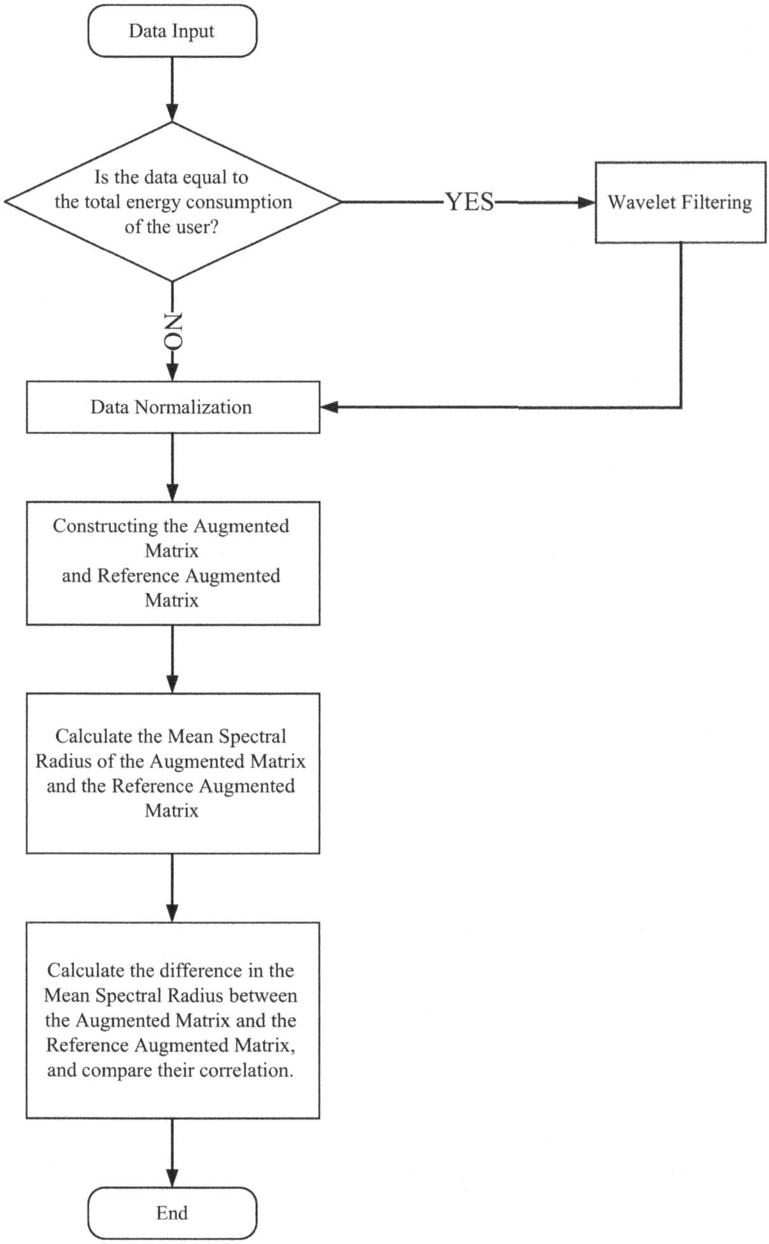

Fig. 2.14 Flowchart of correlation analysis based on random matrices

Table 2.2 Correlation analysis results of energy consumption in an industrial park in guangzhou based on random matrices

Feature factors	K_{MSR}			
	User Type 1	User Type 2	User Type 3	User Type 4
Economic level	0.1986	0.2457	0.2776	0.3075
Response willingness	0.2046	0.1459	0.2336	0.2641
Holidays	0.0612	0.0949	0.1864	0.2053
Temperature	0.0276	0.0413	0.0212	0.0394
Ideal price	0.0705	0.1010	0.1632	0.1348

As an example, the correlation of energy consumption response capabilities of different categories of users in an industrial demonstration park in Guangzhou, China, is considered. The load shifting amount before and after price adjustment is selected as the state variable, and five data sources—economic level, response willingness, holidays, temperature, and ideal price deviation—are chosen as the influencing factors. The resulting data matrix is a 3-dimensional high-dimensional matrix, i.e., the number of influencing factors × the number of users of the same category × the number of sampling points. The correlation of these five influencing factors with the response capabilities of four user categories is analyzed. For each user category, 20 typical users are selected, with each typical user having 8760 sampling points. The correlation analysis of energy consumption behavior for users in the industrial demonstration park is shown in Table 2.2.

From Table 2.2, it can be observed that the economic level and response willingness of users have a significant impact on their response capabilities. Temperature and ideal price have a relatively small effect on the response capabilities of most industrial users. The sensitivity to the holiday factor varies greatly among different user types, with users of Type 3 and Type 4 showing significantly higher sensitivity to holidays compared to those of Type 1 and Type 2.

2.4 Data Processing Technology Based on Cloud-Edge Computing

Distributed energy grid clusters, as an effective form of an active integrated energy system, enable the flexible and efficient utilization of DERs. It integrates multiple energy forms, including electricity, heat, and gas, connecting distributed power sources, controllable loads, ES devices, thermal energy equipment, and gas energy devices through numerous power electronic devices and advanced control technologies. These components are linked to the superior energy grid via one or multiple common connection points and can flexibly switch between grid-connected and islanded operating modes. The construction of distributed energy grid clusters facilitates the efficient utilization of unpredictable energy sources such as solar and wind energy, as well as novel energy forms like waste heat recovery and gas-based distributed energy. This serves as a powerful complement to traditional energy systems, further advancing the transition and transformation of conventional power grids into intelligent integrated energy grids. In such grids, electricity flow, thermal energy flow, gas energy flow, information flow, and business flow are highly integrated. To ensure the flexible and reliable operation of the energy grid and coordinated optimization of multiple energy forms, extensive deployment of information collection devices is required.

With the advancement of automation and informatization in energy systems, many energy enterprises have built or are in the process of constructing integrated EMSs tailored to their needs. Information monitoring systems such as EMSs, comprehensive energy dispatch systems, and energy flow monitoring systems in distributed energy grid clusters enable online collection and monitoring of operational states, equipment conditions, and energy quality data for various energy forms, including electricity, thermal energy, and gas energy. In modern distributed energy grid clusters, monitoring data includes structured data such as remote signaling and telemetry, as well as unstructured data like waveform data, thermal flow data, and gas flow data. These energy grid status data are collected and monitored in real time through data acquisition and supervisory control systems, ensuring efficient and coordinated operation of multiple energy forms. The monitoring data in distributed energy grid clusters exhibit the following characteristics.

(1) Large Data Volume and High Dimensionality

In distributed energy grid clusters, the production, transmission, storage, and load processes of electricity, heat, and gas generate massive amounts of data. As requirements for data analysis accuracy and collection rates increase, sampling precision is also improving. This leads to exponential growth in data volume, placing higher demands on data storage, processing, and analytical capabilities.

(2) Complex Correlations

Monitoring data in distributed energy grid clusters comes from widely distributed multi-energy devices and various data management and collection systems, with complex correlations between different data points. These data exhibit spatial and temporal dispersion as well as heterogeneity. Aggregation technologies are required to integrate the data, extract feature information, and use methods such as correlation analysis and principal component analysis to uncover the intrinsic relationships between energy grid behaviors and characteristic variables. This enables coordinated optimization of multiple energy forms.

(3) Low-Value Density

Despite the large volume of monitoring data in distributed energy grid clusters, the value density of the data is relatively low and is often affected by social, environmental, and technical factors. Data mining techniques are needed to extract information that reflects the operational state and optimization control of the energy grid from the vast amount of data, providing support for decision-making.

Big data refers to "complex datasets that cannot be stored, searched, extracted, shared, analyzed, or processed using existing software tools." The monitoring systems in distributed energy grid clusters generate continuous data streams characterized by enormous volumes, diverse types, and strong real-time requirements, which exemplify the attributes of big data.

Currently, distributed energy grid clusters data acquisition and monitoring systems face three major challenges:

(1) With the intelligent development of energy grids, the data volume generated by monitoring devices is surging. Traditional methods encounter bottlenecks in data collection, storage, and querying for massive datasets.
(2) Coordinated optimization and control scheduling of multi-energy forms require real-time data sharing and interaction across components. However, existing energy dispatch systems struggle to achieve large-scale, secure sharing of data resources.
(3) The real-time computational complexity of massively distributed energy grid clusters monitoring data is high, and traditional methods fail to meet the demands of high-performance analysis.

Traditional data analysis and processing platforms can no longer meet the storage and processing requirements of monitoring data in distributed energy grid clusters. There is an urgent need to establish a secure and efficient big data processing platform for monitoring. A high-performance big data platform can not only enhance the operational control, decision-making, and management strategies of energy grids, but also significantly improve the planning and optimization levels of the energy market, offering substantial research and application value.

2.4.1 Cloud Computing

Cloud computing is an information processing approach that organizes and flexibly utilizes various information and communications technology (ICT) resources through a network to achieve large-scale computation. It leverages resources such as software, storage, networks, and servers via virtualization technology to form an abstract service model. The goal of cloud computing is to provide personalized services that are cost-effective, highly reliable, highly available, and scalable. Achieving this goal requires the support of several key technologies, including data center management, virtualization, massive data processing, and resource management and scheduling. The main advantages of cloud computing include:

(1) Integration of Heterogeneous Data Resources

The heterogeneous distributed file system, which adopts a master–slave backup mechanism, provides a reliable access and storage carrier for heterogeneous data processing. The construction of an ontology-based heterogeneous data integration model under cloud computing offers unified data management and processing methods for various business applications in the cloud environment. This facilitates the efficient and unified retrieval and query of heterogeneous data and establishes substantial correlations and mappings between various types of heterogeneous data processed by business applications.

(2) Dynamic System Scalability

Compared to the vertical expansion model of traditional integrated energy platforms, which relies on upgrading hardware or replacing core components, cloud computing employs horizontal scaling. This reduces the need for single-machine capacity upgrades. When data processing demands increase, system performance can be improved by adding new computing and storage nodes to the resource pool rather than upgrading system hardware. This dynamic scalability reduces hardware upgrade costs, simplifies operations, and enhances storage and computing capabilities for big data.

(3) Efficient Storage and Computing Capabilities

Energy grid monitoring systems generate diverse, voluminous, and widely distributed data, placing high demands on data processing platforms. Cloud computing uses distributed storage technology and parallel computing models to meet the complex computational and analytical requirements. By leveraging virtualization technologies such as host and storage virtualization, cloud computing abstracts various resources into service forms, providing personalized services based on customer data processing needs.

Cloud computing primarily offers three types of services: infrastructure as a service (IaaS), platform as a service (PaaS), and software as a service (SaaS). Hadoop is one of the most mature distributed system frameworks currently available. As a PaaS solution built on infrastructure, Hadoop enables rapid distributed data processing across server clusters. Hadoop consists of many components, with its core including the distributed file system HDFS for data storage and the MapReduce framework for data-parallel computing. Figure 2.15 illustrates the Hadoop ecosystem.

(1) Hadoop Distributed File System (HDFS)

HDFS consists of three core components: the NameNode, Secondary NameNode, and DataNode. It operates on a master–slave architecture, with one NameNode and multiple DataNodes. The NameNode serves as the manager of the distributed file system, acting as the master controller. It manages the file system namespace, cluster configuration information, and block replication. The NameNode maintains the namespace of HDFS and uses an EditLog transaction log to record detailed modifications to the file system metadata. Changes, such as altering the replication factor of a file, are logged in the EditLog, stored locally on the NameNode. Additionally, the NameNode stores a file called FsImage, which contains mappings, attributes, and other essential information.

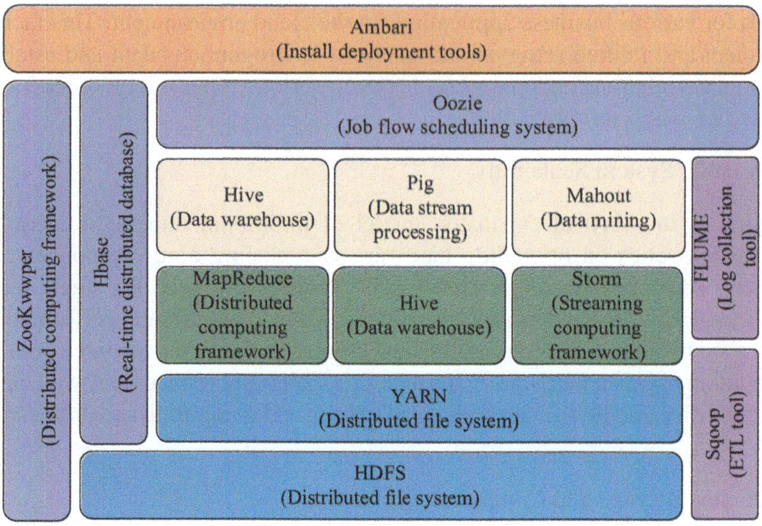

Fig. 2.15 The Hadoop ecosystem

The DataNode is the fundamental storage unit, storing parts of HDFS metadata as blocks in the local file system, with each block stored as a separate file. During file creation, the DataNode calculates the optimal number of files per directory to prevent performance degradation caused by an excessive number of files in a single directory. DataNodes periodically send block status reports to the NameNode. These reports, generated by scanning the local file system, list HDFS data blocks corresponding to the DataNode.

HDFS is a distributed file system designed to handle petabyte-scale datasets. It distributes replicated datasets across different storage nodes, enabling high throughput for target data through a TCP-based streaming data access interface. By splitting data files across the nodes in a cluster, HDFS provides fault tolerance and recovery mechanisms, ensuring high system stability.

HBase, a typical NoSQL database, is suitable for storing sparse, semi-structured, or unstructured data, effectively addressing the diversity of power data structures. HBase allows dynamic addition of columns and does not store empty columns, saving storage space. It also transparently partitions data, enabling horizontal scalability of storage. HBase supports high-performance concurrent read and write operations. When combined with HDFS, HBase enables efficient data reading while maintaining high data throughput.

(2) MapReduce Distributed Computing Framework

MapReduce is a parallel programming paradigm that divides a large computational task into smaller tasks that can be executed in parallel on different machines. It allows parallel processing of large datasets without requiring synchronization of data between nodes. This enables computational tasks to be distributed across multiple machines in a cluster, ensuring reliable and efficient operation on a large scale. Hadoop's MapReduce is an open-source implementation of Google's MapReduce. It performs two main functions:

Map: Decomposes a task into smaller tasks distributed across different cluster nodes.
Reduce: Performs parallel computation on the decomposed tasks, aggregates the results, and outputs the final analysis results to HDFS.

During the Map phase, block data is split into input fragments, typically corresponding to block sizes. Each fragment is represented as a key-value pair. After processing by the map function, a new set of key-value pairs is generated and stored in a circular memory buffer. When the buffer exceeds a threshold, a background thread writes the content to disk. Before writing, the buffer data is partitioned, sorted, and merged to save disk space. The partitioning result is determined by the partition interface method. The system also sorts and merges the Map output records to reduce the amount of data written to disk and lower the network transmission burden.

Once the Map phase is complete, tasks and data are obtained from the Map side via HTTP. Small datasets are stored directly in memory, while larger datasets are merged and written to disk. Tasks and data stored in memory and disk, will group data with the same key together. A custom function processes these key-value pairs, producing a larger ordered file. This merging process iterates until the final result is written to disk as a key-value list. The Reduce phase then aggregates values with the same key and writes the output to HDFS.

Taking the analysis of users' energy consumption patterns as an example, the application of cloud computing in the energy grid is illustrated. MapReduce enables parallel computing through its software processing model. Programs based on this structure can efficiently segment and process data from large-scale distributed intelligent devices. However, it lacks real-time data processing capabilities and is better suited for analyzing data whose value does not diminish over time. For example, in a study of energy consumption patterns in an industrial park in Guangzhou, user information and load curves were analyzed using a simplified parallel computing MapReduce distributed program model. A big data cloud architecture was designed based on MapReduce. Figure 2.16 illustrates a big data clustering scheme for user electricity consumption behavior patterns using the MapReduce framework.

On the Hadoop platform, the HDFS comprises multiple management nodes and data nodes, which are intelligent measurement terminals or computing devices. The system database is composed of the HBase database and HDFS. Hive maps structured data into data tables, enabling the transformation of structured query language (SQL) database queries into MapReduce tasks. Each stage transmits data in the form of a <key, value> structure, where the key represents the sample number, and the value represents the coordinate values of various dimensions of electricity usage data.

By setting up four clustering centers and four MapReduce parallel nodes, a comparison was made between the computational efficiency of a traditional serial clustering algorithm and a clustering algorithm based on the MapReduce framework

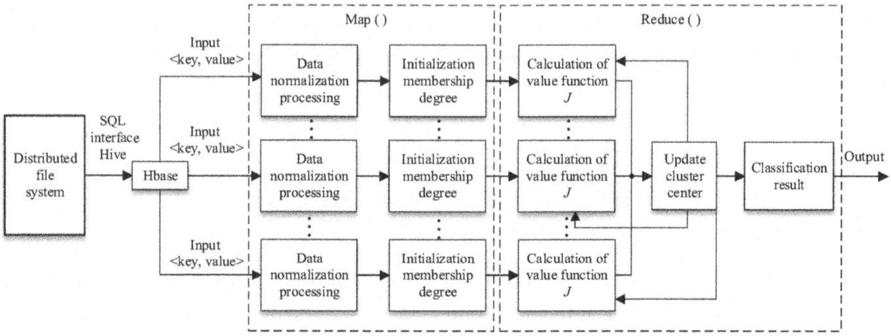

Fig. 2.16 User energy consumption pattern clustering process based on MapReduce architecture

Table 2.3 Comparison of efficiency between MapReduce and traditional serial computing

Number of experiments	Record number	Computing time/s	
		MapReduce	Traditional serial computing
1	7300	0.890	1.87
2	14,600	1.065	2.45
3	29,200	2.791	4.62
4	58,400	3.259	8.84
5	175,200	3.729	11.5

in the Java environment. The results, shown in Table 2.3, indicate that the clustering algorithm based on the MapReduce framework demonstrates significantly higher computational efficiency. Moreover, the advantages of the algorithm become increasingly evident as the data volume grows (Hao et al. 2017).

2.4.2 *Edge Computing*

Edge Computing refers to a technology that provides services by integrating network, computing, storage, and information technologies near the source of data or devices. It enables rapid responses to network service demands, making it particularly suitable for scenarios in distributed energy grid clusters that require massive real-time data transmission, computation, storage, and edge-based security and privacy protection. By deploying devices on the user side, edge computing collects, stores, and processes data from various energy sources such as electricity, heat, and gas. It can analyze and label data in real time, transmitting processed results or reduced data to cloud computing centers to work collaboratively, forming a complementary relationship. The application of edge computing in distributed energy grid clusters offers several significant advantages.

(1) Fast Response Time

Edge computing complements and extends cloud computing. In cloud platforms, most tasks are performed through centralized computation. However, for complex computation tasks like multi-energy collaborative optimization in distributed energy grid clusters, cloud computing may face high latency and fail to meet real-time requirements. By offloading some computation tasks to edge nodes near the data source, the distributed architecture of edge computing significantly reduces the load on cloud computing. In distributed energy grid clusters, edge computing can process multi-energy flow data (electricity, heat, gas) in real time, providing ultra-low latency responses (<10 ms). This makes it ideal for scenarios requiring quick responses, such as multi-energy real-time scheduling, anomaly detection, and islanding control.

(2) Enhanced Security

Edge computing protects local data privacy through regional isolation, avoiding the direct transmission of sensitive data to the cloud. For energy data involving electricity, heat, and gas in distributed energy grid clusters, edge computing performs preliminary processing locally, effectively reducing security risks during data transmission. Additionally, the distributed architecture prevents the cloud center from becoming a single point of attack, addressing privacy protection challenges in cloud computing and further enhancing the information security of distributed energy grid clusters.

(3) Low Energy Consumption

Edge computing brings computing resources and efficient services to the network edge, leveraging local storage, network, and computing resources. In distributed energy grid clusters, edge computing can filter, analyze, and process electricity, heat, and gas data at edge nodes and dynamically adjust the data flow uploaded to the cloud based on security policies. This mechanism reduces data transmission volume, network bandwidth usage, overall data processing costs, and device energy consumption, thereby improving the operational efficiency of the energy system.

(4) Intelligence

In distributed energy grid clusters, edge computing, combined with deep learning technologies, can offload some model inference tasks to nearby edge nodes, enabling collaborative work between terminal devices and edge servers. This collaboration reduces the resource and energy consumption of terminal devices and minimizes response latency. Additionally, artificial intelligence technologies can automatically extract optimal scheduling strategies based on historical data and quickly adapt to real-time changes in user demands and energy states. This optimizes the efficiency of multi-energy collaborative operations.

(5) Reliability

The distributed architecture of edge computing reduces the impact of single-point failures on cloud computing platforms. In distributed energy grid clusters, proper task load balancing and resource scheduling mechanisms ensure the efficient operation of data processing tasks for electricity, heat, and gas. Even if cloud computing services are disrupted due to network failures, edge computing can independently handle critical tasks, ensuring the stable operation of multi-energy systems and uninterrupted user services.

As the business of distributed energy grid clusters grows rapidly, the increasing number of demand-side terminals and systems imposes higher requirements on centralized data processing platforms. Edge computing effectively addresses the challenges of cloud computing, providing strong support for the construction and development of distributed energy grid clusters. By being deployed close to the user-side edge, edge computing enhances the information interaction capabilities between users and the energy grid. Its low-latency characteristics enable quick responses to anomalies, implementing islanding control for abnormal regions and eliminating anomalies through adjustment mechanisms, thereby improving the stability of the energy grids.

Moreover, distributed energy grid clusters extensively adopt new energy sources such as wind, solar, waste heat recovery, and gas energy. The real-time response mechanism of edge computing facilitates the efficient integration and resource scheduling of new energy sources, providing a basis for multi-energy collaborative optimization and seamless switching between grid-connected and islanding modes. This significantly enhances the operational efficiency and flexibility of distributed energy grid clusters.

2.4.3 Data Processing Technology Based on Cloud-Edge Computing

Cloud computing and edge computing form a complementary and collaborative relationship. The close coordination between the two can better meet various big data processing needs. Edge computing primarily handles real-time, short-cycle data locally, providing high-value data to the cloud; cloud computing, on the other hand, undertakes computational tasks beyond the capacity of edge nodes, focusing on non-real-time, long-cycle data processing, optimizing business rules or models, and deploying them back to the edge to better meet local requirements. This cloud-edge collaboration provides a platform and foundation for big data processing in power grids. Currently, the data processing technologies applied in smart energy grid analysis include the following.

(1) Big Data Storage and Processing

The distributed energy grid generates massive amounts of data, encompassing both structured and unstructured formats. The transmission and storage of such vast data, produced by the collaborative operation of electricity, heat, and gas systems, are critical challenges for efficient big data processing in distributed energy grid clusters.

Data compression techniques, where data is compressed before transmission and decompressed at the monitoring center, can effectively reduce network transmission volume and improve efficiency.

In terms of big data storage and processing, two primary techniques are used: stream processing and batch processing. Stream processing enables millisecond-level rapid processing of continuously incoming data streams, suitable for scenarios requiring high real-time performance. In distributed energy grid clusters, stream processing can be used for online evaluation of heterogeneous data, joint scheduling of electricity, heat, and gas loads, and online equipment monitoring. Popular frameworks like Storm, Spark, and Samza support real-time energy grid control and optimization. Batch processing temporarily stores data before computation, making it suitable for tasks with large, complex datasets but low real-time requirements. In distributed energy grid clusters, batch processing can analyze historical multi-energy operation data, optimize long-term planning, and identify equipment failure patterns. Typical batch-processing models include frameworks like MapReduce.

Cloud computing serves as the foundational platform for big data storage and processing in distributed energy grid clusters. By integrating distributed file systems, data processing systems, and database technologies, cloud computing provides robust data storage and computation capabilities. It supports unified management and analysis of electricity, heat, and gas data, offering technical support for multi-energy collaborative optimization, resource scheduling, and operational mode switching.

(2) Data Visualization

The core idea of data visualization is to represent each data item in a database as an independent graphical element, forming a multidimensional data image. By observing data from various perspectives, graphical representations make complex information easier to understand and analyze. In distributed energy grid clusters, data visualization plays a crucial role, especially in scenarios involving electricity, heat, and gas collaboration. It provides researchers with intuitive analysis tools and presentation methods.

(a) Panoramic Display of Multi-energy System Operations

Data visualization provides a comprehensive view of the operation status of electricity, heat, and gas systems in production, operation, scheduling, and management. It reveals energy consumption patterns and their correlation with economic development, highlighting dynamic characteristics and predicting future energy demand trends.

(b) Focus on Key Data of Interest

By amplifying critical information and filtering unnecessary data, visualization tools can provide targeted displays. For example, they can detail energy consumption patterns for different user types, energy efficiency levels, and the distribution of heat and gas energy.

(c) Visualization of Trends and Anomalies

Visualization tools can predict and display uncertain development points, such as spatial load growth trends, energy grid expansion, and potential anomalies or faults. This enables operators to take preemptive optimization measures or emergency responses.

(3) Heterogeneous Multi-source Data Processing

Distributed energy grid clusters involve heterogeneous data from electricity, heat, and gas systems. Efficient integration, querying, transmission, and processing of such data are critical for collaborative operations and intelligent development. Traditional centralized storage models face scalability and reliability challenges. By introducing Hadoop-based distributed storage and processing technologies, these bottlenecks can be addressed.

For example, a framework based on ontology integrates heterogeneous datasets, offering unified interfaces for data acquisition, analysis, and application. This includes a customized SQL-like language for managing and accessing heterogeneous data sources, combined with cloud computing and semantic mapping techniques to enhance multi-source data integration.

(4) Data Mining

Data mining involves extracting hidden, previously unknown but potentially valuable information from massive, dynamic, and heterogeneous datasets in distributed energy grid clusters. It uncovers patterns in energy system data, identifying commonalities and summarizing characteristics across electricity, heat, and gas data.

The process includes:

- Business analysis
- Data collection and organization
- Data analysis and processing
- Indicator display
- Data mining modeling
- Interpretation and presentation of results

Traditional centralized data mining methods face limitations, such as poor scalability for heterogeneous data and labor-intensive preprocessing. However, advancements in cloud computing, edge computing, IoT, and intelligent terminals have revolutionized data mining approaches.

For instance, Hadoop-based platforms integrate diverse computing and storage modes, enabling efficient handling of real-time, mixed-structure data from electricity, heat, and gas systems. Such platforms significantly enhance the processing speed of distributed energy grid data, providing robust support for system optimization and multi-energy collaboration.

References

Bedi G, Venayagamoorthy GK, Singh R, Brooks RR, Wang KC (2018) Review of Internet of Things (IoT) in electric power and energy systems. IEEE Internet Things J 5(2):847–870
Chen J, Ding J, Tian S et al (2018) Clustering analysis of electric load curves based on an improved fast density peaks algorithm. Power Syst Protect Control 46(20):91–99
Čolaković A, Hadžialić M (2018) Internet of Things (IoT): a review of enabling technologies, challenges, and open research issues. Comput Netw 144:17–39
CPS Steering Group (2008) Cyber-physical system executive summary
Han J, Cai Y, Cercone N (1993) Data-driven discovery of quantitative rules in relational databases. IEEE Trans Knowl Data Eng 5(1):29–40
Hao R, Ai Q, Xiao F (2017) Architecture based on multivariate big data platformfor analyzing electricity consumption behavior. Electr Power Autom Equip 37(08):20–27
Jieyang P, Kimmig A, Dongkun W, Niu Z, Zhi F, Jiahai W et al (2023) A systematic review of data-driven approaches to fault diagnosis and early warning. J Intell Manuf 34(8):3277–3304
Koeppel GA (2007) Reliability considerations of future energysystems: multi-carrier systems and the effect of energy storage. Federal Institute of Technology, Zurich, Switzerland, Swiss
Koeppel G, Andersson G (2006) The influence of combined power, gas, and thermal networks on the reliability of supply. In: The sixth world energy system conference, Torino, Italy, pp 10–12
Liu et al (2008) Cyber physical systems: a new frontier. In: IEEE international conference on sensor networks, ubiquitous and trustworthy computing
McGrangahan M, Goodman F (2005) Technical and system requirements of advanced distribution automation. In: 18th International conference and exhibition on electricity distribution, CIRED, Turin, Italy
NSF Cyber-Physical System Program Solicitaion [EB.OB] (2011) http://www.nsf.gov/
NSF Workshop on Cyber-Physical Systems [EB/OB] (2006) Carngie Mellon University, Austin, Tex. http://varma.eee.cmu.edu/CPS/
Rajkumar R, Lee I, Sha L et al (2010) Cyber physical systems: the next computing revolution. In: Proceedings of DAC. ACM, New York, pp 731–736
Revoda J, Lysecky S et al (2011) Interface model based on cyber-physical energy system designed for smart grid. In: IEEE 19th conference on VLSI and system-on-chip, pp 368–373
Wang J, Shi Q (2013) Short-term traffic speed forecasting hybrid model based on chaos–wavelet analysis-support vector machine theory. Transp Res Part C Emerg Technol 27:219–232
Wang D, Zhou F (2018) Electricity load pattern extraction based on unsupervised extreme learning machine. Grid Technol 42(10)
Wang Y, Chen Q, Kang C et al (2017) Clustering of electricity consumption behavior dynamics toward big data applications. IEEE Trans Smart Grid 7(5):2437–2447
Xiong Y, Yao W, Chen W, Fang J, Ai X, Wen J (2020) A data-driven approach for fault time determination and fault area location using random matrix theory. Int J Electr Power Energy Syst 116:105566
Yi S, Yun F, Can C et al (2017) Power user load coding and behavior analysis based on dynamic adaptive K-means clustering. J Electr Power Sci Technol (3)

Chapter 3
Electric-Thermal-Gas Coordinated Control Technology of Distributed Energy Grid Clusters

The coordinated control of multiple energy forms, including electricity, thermal energy, and natural gas, will play a pivotal role in the evolving energy landscape. As the integration of multi-energy flows becomes increasingly essential, establishing an efficient, reliable, and flexible distributed energy grid has become a critical priority. This chapter explores the coordinated control technologies for electricity, heat, and gas, emphasizing the organic interaction and optimization of energy generation, distribution, conversion, and consumption. By analyzing the dynamic characteristics and coupling relationships of multi-energy systems, this work discusses strategies to enhance energy efficiency and economic benefits through distributed coordinated control, providing both a theoretical foundation and practical guidance for the future management of energy grids.

3.1 Coordinated Control Scheme Based on Electric, Thermal, and Gas Energy Synergy

In the future, the interconnection of energy systems will extend beyond the interconnection of electricity and will encompass the integration of multi-energy flows such as electricity, heat, and gas (Ma et al. 2021). The electric-thermal-gas interconnected system is based on a tightly coupled multi-dimensional energy physical system. By organically interacting and coordinating the optimization of energy generation, distribution, conversion, and consumption, it enables bidirectional on-demand transmission and dynamically balanced use of energy, thus forming an integrated system for energy production, supply, and consumption.

T. Lu et al., *Operation Modeling and Collaborative Regulation of Distributed Energy Grid Clusters*, Smart Energy Systems, https://doi.org/10.1007/978-981-96-9942-1_3

3.1.1 Objectives and Characteristics of Coordinated Control

Currently, the energy flow analysis models of power grids, gas networks, and thermal networks, along with their corresponding energy conversion models for various devices, are not uniform. The operation of each energy system is relatively independent, with a lack of necessary communication and interaction between systems. Multi-energy devices struggle to cooperate with one another, leading to suboptimal energy utilization. Therefore, it is necessary to propose a distributed coordinated control method that enhances energy efficiency and economic benefits, based on the dynamic characteristics and coupling relationships of multi-energy systems.

The operational constraints and control variables differ for different types of systems. For example, in natural gas transmission networks and electrical hybrid networks, the same variables may be subjected to different constraints. To achieve the coordinated optimization of multi-energy systems, it is essential to establish a stable analysis model that can quickly and accurately integrate different system state variables (Huang et al. 2022). Employing coordinated control strategies to improve the energy utilization efficiency of DEGs is a key research focus. The staged use of fossil energy and the establishment of multi-objective optimized energy systems can effectively enhance the energy efficiency of the system. In distributed energy grid clusters, with the aim of enhancing the economic and reliability performance of the power grid, hierarchical models, control architectures, optimization objectives, and coordination strategies can be designed to achieve coordinated optimization control of complex power grids.

For distributed energy grid clusters with electric-thermal-gas synergy, optimization should consider the staged use of energy, making comprehensive use of input energy and internal energy to achieve higher energy efficiency. For systems with ES, the scheduling of ES can save energy costs, extend the lifespan of storage, and perform decoupled optimization of energy flows without compromising the complementary benefits of multi-energy sources, thereby maximizing the synergistic effects between different energy flows.

The objective of coordinated control in energy grid clusters is to enhance overall energy efficiency by optimizing the collaborative operation of electricity, heat, and natural gas systems while ensuring the stability and reliability of the system. Through intelligent scheduling and real-time response, the system can dynamically balance energy demands, quickly adapt to load fluctuations and external changes, reduce energy and operational costs, and prolong the lifespan of equipment. Moreover, by reducing carbon emissions, optimizing the energy utilization structure, and increasing the proportion of low-carbon energy use, the promotion of green development is also a focal point of current research.

The coordinated control of energy grid clusters is a complex engineering task, with its main characteristics as shown in Table 3.1.

Table 3.1 Characteristics of coordinated control in electric-thermal-gas synergy energy grid clusters

Characteristics	Description
Dynamic response and real-time adjustment	The system involves multiple types of energy and loads. The coordinated control can respond in real-time to changes in the external environment, load fluctuations, and system state variations
Flexibility and expandability	The system can flexibly adjust the distribution of electricity, thermal energy, and natural gas according to demand and environmental changes. It also facilitates the integration of renewable energy sources
High reliability and robustness	Through redundancy design and fault-tolerant control, the system can self-regulate to ensure overall reliability and continuous operation, even when a single device fails
User DR and comfort	Users can flexibly adjust energy plans based on their individual needs and energy pricing to improve adaptability
Efficient utilization of renewable energy	Dynamic coordinated control can effectively address the variability and uncertainty of PV and WP output

3.1.2 Energy Routers, Converters, and Hubs

The Energy Router (ER), as one of the key energy conversion devices, plays a crucial role in enabling the aforementioned energy internet technologies. In addition to functioning as a multi-level converter system, it must also be closely coupled with information networks. ERs provide flexible AC/DC ports within the transmission network, thereby enabling AC/DC grid interconnection. This serves as the foundation for developing "plug-and-play" terminal interfaces for distributed energy sources, ES devices, electric vehicles, and charging piles. Moreover, due to the close integration between ERs and information networks, they can monitor, collect, and control electrical quantities at each port in real-time. This provides comprehensive operational data for the entire system to meet the management and scheduling requirements of various networks.

Unlike ERs, Energy Switches (ESW) and Energy Hubs (EHs) are core components on the distribution side (or distributed energy grid side) (Zhang et al. 2021). The energy switch, based on power electronics and information technologies, is capable of altering grid voltage levels and converting energy forms. This enables functions such as power isolation, transmission, and power quality control.

Structurally, energy converters and ERs are similar, as both can perform AC/DC conversions. The EH unit, on the other hand, serves as an interface between various energy infrastructures and/or loads. Within an EH, energy can be converted through a series of devices such as CHP systems, gas compressors, transformers, and power electronics equipment. Each energy grid can utilize other networks to fulfill part of its own system's load demands. This highlights how EHs can effectively address network congestion issues, reduce the probability of congestion occurrences, and enhance production efficiency and energy utilization (Dezhdar et al. 2023).

Another major advantage of EHs is their ability to simplify management processes. In distributed energy grid clusters, the energy grid operator only needs to manage the input to the grid where the EH is located, without overseeing the output from the hub. The EH itself can support demand-side management across electricity, heat, and gas, enabling a more efficient and streamlined overall management procedure for the multi-energy system.

It is evident that ERs, converters, and hubs are the core devices for realizing multi-energy complementarity.

3.1.3 Hierarchical Coordinated Control of Electricity-Thermal-Gas Systems Based on EHs

The electricity-thermal-gas coordinated energy grid system utilizes EHs to achieve the conversion between different energy types, forming a hybrid network that integrates electricity, heat, and other energy sources. This represents an important development direction for future energy interconnection.

EHs enable energy conversion between various sources and serve as regulators for economic dispatch and optimal power flow in distributed energy grid clusters. Acting as energy conversion stations, EHs achieve energy transformation through the internal coordination of various devices and components (Kim et al. 2024). With the introduction of EHs, the different energy carriers within the grid become interdependent and complementary, enhancing the overall economic efficiency of the system.

(1) Functional Characteristics of EHs

The EH, as an energy conversion unit capable of meeting multiple energy demands, provides interfaces for the input and output of various energy types. Its specific functions are as follows.

(a) Plug-and-Play Capability

EHs are equipped with universal standardized interfaces, ensuring simplicity and ease of use. These interfaces integrate both power conversion and communication functions. The standardization of power interfaces is achieved through power electronic converter devices, while the communication system can quickly identify the electrical characteristics of different energy types and facilitate energy exchange based on user terminal requests.

(b) Bidirectionality

From the perspective of the energy grid, user terminals can act either as energy consumers or energy suppliers. EH interfaces are bidirectional to accommodate the dual-directional flow of energy. For economic and practical considerations, interfaces connected to loads can be output-only, those connected to power sources can be input-only, and those connected to ES devices must allow bidirectional flow.

(c) Energy Management Functionality

EHs collect energy data and device status information to coordinate power distribution and optimize energy scheduling between DEGs and distribution networks. With the support of ES devices, they achieve a balance between energy supply and demand within DEGs.

(d) Real-Time Communication Capability

EHs rely on real-time communication to acquire accurate data for precise control and management functions.

EHs establish corresponding coupling models for power conversion between different energy carriers. From a system perspective, coupling various energy carriers offers numerous potential advantages compared to traditional decoupled energy supply networks. The redundant energy flow paths provide a degree of freedom, creating opportunities for multi-energy coordinated optimization.

As shown in Fig. 3.1, the dashed box in the diagram represents an EH. Its inputs and outputs are of different energy forms, and it participates in both the feedback and feedforward regulation of the system. In the diagram, VP_g represents the efficiency of natural gas input into the CHP, and $(1-V)P_g$ represents the efficiency of natural gas input into the GBs. $G_{DR}(t)$ and $T_{DR}(t)$ correspond to the demands for energy substitution and electric energy conversion, respectively.

An EH consists of two basic components: direct connections and converter connections (Ding et al. 2022). The EH uses a conversion matrix T to link the

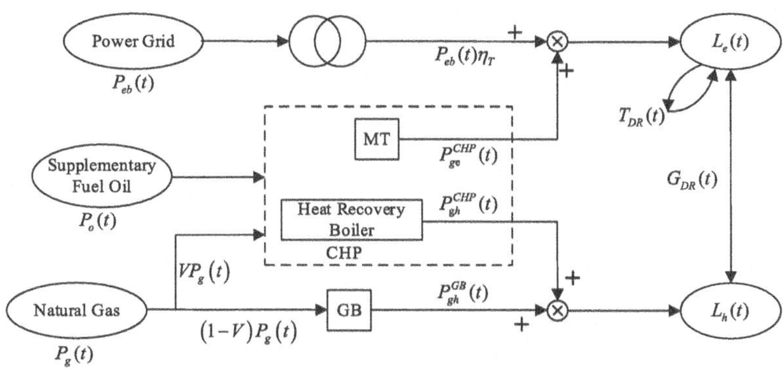

Fig. 3.1 Hierarchical system framework of electricity-thermal-gas based on EHs

input energy matrix P with the output energy $L = TP$. The matrix T represents the conversion efficiency between all input and output energy types, where $T = 0$ indicates no conversion relationship between certain energy types.

The fundamental problem in operating a distributed energy grid based on the EH is to find the optimal conversion matrix T under certain constraints.

The heat-electricity decomposition matrix is as follows:

$$\begin{bmatrix} L_{\mathrm{e}}(t) \\ L_{\mathrm{h}}(t) \end{bmatrix} = \mathbf{AB}$$
$$\mathbf{A} = \begin{bmatrix} \eta^{T} & \frac{1}{1+V_{\mathrm{CHP}}(t)} V(t)\eta_{\mathrm{ge}}^{\mathrm{CHP}} & \eta_{\mathrm{oe}}^{\mathrm{CHP}} \\ 0 & (1 - V(t))\eta_{\mathrm{gh}}^{\mathrm{GB}} \frac{V_{\mathrm{CHP}}(t)}{1+V_{\mathrm{CHP}}(t)} V(t)\eta_{\mathrm{gh}}^{\mathrm{CHP}} & \eta_{\mathrm{oh}}^{\mathrm{CHP}} \end{bmatrix} \tag{3.1}$$
$$\mathbf{B} = \begin{bmatrix} P_{\mathrm{e}}(t) \\ P_{\mathrm{g}}(t) \\ P_{\mathrm{o}}(t) \end{bmatrix}$$

where, $L_{\mathrm{e}}(t)$ and $L_{\mathrm{h}}(t)$ represent the electric and thermal load for transfer, respectively. $P_{\mathrm{e}}(t)$, $P_{\mathrm{g}}(t)$, and $P_{\mathrm{o}}(t)$ represent the total energy consumption of electricity, natural gas, and oil within the regional energy grid. V represents the proportion of natural gas input to the CHP system relative to the total natural gas purchase, satisfying $0 \leq \mathrm{V} \leq 1$. The heat-to-power ratio, $V_{\mathrm{CHP}}(t) = P_{\mathrm{gh}}^{\mathrm{CHP}}/P_{\mathrm{ge}}^{\mathrm{CHP}}$, represents the ratio of CHP thermal output power to electric output power. $\eta_{\mathrm{gh}}^{\mathrm{GB}}$ refers to the heating efficiency of the GBs. η^{T} represents the efficiency of the transformer. $\eta_{\mathrm{ge}}^{\mathrm{CHP}}$ and $\eta_{\mathrm{gh}}^{\mathrm{CHP}}$ represent the efficiency of natural gas converting to electricity and heat through CHP, respectively, where $\eta_{\mathrm{oe}}^{\mathrm{CHP}}$ and $\eta_{\mathrm{oh}}^{\mathrm{CHP}}$ denote the electrical and thermal conversion efficiencies of CHP.

(2) Coordinated Electricity-Thermal-Gas Energy Grid System Model

(a) Thermal System Model

The thermal system typically consists of thermal sources, thermal networks, and thermal loads. The thermal energy transmitted or consumed in a thermal system is determined by the water flow rate and temperature, as expressed in Eq. (3.2). The thermal loss during transmission in thermal pipelines can be described by the temperature drop of the water flow within the pipelines, as shown in Eq. (3.3). At pipeline junction nodes, hot water satisfies the law of energy conservation, and the inflow-outflow relationship at the nodes is illustrated in Eq. (3.4).

$$H_i(t) = \kappa \cdot m_i(t) \cdot \left(T_i^{\mathrm{g}}(t) - T_i^{\mathrm{r}}(t)\right) \tag{3.2}$$

$$T_p^{\mathrm{end}}(t) - T_p^{\mathrm{a}}(t) = \left(T_p^{\mathrm{start}}(t) - T_p^{\mathrm{a}}(t)\right) \cdot \mathrm{e}^{-\frac{\delta_p \cdot L_p}{\kappa \cdot m_p(t)}} \tag{3.3}$$

$$\left(\sum m_i^{\text{out}}(t)\right) \cdot T_i^{\text{out}}(t) = \sum\left(m_i^{\text{in}}(t) \cdot T_i^{\text{in}}(t)\right) \tag{3.4}$$

where, H_i represents the thermal supply power or thermal load demand power at node i, while κ is the specific thermal capacity of water. m_i denotes the mass flow rate of water entering or leaving the thermal source or thermal load node i, and T_i^g and T_i^r are the supply and return water temperatures at the thermal source or thermal load node i, respectively. T_p^a, T_p^{start} and T_p^{end} represent the ambient temperatures at the start and end of pipeline p, respectively, while δ_p, L_p and m_p denote the thermal transfer coefficient, pipeline length, and mass flow rate of pipeline p, respectively. Additionally, T_i^{in} and m_i^{in} are the temperature and mass flow rate of water flowing into node iii from the connecting pipe, T_i^{out} and m_i^{out} are the temperature and mass flow rate of water flowing out of node iii to the connecting pipe.

(b) Natural Gas System Model

The natural gas system primarily consists of natural gas pipelines, compressor stations, gas loads, pressure regulating valves, and other components (Zhang et al. 2022). The flow of natural gas through the pipeline can be effectively controlled by adjusting pressure regulation valves and similar devices. The gas flow rate within the pipeline is closely related to the pressures at the nodes on either side of the pipeline, as well as the transmission conditions of the pipeline itself, i.e.,

$$f_{ij}(t) = \sigma_{ij} \cdot \upsilon_{ij}(t)\sqrt{\upsilon_{ij}(t) \cdot \left[(\pi_i(t))^2 - \left(\pi_j(t)\right)^2\right]} \tag{3.5}$$

where, f, σ, υ and π represent the natural gas flow rate, pipeline characteristic parameters, gas flow direction variables, and pressure, respectively. The subscripts i, j and ij refer to the pipeline's starting node, ending node, and pipeline itself, respectively. Specifically, when $\pi_i \leq \pi_j$, $\upsilon_{ij} = -1$, the flow is directed from the starting node to the ending node; otherwise, $\upsilon_{ij} = +1$, the flow direction is reversed.

The compressor station consists of a GT, an engine, and a compression mechanism. Its function is to increase the pressure of the natural gas within the pipeline to counteract the pressure loss during transportation. The GT draws natural gas from the compressor station to provide the necessary electrical energy for the operation of the compressor. The compressor model is represented by:

$$\begin{cases} P_{\text{com}}(t) = o_{\text{com}} \cdot f_{\text{com}}\left[\left(\frac{\pi_{\text{com}}^{\text{out}}(t)}{\pi_{\text{com}}^{\text{in}}(t)}\right)^Z - 1\right] \\ \tau_{\text{com}} = \omega_{\text{com}} + \theta_{\text{com}} \cdot P_{\text{com}}(t) + \rho_{\text{com}} \cdot \left(P_{\text{com}}(t)\right)^2 \end{cases} \tag{3.6}$$

where, o, τ and Z represent the compressor characteristic parameters, the flow rate consumed by the compressor, and the parameters related to the compressor factor, respectively. The subscript "com" denotes the compressor. ω, θ and ρ correspond to the characteristic curve parameters associated with the compressor's energy consumption.

(c) Electricity System Model

In this paper, the power system uses a DC power flow model. The node power balance model is given by Eq. (3.7), and the branch power flow model is provided by Eq. (3.8).

$$P_i(t) = \sum_{j \in i} \left(B_{ij} \cdot \theta_j(t) \right) \tag{3.7}$$

$$P_{ij}(t) = -B_{ij} \cdot \Delta\theta_{ij}(t) = \frac{\theta_i(t) - \theta_j(t)}{x_{ij}} \tag{3.8}$$

where, B, θ, x and Δ represent the branch admittance, node phase angle, branch reactance, and the difference, respectively. The subscripts i and j denote the different node numbers in the power system. The branch with starting and ending nodes ij is denoted as the branch between node i and node j. Specifically, when $B_{ij} = -1/x_{ij}$, it implies there is no difference in the node phase angle, $B_{ii} = \sum_{j \in i, j \neq i} 1/x_{ij}$.

(3) Hierarchical Optimization Model for the Integrated Electric-Thermal-Gas Synergy System

The main difference between CHP systems and conventional power generation units lies in their ability to collect excess thermal from power generation through a waste thermal boiler, thereby improving fuel utilization efficiency. Waste thermal boilers can be classified into two types: those without supplementary combustion devices (non-supplementary waste thermal boilers) and those with supplementary combustion devices (supplementary combustion waste thermal boilers). The thermal-to-power ratio of non-supplementary waste thermal boilers is generally considered unchangeable, while units with supplementary combustion devices can adjust the thermal-to-power ratio within a certain range by burning additional fuel. In most cases, the additional fuel is used solely to regulate the thermal-to-power ratio and does not generate any net thermal or electrical power. Therefore, the thermal-to-power conversion efficiencies, $\eta_{\text{oe}}^{\text{CHP}}$ and $\eta_{\text{oh}}^{\text{CHP}}$, the supplementary fuel are both considered to be zero.

As shown in Fig. 3.2, the overall optimization model of the integrated electric-thermal-gas synergy system can be divided into three layers: the multi-energy dispatch layer, the energy efficiency optimization layer, and the unit allocation layer. These layers are described as follows:

(a) Multi-energy Dispatch Layer

The objective of this layer is to minimize the overall energy cost of the distributed energy grid. The CHP obtained from the lower layer optimization is substituted into the optimization calculations. The CHP unit is a power-thermal coupling element,

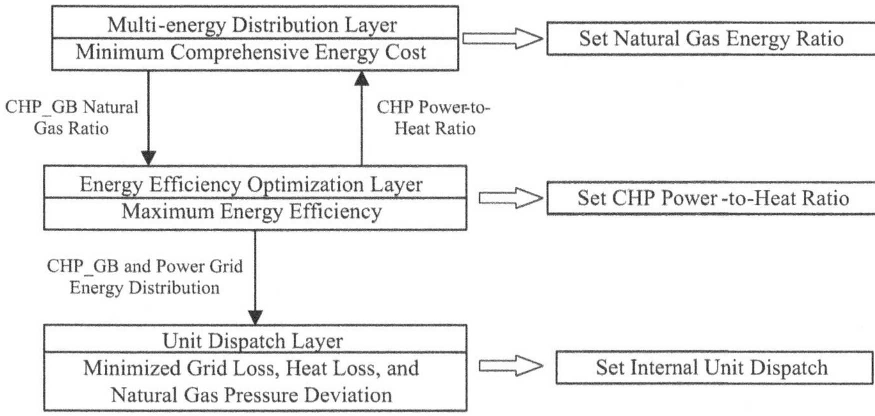

Fig. 3.2 Hierarchical optimization model of electric-thermal-gas interconnected energy system

and when the electricity and thermal prices differ, the system's overall energy cost is most sensitive to the real-time natural gas consumption ratio of CHP and GB. The real-time natural gas consumption ratio for CHP and GB is set as the output of this layer, and the optimization results are then used as known parameters in the lower-layer optimization model.

(b) Energy Efficiency Optimization Layer

Based on the natural gas consumption ratio for CHP and GB provided by the upper layer optimization, this layer maximizes the energy efficiency of the integrated energy system. Transformers, waste thermal recovery systems, and GBs have relatively high energy conversion efficiencies (around 90%), while the energy conversion efficiency of CHP units for power generation is lower (around 36%). The system's energy efficiency is highly sensitive to the real-time thermal-to-power ratio of the CHP units. The CHP thermal-to-power ratio is set as the output of this layer, and the optimization results are then used as known parameters in the upper-layer optimization model.

(c) Unit Allocation Layer

Based on the optimization results from the previous two layers, this layer establishes a dynamic energy connector model. The optimization quantity is the power distribution of similar units, and it selects multi-energy grid performance evaluation indicators (such as line losses, thermal losses, and gas network node pressure deviations) to optimize the real-time operation state of the integrated energy grid.

(4) Hierarchical Optimization of Electricity-Thermal-Gas Energy Flow

(a) Objective Function

The upper-layer optimization aims to minimize the total energy cost, which includes both the **comprehensive operational cost** $C(t)$ and the **pollutant emission cost**

$P(t)$. The decision variable $V(t)$ in this layer is the ratio of the natural gas power input to the total purchased gas power, which is used for the **CHP** system.

$$\min F(V(t)) = \sum_t C(t) + \sum_t P(t) \tag{3.9}$$

where, C denotes the total operational cost, and P denotes the pollutant emission cost.

The total operational cost considers the energy procurement cost, equipment depreciation cost, and demand-side response revenue, as shown in Eq. (3.10).

$$C_1(t) = \frac{P_{\text{g}}(t)}{Q_{\text{gas}}}\lambda_t^{\text{g}} + P_{\text{eb}}(t)\lambda_t^{\text{eb}} + \frac{P_{\text{o}}(t)}{Q_{\text{o}}}\lambda_t^{\text{o}} \tag{3.10}$$

where, $C_1(t)$ represents the energy procurement cost, which consists of three components on the right-hand side of the equation: the natural gas procurement cost, the electricity price cost, and the supplementary fuel cost. Specifically, λ_t^{g}, λ_t^{eb} and λ_t^{o} denote the prices of natural gas, real-time electricity prices from the grid, and supplementary fuel, respectively. Q_{gas} refers to the lower heating value of natural gas, which is taken as 9.97 kW h/m^3, and Q_{o} is the lower heating value of fuel, taken as 11.917 kW h/kg. Finally, $P_{\text{o}}(t)$ represents the response fuel consumption.

$$C_2(t) = \frac{C_{\text{CHPIns}}\delta_{\text{FRCHP}}}{8760P_{\text{N,CHP}}}(P_{\text{ge}}^{\text{CHP}}(t) + P_{\text{gh}}^{\text{CHP}}(t)) + \frac{C_{\text{GBIns}}\delta_{\text{FRGB}}}{8760P_{\text{N,GB}}}P_{\text{gh}}^{\text{GB}}(t) \tag{3.11}$$

where, $d_i(i = CHP, GB)$ represents the annual depreciation rate of power generation type i. The annual depreciation rates for CHP and GB are both set to 0.1. Additionally, L_i denotes the service life limit for the power generation type i.

$$\delta_{\text{FR}i} = \frac{d_i(1+d_i)^{L_i}}{(1+d_i)^{L_i} - 1} \tag{3.12}$$

where, $d_i(i = CHP, GB)$ represents the annual depreciation rate for the corresponding power generation type i. The annual depreciation rates for CHP and GB can both be set to 0.1. L_i denotes the depreciation lifespan for power generation type i.

$$C_{\text{DR}}(t) = P_{\text{ge}}^{\text{CHP}}(t)\lambda^{\text{IDR}} + G_{\text{DR}}(t)\left(\lambda_t^{\text{eb}} - \frac{\lambda_t^{\text{g}}}{Q_{\text{gas}}}\right) + T_{\text{DR}}(t)\lambda_t^{\text{eb}} \tag{3.13}$$

where, $C_{\text{DR}}(t)$ represents the benefits from the demand-side response, including CHP power generation revenue, energy substitution benefits, and electricity peak-shifting benefits. λ^{IDR} refers to the portion of the CHP power generation feed-in tariff that exceeds the standard electricity price.

$$\begin{aligned} P_{\mathrm{p}}(t) &= P_{\mathrm{gh}}^{\mathrm{GB}}(t)\sum \lambda_i^{\mathrm{GB}} G_i^{\mathrm{GB}} + P_{\mathrm{o}}(t)\sum \lambda_i^{\mathrm{oil}} G_i^{\mathrm{oil}} \\ &\quad + P_{\mathrm{ge}}^{\mathrm{CHP}}(t)\lambda_i^{\mathrm{CHP}} G_i^{\mathrm{CHP}} + P_{\mathrm{eb}}(t)\sum \lambda_i^{\mathrm{eb}} G_i^{\mathrm{eb}} \\ &= P_{\mathrm{gh}}^{\mathrm{GB}}(t)C_{\mathrm{GB}} + P_{\mathrm{o}}(t)C_{\mathrm{o}} + P_{\mathrm{ge}}^{\mathrm{CHP}}(t)C_{\mathrm{CHP}} + P_{\mathrm{eb}}(t)C_{\mathrm{eb}} \end{aligned} \tag{3.14}$$

where, $P_{\mathrm{p}}(t)$ represents the cost of pollutant treatment; λ_i^{GB}, λ_i^{oil}, λ_i^{CHP} and λ_i^{eb}, as well as G_i^{GB}, G_i^{oil}, G_i^{CHP} and G_i^{eb}, respectively denote the environmental costs of pollution and the pollutant emission intensities (in units of kg/(kW h)) associated with GB, supplementary fuel equipment, CHP, and the power grid. Here, iii represents the type of pollutant, including CO, SO_2, and NO_x. The pollutant treatment costs per unit of power generation for GB, supplementary fuel equipment, CHP, and the power grid are denoted by C_{GB}, C_{o}, C_{CHP} and C_{eb}, respectively.

(b) Objective Function

For distributed energy grid clusters, user participation in demand-side response is no longer limited to traditional demand-side response mechanisms. Regarding conventional electricity elastic loads, user response behaviors are specifically reflected in two ways: reducing unnecessary loads during peak electricity consumption periods and shifting part of the load from peak periods to off-peak periods.

Conventional electricity elastic loads are characterized by user responses to electrical loads on a temporal scale, representing one form of demand-side response for regional integrated energy users. Another form occurs when users shift some electrical loads to thermal energy during periods of high electricity prices or convert part of their thermal loads to electricity during periods of low electricity prices. In this case, elastic loads are no longer confined to electrical loads.

This chapter refers to such demand-side responses as generalized demand-side responses, as detailed in Fig. 3.3.

Based on the characteristics of integrated demand-side response, a demand-side response model for the distributed energy grid clusters is established. Let $P_{\mathrm{GDR}}(t)$ denote the power at time t when electricity is substituted for thermal energy, and

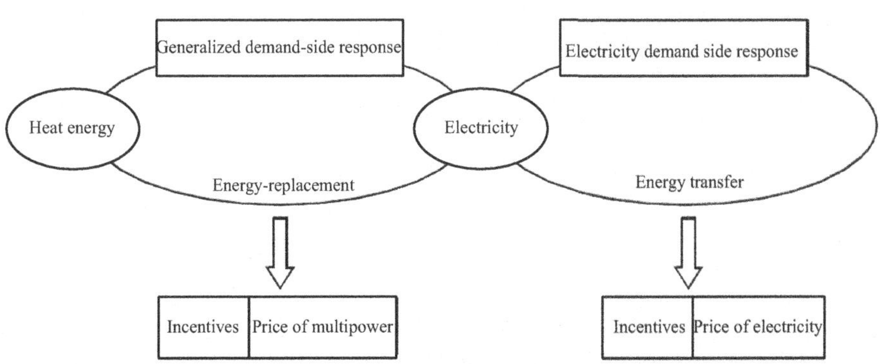

Fig. 3.3 Illustration of integrated demand-side response

$P_{\mathrm{TDR}}(t)$ denote the power reduction by users at time t. The specific response model is given as follows:

$$\begin{cases} L_{\mathrm{e}}(t) = \overline{L}_{\mathrm{e}}(t) - G_{\mathrm{DR}}(t) + T_{\mathrm{DR}}(t) \\ L_{\mathrm{h}}(t) = \overline{L}_{\mathrm{h}}(t) + G_{\mathrm{DR}}(t) \end{cases} \tag{3.15}$$

where, $\overline{L}_{\mathrm{e}}(t)$ and $\overline{L}_{\mathrm{h}}(t)$ represent the predicted values of electrical and thermal loads, respectively.

To minimize the impact of the distributed energy grid clusters on the local power grid, the electricity market imposes regulations on the range of regional electricity purchases, setting upper and lower fluctuation limits on the amount of electricity procured within the region.

$$0 \le P_{\mathrm{eb}}(t) < P_{\mathrm{eb,max}} \tag{3.16}$$

Based on the characteristics of the generation units, the upper and lower bounds of power output for CHP and GTs are constrained as follows:

$$\begin{aligned} 0 \le P_{\mathrm{ge}}^{\mathrm{CHP}}(t) < P_{\mathrm{ge,max}}^{\mathrm{CHP}} \\ 0 \le P_{\mathrm{gh}}^{\mathrm{GB}}(t) < P_{\mathrm{GB,max}} \end{aligned} \tag{3.17}$$

A linear model is used to fit the supplementary combustion rate A(%), the thermal-to-electricity ratio V_{CHP}, and the corresponding fuel consumption $P_{\mathrm{o}}(t)$ of the supplementary combustion units in CHP systems. The fitted expressions are given as follows:

$$\begin{aligned} V_{\mathrm{CHP}}(t) = K_{\mathrm{a1}}A(t) + K_{\mathrm{a2}} \\ A(t) = K_{\mathrm{p1}}P_{\mathrm{o}}(t) + K_{\mathrm{p2}} \end{aligned} \tag{3.18}$$

where, K_{a1}, K_{a2}, K_{p1} and K_{p2} are the model parameters fitted based on the characteristics of the supplementary combustion unit.

Substituting the above equation into the thermoelectric decoupling matrix and applying the definition of the thermal-to-electricity ratio, we obtain:

$$\eta_{\mathrm{gh}}^{\mathrm{CHP}} = \eta_{\mathrm{ge}}^{\mathrm{CHP}} V_{\mathrm{CHP}} = \eta_{\mathrm{ge}}^{\mathrm{CHP}}[K_{\mathrm{a1}}(K_{\mathrm{a1}}P_{\mathrm{o}}(t) + K_{\mathrm{p2}}) + K_{\mathrm{a2}}] \tag{3.19}$$

Maximum supplementary combustion rate constraint:

$$0 \le A \le A_{\mathrm{max}} \tag{3.20}$$

where, A_{max} represents the maximum supplementary combustion rate of the internal supplementary combustion unit, expressed as a percentage (%).

The thermal system satisfies the nodal mass flow balance equation, i.e.,

$$A_{\mathrm{DHS}}^{\mathrm{w}} \cdot \dot{m}_{\mathrm{p}} = \dot{m}_0 \tag{3.21}$$

where, $A_{\mathrm{DHS}}^{\mathrm{w}}$ is the fully connected incidence matrix of the nodes and pipelines in the thermal system; $\dot{m}_{\mathrm{p}}$ represents the mass flow vector of the thermal system pipelines; and $\dot{m}_0$ denotes the net load vector of the nodes in the thermal system.

In the natural gas system, the nodal mass flow must satisfy the flow conservation law, i.e.,

$$A_{\mathrm{NGS}}^{\mathrm{w}} \cdot \dot{\boldsymbol{f}}_{\mathrm{pipe}} + A_{\mathrm{NGS}}^{\mathrm{s}} \cdot \dot{\boldsymbol{\tau}}_{\mathrm{com}} = \dot{\boldsymbol{f}}_0 \tag{3.22}$$

where, $A_{\mathrm{NGS}}^{\mathrm{w}}$ is the fully connected incidence matrix of the nodes and pipelines in the natural gas system; $\dot{\boldsymbol{f}}_{\mathrm{pipe}}$ represents the flow vector of the natural gas pipelines; $A_{\mathrm{NGS}}^{\mathrm{s}}$ is the incidence matrix of the starting points in the natural gas system; $\dot{\tau}_{\mathrm{com}}$ denotes the flow vector of the compressors in the natural gas system; and $\dot{\boldsymbol{f}}_0$ the net load vector of the nodes in the natural gas system.

The power balance constraint of the electrical system is primarily determined by the nodal power injection and the nodal phase angle, i.e.,

$$\dot{P}_{\mathrm{PS}} = B \cdot \dot{\theta} \tag{3.23}$$

where, $\dot{P}_{\mathrm{PS}}$ represents the nodal power injection vector of the electrical system; B is the branch susceptance matrix; and $\dot{\boldsymbol{\theta}}$ is the nodal voltage phase angle vector of the electrical system.

(5) Efficiency Optimization Layer

The thermal efficiency $\eta_{\mathrm{gh}}^{\mathrm{CHP}}$ and power generation efficiency $\eta_{\mathrm{ge}}^{\mathrm{CHP}}$ of CHP differ significantly. The distributed energy grid clusters' efficiency is most sensitive to the CHP thermal-to-electricity ratio (Mallier et al. 2021). The optimization objective of the efficiency optimization layer is to maximize the utilization rate of energy resources, with the decision variable being the CHP thermal-to-electricity ratio. The objective function is defined as:

$$\max F(V_{\mathrm{CHP}}) = \frac{\sum L}{\sum P_{\mathrm{in}}} = \frac{\sum [L_{\mathrm{e}}(t) + L_{\mathrm{h}}(t) + L_{\mathrm{c}}(t)]}{\sum [P_{\mathrm{e}}(t) + P_{\mathrm{g}}(t) + P_{\mathrm{o}}(t)]} \tag{3.24}$$

where, F represents the ratio of total energy output to energy input. The decision variables of the multi-energy flow layer and the efficiency optimization layer are mutually dependent. Solving the optimization problem of the multi-energy flow layer and the efficiency optimization layer is a typical NBLP (Non-deterministic Polynomial Linear Programming) problem.

NBLP is an NP (Non-deterministic Polynomial) problem because the lower-level model has the characteristics of continuity, non-linearity, and convexity, and the lower-level model is feasible and differentiable. According to the KKT optimization conditions, the extreme value of the lower-level model is treated as 0. The KKT model can be used to establish Lagrange functions for the objective function and constraints of the lower-level model. Assuming there are N constraints in the lower-level model, the specific optimization model is:

$$\begin{aligned} &\max F(V, V_{\mathrm{CHP}}) \\ &\text{s.t.} \quad \mathbf{g}_i(V, V_{\mathrm{CHP}}) = 0 \quad i = 1, 2, \ldots, N \end{aligned} \tag{3.25}$$

Let $(\overline{V}, \overline{V}_{CHP})$ represent the feasible solution for the optimization of the multi-energy flow layer, and $\boldsymbol{\lambda}^{\mathrm{T}}$ denote the Lagrange multiplier vector. For a fixed $\overline{V}$, based on the optimality conditions, the efficiency optimization problem is equivalent to solving the KKT point problem. The KKT conditions can be expressed as follows:

$$\begin{aligned} \min L = &\left\| \nabla_{V_{\mathrm{CHP}}} F(V, V_{\mathrm{CHP}}) + \boldsymbol{\lambda}^{\mathrm{T}} \nabla_{V_{\mathrm{CHP}}} g_i(V, V_{\mathrm{CHP}}) \right\|^2 \\ &+ \left\| \boldsymbol{\lambda}^{\mathrm{T}} \nabla_{V_{\mathrm{CHP}}} g_i(V, V_{\mathrm{CHP}}) \right\|^2 \\ \text{s.t.} \quad &\lambda_i \geq 0 \end{aligned} \tag{3.26}$$

where, $\boldsymbol{\lambda}^{\mathrm{T}} = (\lambda_1, \lambda_2, \ldots, \lambda_N)$ represents the Lagrange multiplier vector. The specific expression for the partial derivative with respect to the lower-level decision variable V_{CHP} is given as:

$$\frac{\partial L(V, V_{\mathrm{CHP}})}{\partial V_{\mathrm{CHP}}} = \frac{-[L_{\mathrm{e}}(t) + L_{\mathrm{h}}(t)]K_1}{(K_1 V_{\mathrm{CHP}} + K_2)^2} + (a_2 - a_1) = 0 \tag{3.27}$$

where,

$$\begin{gathered} K_1 = 1/(K_{\mathrm{a1}} K_{\mathrm{p1}}) \\ K_2 = (L_{\mathrm{c}}(t) - P_{\mathrm{ge}}^{\mathrm{CHP}}(t))/\eta_{\mathrm{T}} + P_{\mathrm{ge}}^{\mathrm{CHP}}(t)/\eta_{\mathrm{ge}}^{\mathrm{CHP}} - (K_{\mathrm{a2}}/K_{\mathrm{a1}} - K_{\mathrm{p2}})/K_{\mathrm{p1}} \\ a_1(V_{\mathrm{CHP}} - K_{\mathrm{a2}} - K_{\mathrm{a1}} A_{\max}) = 0 \\ a_2(-V_{\mathrm{CHP}} + K_{\mathrm{a2}}) = 0 \\ a_1, a_1 \geq 0 \end{gathered} \tag{3.28}$$

The Lagrange function is used as a feasibility metric for the upper-level multi-energy flow. The feasible region is defined as the set of all unknown variables that satisfy $L = 0$.

(6) Unit Optimization Allocation Layer

Under the framework of the distributed energy grid clusters, the close coupling of various energy conversion devices introduces challenges in optimal power flow and operational decision-making. The power network directly transmits electrical energy to the load side, with some generation nodes, such as CHP, coupled with the thermal system. The thermal system includes thermal sources, thermal loads, heating loops, and heat recovery loops. Thermal sources are regarded as load nodes of the natural gas network, which itself comprises gas sources, pipelines, compressors, and natural gas loads.

Traditional power flow analysis methods are only applicable to pure power systems, presenting significant limitations when applied to distributed energy grid clusters. The steady-state Multi-Carrier Power Flow (MCPF) model for electrothermal coupled systems offers a new perspective for scheduling decisions in distributed energy grid clusters.

3.2 Coordination Controller of Distributed Energy Grid

3.2.1 Structure of the Coordination Controller

Taking into account the operational interaction between multiple energy sources, such as electricity, heating, and gas, a multi-dimensional information model is developed. This model integrates operational data from the distribution network, various distributed energy sources, and flexible loads. The aim is to enhance resource utilization and precise control of power flow across the interconnecting lines while ensuring the power quality and reliability of the distribution network (Stennikov et al. 2022). This system enables functions such as energy management, operational control, and optimal scheduling.

In light of the future ubiquity of the IoT, this section designs an intelligent hierarchical regulation framework based on dynamic multi-agent technology (Laghari et al. 2021). The proposed architecture consists of a layered energy grid system with four layers: the perception layer, the network layer, the platform layer, and the application layer. The platform design architecture is shown in Fig. 3.4.

(1) **Perception Layer**: This includes the information collection and detection of various types of distributed energy sources and multi-flexible loads. The distributed energy source intelligent agents have modules such as SA, command execution, and data communication. They are capable of interacting with multiple energy forms (electricity, thermal, gas) based on instructions from the system control center and their own operational constraints while enabling real-time information exchange and data uploading between intelligent agents.

    The coordination controller has functions such as forecasting, communication, load control, and DR. It can upload load information in real-time, control the load, and provide feedback on the load status.

(2) **Network Layer**: This includes the physical network and information network of multi-energy flows such as electricity, thermal, and gas. The physical network receives control signals from the regionally distributed energy coordination controllers, while the information network is responsible for transmitting these control signals.

    The coordination controller features regional cooperative regulation and real-time communication modules. It can accept control instructions from

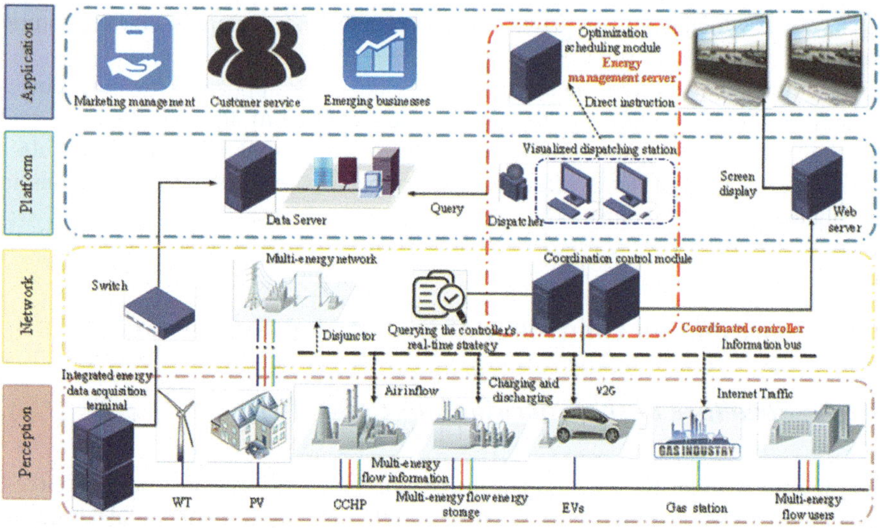

Fig. 3.4 Hierarchical control framework design

higher-level systems and, considering the system's energy efficiency constraints, perform cooperative regulation of distributed energy systems between different regions. This ensures the real-time dynamic adjustment requirements of the distributed energy systems are met. The controller issues instructions based on the priority of load and distributed energy.

Based on a multi-agent marginal consistency control algorithm, the regional distributed energy regulation system ensures "centralized management, decentralized control" of the local perception layers within the region, guaranteeing voltage and frequency stability across all regions.

(3) **Platform Layer**: This includes the visualization of the EMS in the distributed energy grid clusters' data center and manual operation functionalities. The platform layer is typically required to be versatile, and capable of collecting operational data from different systems and storing it in a standardized format. The platform must be able to perform in-depth analysis of data with the same data structure. Visualization can display operational data in graphical and chart formats, providing an intuitive view of the platform's real-time operating status and long-term operational results.

(4) **Application Layer**: This is a collection of various advanced applications and functions that can be customized and extended according to user needs. It generally includes functions such as optimization scheduling, marketing management, and customer service. The main task of optimization scheduling is to perform optimized scheduling and decision-making. By integrating operational information from lower-level intelligent agents and the operational states of multiple agents, it coordinates and optimizes scheduling across multiple energy types, diverse users, and time scales. It also coordinates controllable resources

with different control response rates, gradually eliminating the impact of forecast errors and disturbances. Marketing management and customer service enable optimized interaction in demonstration parks and with diverse users.

3.2.2 Coordination Controller Functional Design

The coordination controller implements three main functions: information collection and issuance, adaptive regulation, and subsystem combination (Pinto et al. 2022). The information collection and distribution function is the primary function, and the self optimization and market trading functions can be expanded on this basis.

The controller lies between the main unit, subsystems, and users, with the overall goal of information relay and real-time self-regulation. This section will base the design of the regional coordination controller architecture on hierarchical control, as shown in Fig. 3.5.

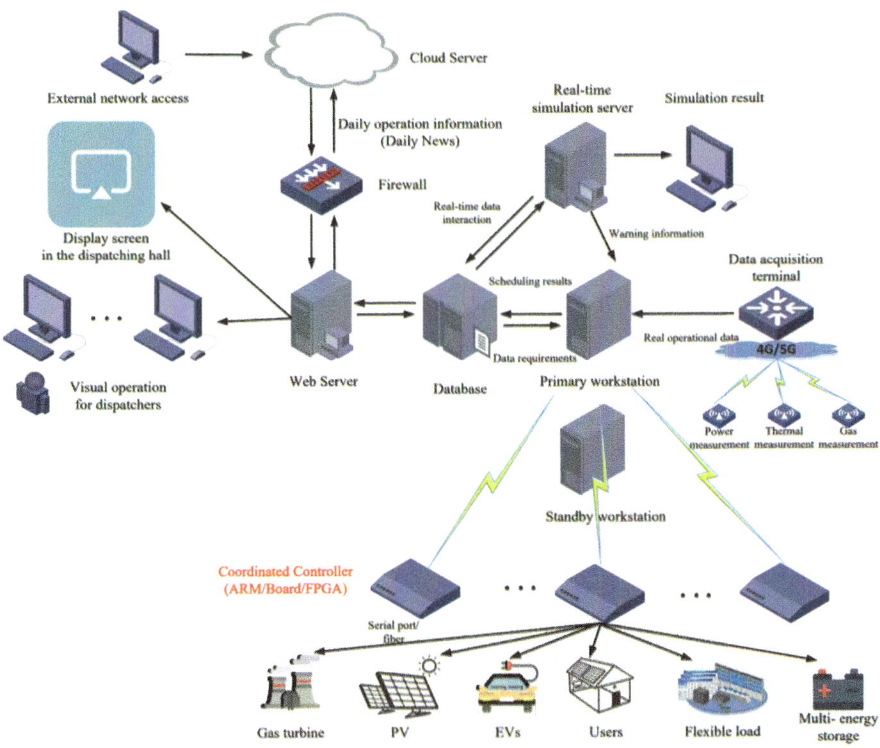

Fig. 3.5 Distributed energy grid control architecture based on regional coordination controller

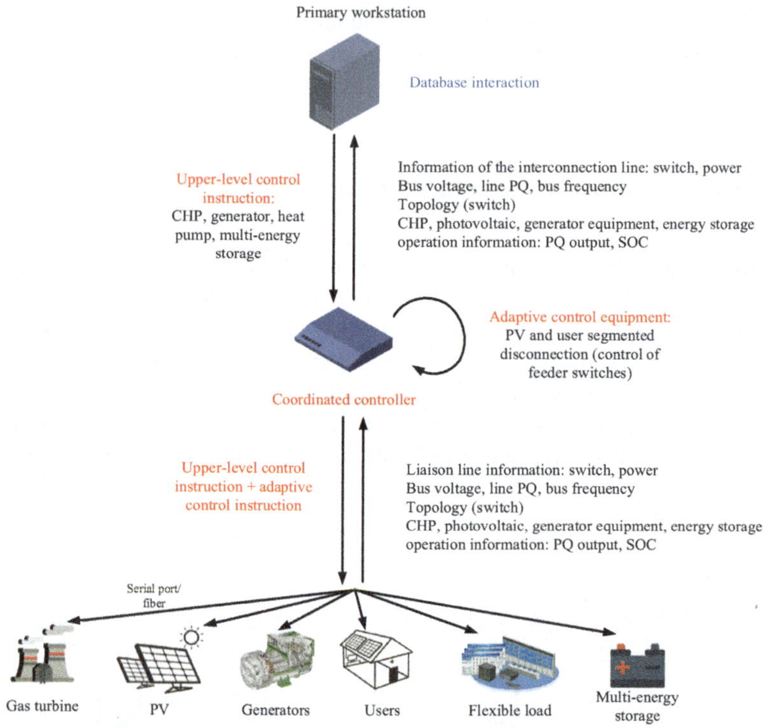

Fig. 3.6 Schematic of information flow in the control process

(1) Information Collection and Distribution Function

The coordination controller is responsible for the collection and distribution of information. The content and relationships of the information flow collection and distribution are shown in Fig. 3.6.

The controller interacts with the upper layer through the database, collecting equipment information and uploading it to the central database. Similarly, it accesses the intraday scheduling results in the central database every five minutes. After data collection, adaptive control commands are issued to each unit based on local conditions.

(2) Adaptive Adjustment Function

When the underlying equipment of the controller encounters abnormal situations that cannot be addressed in time, and there is insufficient time to report to the superior energy grid, the adaptive adjustment function of the coordination controller should be triggered.

Generally, three situations will trigger the adaptive adjustment function. These situations and the corresponding adjustment methods are as follows:

(a) Grid-Connected and Off-Grid Transitions. In the case of a transition from grid-connected to off-grid, the ES control mode changes from **PQ mode → VF mode**. In the case of a transition from off-grid to grid-connected, the ES control mode changes from **VF mode → PQ mode**.

(b) Power Quality Exceeding Limits (Failure to meet distributed energy grid operation standards). If frequency or voltage exceeds the upper limit, PV generation is curtailed in stages with a 10-s interval between actions. If the limit violation persists, an additional stage is curtailed. If frequency or voltage falls below the lower limit, loads are curtailed in stages with a 10-s interval between actions. If the limit violation persists, an additional stage is curtailed.

(c) Tie-Line Deviation Greater Than 4%. When the tie-line power deviates significantly from its reference value, real-time active power from the tie-line is collected, and the reference value is retrieved from the upper layer (with a 5-min step size). The deviation is then calculated. If the power injected by the grid into the distributed energy grid system is excessive and the tie-line power deviation exceeds 4%, the power deviation is calculated, and the number of load stages to curtail is determined to control the deviation within 3% (no action is taken if no feasible solution exists). Loads are curtailed in stages with a 10-s interval between actions. If the limit violation persists, an additional stage is curtailed. If the distributed energy grid's output to the grid is excessive and the deviation exceeds 4%, the power deviation is calculated, and the number of PV stages to curtail is determined to control the deviation within 3% (no action is taken if no feasible solution exists). PV generation is curtailed in stages with a 10-s interval between actions. If the limit violation persists, an additional stage is curtailed.

Based on **GB/T 34930-2017**, the abnormal response characteristics of voltage and frequency at the grid connection point are shown in Tables 3.2 and 3.3. These tables can be used to design the power quality action thresholds. According to the tables, the reference voltage and frequency range for autonomous adjustment is: Voltage: <90% U_N or >110% U_N. Frequency: f <49.5 Hz or >50.2 Hz.

(3) Subunit Power Distribution Function

In output units that include multiple controllable subunits, the upper-level optimization provides the overall optimal output strategy, delivering total power output instructions for each scheduling period. Therefore, the coordination controller must determine the operating status (on/off) of the subunits and implement the power distribution among them.

The specific process is illustrated in the following flowchart (Fig. 3.7):

Step 1: Determine the number of operating subunits during the scheduling period.

Assume that the output unit of the controllable subunits consists of NNN identical subunits, each with active power and reactive power limits denoted by $P_{\max}$, $P_{\min}$ and $Q_{\max}$, $Q_{\min}$, respectively. The total power output command from the upper level is denoted as P_{all} and Q_{all}.

Table 3.2 Voltage abnormal response characteristics requirements for DEGs

Grid connection point voltage	Grid connection point voltage
$U < 50\% \ U_N$	If the grid connection point voltage is less than 50% U_N and the duration is 0.2 s, the distributed energy grid should disconnect from the grid, transitioning from grid-connected mode to off-grid mode
$50\% \ U_N < U < 90\% \ U_N$	The distributed energy grid should avoid drawing power from the grid and instead supply power to support the grid connection point voltage. If the grid connection point voltage is greater than or equal to 50% U_N but less than 90% U_N for 2 s, the distributed energy grid should disconnect from the grid, transitioning from grid-connected mode to off-grid mode
$90\% \ U_N < U < 110\% \ U_N$	Normal grid-connected operation
$110\% \ U_N < U < 120\% \ U_N$	The distributed energy grid should avoid supplying power to the grid and instead absorb power to reduce the grid connection point voltage. If the grid connection point voltage is greater than 110% U_N but less than or equal to 120% U_N for 2 s, the distributed energy grid should disconnect from the grid, transitioning from grid-connected mode to off-grid mode
$120\% \ U_N < U$	If the grid connection point voltage exceeds 120% U_N for 0.2 s, the distributed energy grid should disconnect from the grid, transitioning from grid-connected mode to off-grid mode

Note U represents the distributed energy grid connection point voltage, and U_N is the nominal voltage at the grid connection point

Table 3.3 Frequency response characteristics requirements for energy grids connected to 10 kV and above voltage level distribution networks

Frequency range (Hz)	Frequency range (Hz)
$f < 48.0$	The distributed energy grid should immediately switch from grid-connected mode to off-grid mode
$48.0 < f < 49.5$	Each time the frequency drops below 49.5 Hz, it is required to maintain a minimum energy supply
$49.5 < f < 50.2$	Continuous operation
$50.2 < f < 50.5$	When the frequency exceeds 50.2 Hz, the distributed energy grid should stop supplying active power to the grid while absorbing as much active power as possible
$f > 50.5$	The distributed energy grid should immediately switch from grid-connected mode to off-grid mode

Note f represents the frequency of the grid to which the distributed energy grid is connected

Fig. 3.7 Power distribution calculation process for subunits

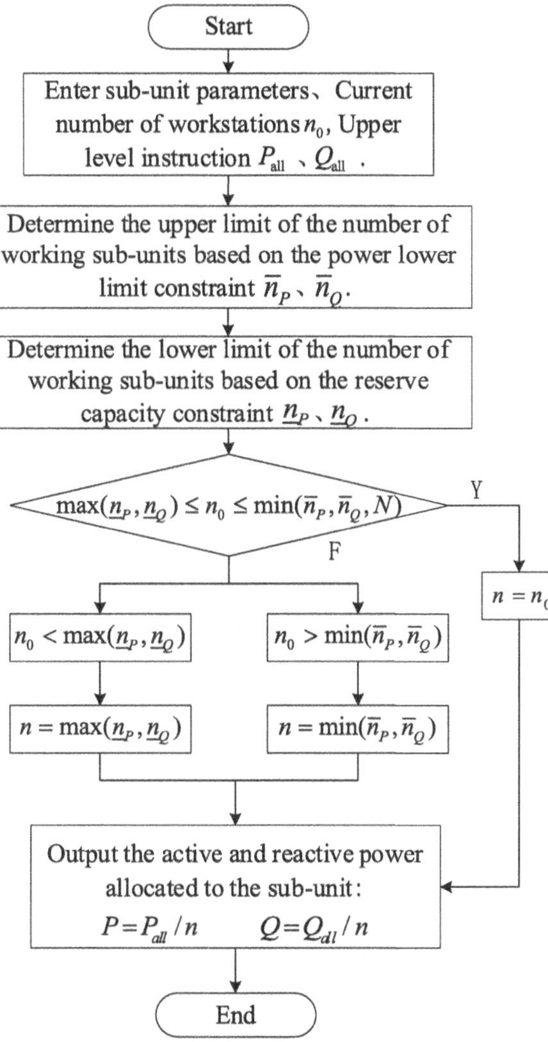

(a) Lower Power Limit Constraint: The maximum operating subunit count index is given by:

(b) Equipment Capacity Constraint: The minimum operating subunit count index is given by:

$$\overline{n}_P = fix(P_{\text{all}}/P_{\text{min}}); \quad \overline{n}_Q = fix(Q_{\text{all}}/Q_{\text{min}}) \tag{3.29}$$

$$\underline{n}_P = ceil((P_{\text{all}} + r_P)/P_{\text{max}}); \quad \underline{n}_Q = ceil((Q_{\text{all}} + r_Q)/Q_{\text{max}}) \tag{3.30}$$

where, $ceil(\cdot)$ represents the ceiling function, and r_P and r_Q represent the system's active and reactive power reserve requirements, respectively.

(c) Determination of the Specific Number of Operating Subunits: If the current number of operating subunits n_0 satisfies: $\max(\underline{n}_P, \underline{n}_Q) \leq n_0 \leq \min(\overline{n}_P, \overline{n}_Q, N)$, then the current number of operating subunits remains unchanged. When $n_0 < \max(\underline{n}_P, \underline{n}_Q)$, new subunits need to be activated to ensure the number of operating subunits reaches:

$$n = \max(\underline{n}_P, \underline{n}_Q) \tag{3.31}$$

When $n_0 > \min(\overline{n}_P, \overline{n}_Q)$, some subunits need to be shut down to ensure the number of operating subunits reaches:

$$n = \min(\overline{n}_P, \overline{n}_Q) \tag{3.32}$$

Step 2: Subunit Power Distribution

After determining the number of operating subunits, the total power command issued by the upper level needs to be distributed among the subunits. Since the subunits are of the same type, it can be concluded, based on the principle of equal incremental rates driven by economic benefits, that having the same output for all subunits ensures optimal economic efficiency. Therefore, the active and reactive power allocated to the NNN operating subunits is given as follows:

$$\begin{aligned} P &= P_{\text{all}}/n \\ Q &= Q_{\text{all}}/n \end{aligned} \tag{3.33}$$

3.3 Coordinated Controller for Energy Grid Clusters

The energy grid group coordination controller aims to optimize the collaborative operation between DEGs and higher-level energy grids, and achieve efficient management and resource allocation through the construction of a proxy module system (Hui et al. 2022). The distributed energy grid proxy module in the system is responsible for internal scheduling, managing tasks including grid connection, disconnection operations, data processing and transmission, and ensuring that appropriate control strategies can still be provided based on component status even in a disconnected state. When restoring grid connection, it strengthens interaction with higher-level agents and adjusts its operating mode to meet the overall demand of the power grid. The higher-level energy grid agent coordinates the electricity market, natural gas market, and thermal market with various agents, processes data from multiple distributed grids based on multi-objective optimization principles, determines energy allocation and control strategies, and ensures system stability and efficiency (Malla et al. 2022). In addition, communication mechanisms such as Federal Communications and Blackboard System ensure the effective flow of information between different agents, supporting the development and application of agent modules in the entire system, thereby promoting harmonious cooperation between DEGs and main power grids, and improving the overall efficiency of energy utilization.

3.3.1 Distributed Energy Grid Proxy Module

The distributed energy grid module contains subunit components and information communication components, which control the underlying components while also controlling the distributed energy grid.

(1) Subunit Components

The subunit component proxy module includes a first grid-connected proxy module, a first data processing module, a first data transmission module, a first data storage module, and an internal control module. Among them, the first grid connection proxy module is responsible for controlling the grid connection and disconnection functions of the components to the distributed energy grid. It receives grid connection and disconnection information provided by the first data transmission module or the distributed energy grid proxy module, and gives high priority to the grid connection commands transmitted by the distributed energy grid module; The grid connection command transmitted by the first data transmission module is of low priority for grid connection; The first data processing module and the first data storage module are responsible for processing and storing data of the represented components, so as to timely process and update the operating status of the components; The first data transmission module is responsible for transmitting component operating status information to the distributed energy grid proxy module and receiving commands from the distributed energy grid proxy module; The internal control module is a specific component control program that inputs corresponding control programs to control the operation of different components, ensuring the safe and stable operation of components controlled by a single component agent. It only accepts commands transmitted by the first data processing module. This component proxy module has an anti-interference ability to the outside world and can make corresponding adjustments to small changes in the outside world.

(2) Information and Communication Components

The distributed energy grid proxy module is designed to manage the scheduling and operation of electricity, thermal, gas, and other proxies within the distributed energy grid. Upon receiving information from the component proxy module, the module provides control strategies to the module, which then change the operating status of the corresponding control components on its own. In addition, the distributed energy grid proxy module provides the operational status of its respective distributed energy grid to the higher-level energy grid proxy module, and coordinates with the higher-level energy grid proxy module through communication to solve the task division and resource allocation among the agents. The distributed energy grid proxy module includes a second grid connection proxy module, a second data processing module, a second data transmission module, and a second data storage module. The second grid connection proxy module is responsible for the grid connection and disconnection functions of the controlled distributed energy grid to the upper-level energy grid. It receives grid connection and disconnection information provided by the second data transmission module or the upper level energy grid proxy module, and takes the grid connection command transmitted by the upper level energy grid proxy module as high priority, and the grid connection command transmitted by the second data transmission module as low priority for grid connection; The second data processing module is responsible for processing the information provided by the component proxy module when the network is disconnected, and providing corresponding control strategies to the corresponding component proxy through the second data transmission module based on the results of data processing; When connected to the grid, the second data processing module will increase its connection with the higher-level energy grid proxy module, receive data transmitted by the communication module between the component proxy module and the higher-level energy grid proxy module, and provide corresponding control strategies for the corresponding component proxy through the second data transmission module based on the data processing results. At the same time, the operating status of the distributed energy grid will be communicated to the higher-level energy grid proxy module again through the second data transmission module for incoming control; The second data storage module stores the results of the second data processing module for other proxy modules to query. The interactive design framework of the distributed energy grid coordination controller is shown in Fig. 3.8.

3.3.2 *Superior Energy Grid Proxy Module*

The superior energy grid is composed of the superior power grid, superior gas network, and superior thermal network, as shown in Fig. 3.9. The higher-level energy grid proxy module is responsible for coordinating and scheduling the electricity market and various agents, and making major decisions based on the comprehensive information of distributed energy grid agents. The higher-level energy grid proxy module includes a third data processing module, a third data transmission module, and a data storage module. The third data processing module is responsible for comprehensively processing the data transmitted by multiple distributed energy grid proxy modules. While ensuring the stability of the distributed energy grid, it calculates the allocation of energy savings according to multi-objective optimization theory and determines whether the energy control strategy is feasible (Kumar et al. 2023). If feasible, it sends instructions to the distributed energy grid proxy module through the third data transmission module, without requiring each distributed energy grid proxy

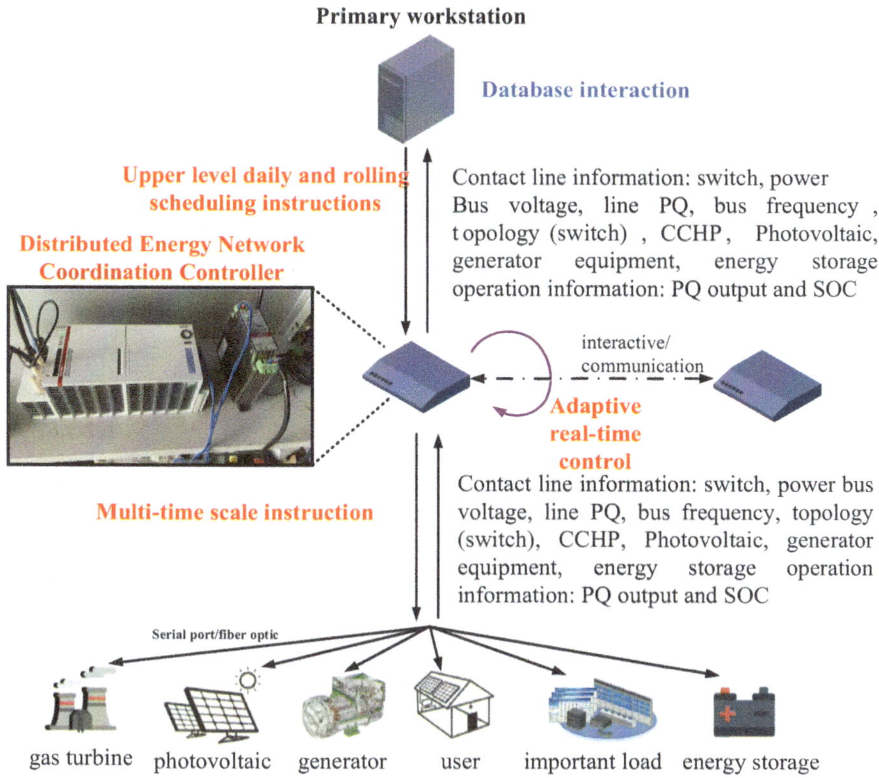

Fig. 3.8 Interactive design of distributed energy grid coordination controller

module to provide a modified operating directory. The third data storage module is responsible for storing the data transmitted by each distributed energy grid proxy module and the control strategy data is processed by the third data processing module, for query by the distributed energy grid proxy module and administrators.

(1) Superior Power Grid

As an important component of the energy grid group coordination and control system, the higher-level power grid achieves efficient management and scheduling of the electricity market through a proxy module system. Under this framework, the higher-level power grid agent module is responsible for receiving information from distributed energy grid agents, conducting comprehensive analysis and processing to ensure the stable operation of the entire power system. The third data processing module plays a core role in this process. It not only needs to process the data transmitted from multiple distributed energy grid proxy modules, but also calculates the allocation of energy savings and evaluates the feasibility of energy control strategies based on multi-objective optimization theory. For the higher-level power grid, its proxy module must also have the ability to handle complex electricity market transactions. This involves interaction with electricity suppliers, consumers, and other

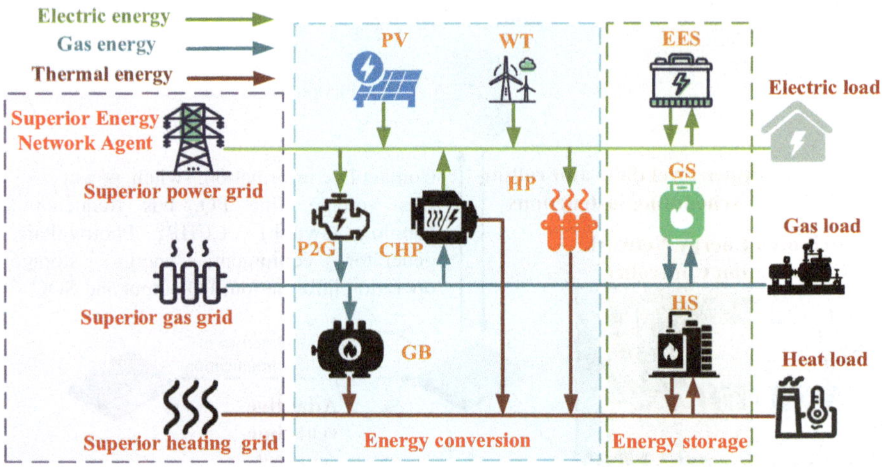

Fig. 3.9 Superior energy grid proxy module

stakeholders in different regions, including real-time pricing determination, execution of electricity trading, and so on. In order to ensure the transparency and efficiency of these processes, the proxy module will use advanced algorithms for simulation and prediction, providing accurate price signals and decision support for market participants. In addition, in emergency situations such as power outages or supply–demand imbalances caused by natural disasters, the proxy module also needs to respond quickly and take necessary measures to restore system stability. In addition, the proxy module in the higher-level power grid also needs to consider the integration of renewable energy. With the increasing proportion of intermittent new energy sources such as wind and solar, balancing their volatility and uncertainty has become a key challenge (Li et al. 2021). For this purpose, the proxy module needs to combine weather forecast data and other relevant information to plan the power generation plan in advance, and reserve sufficient adjustment capacity to cope with possible power deviations. At the same time, in order to promote the effective utilization of distributed energy resources, the proxy module also needs to establish a flexible grid connection mechanism, allowing small power generation facilities to directly participate in the electricity market, thereby improving the overall security and economy of energy supply.

(2) Superior Thermal Network

As a core component of urban heating systems, the higher-level heating network not only connects heat sources (such as thermal power plants and cogeneration units) with end-users, but also constructs a complex and precise energy transfer network. The higher-level heating network aims to coordinate every link from heat production to transportation and consumption, ensuring that heat can be accurately and precisely delivered to each end user according to time and quantity requirements. In order to achieve this goal, the third data processing module plays a crucial role by collecting data streams from various distributed energy grid proxy modules, which cover key parameters such as temperature, flow rate, and pressure. Through artificial intelligence such as temporal neural networks, transfer learning, and reinforcement learning, the massive information is deeply analyzed, providing a reliable basis for optimizing energy allocation strategies. In order to achieve precise energy scheduling, the higher-level heat network proxy module must have a high sensitivity to future changes in heat load demand trends. This requires it to not only understand the current heat consumption patterns, but also establish accurate prediction models based on long-term accumulated historical data. Given the significant differences in heat consumption characteristics between different regions such as residential areas, commercial areas, and industrial areas, the proxy module needs to develop a highly adaptable and flexible heating planning scheme. Specifically, during the severe winter season when heating demand sharply increases, the proxy module may initiate a series of measures to enhance the working intensity of the cogeneration unit and ensure sufficient heat supply; In summer, as the demand for heating decreases, the proxy module will adjust its strategy and focus more on the efficiency of cooling services or electricity production. In addition, considering the increasing frequency of extreme weather events, modern heating systems should also have the ability to respond to unexpected situations, such as quickly responding to additional load challenges caused by sudden temperature changes through intelligent control systems. In order to further enhance the reliability and flexibility of the system, future research directions may include but are not limited to: introducing advanced machine learning techniques to improve load forecasting accuracy; Explore the application of new ES materials and technologies to more effectively store excess heat energy; And develop more complex multi energy coupling systems, integrate multiple forms of renewable energy, and jointly serve the comprehensive energy needs of cities.

(3) Superior Gas Network

As a key node in the natural gas supply chain, the higher-level gas network not only covers all links from upstream extraction, and transportation to downstream distribution, but also plays an indispensable role in the design of the energy grid group coordination controller. It ensures the safe and reliable supply of natural gas through a proxy module system, ensuring that this important energy can be delivered to every place in the shortest possible time and at the lowest cost. This module is mainly composed of a third data processing module, a third data transmission module, and a third data storage module. These components work together to form a complex and efficient intelligent system.

Maintaining a balance between natural gas production and consumption is one of the core tasks of the higher-level gas network proxy module. Due to the fact that natural gas is a commodity that is relatively difficult to store on a large scale, the dynamic adjustment between its supply and demand is particularly crucial. The third data processing module integrates data from multiple distributed energy grid proxy modules, including multi-source information such as user demand forecasting, pipeline transportation capacity, and gas storage inventory levels. In order to achieve the optimal gas source allocation scheme, this module adopts various advanced artificial intelligence algorithms, such as commercial solvers, heuristic algorithms, and reinforcement learning. For example, during the peak winter season, when the demand for urban heating sharply increases, the proxy module may prioritize increasing production from gas fields near major consumption areas and simultaneously activate underground gas storage to release reserve gas, in order to alleviate the supply shortage in local areas. In addition, the proxy module will adjust production plans based on real-time weather forecasts and respond in advance to extreme weather conditions that may affect gas supply stability.

In addition to daily operations, the higher-level gas network proxy module also needs to have strong emergency response capabilities. In the face of emergencies such as pipeline leaks or extreme weather, the proxy module can quickly activate emergency plans and take a series of measures to minimize accident losses. This includes but is not limited to switching backup lines, limiting the gas consumption of non-essential users, and coordinating emergency repair teams to quickly arrive at the scene. By developing detailed emergency plans in advance and working closely with local governments and other relevant departments, the proxy module can effectively manage and mitigate potential risks, ensuring the continuity and reliability of the natural gas supply. At the same time, the proxy module also utilizes big data analysis and machine learning techniques to conduct retrospective research on historical events, continuously improving and perfecting emergency response mechanisms, and enhancing the overall resilience of the system.

It is worth noting that the higher-level gas network does not operate in isolation, but forms a closely linked relationship with the higher-level heating network and the higher-level power grid. This linkage is not only reflected in direct physical connections, but also in information sharing and technological collaboration at the logical level. For example, in the cold season, when the higher-level heating network needs more heat for urban heating, the higher-level gas network can support the efficient operation of the cogeneration device by adjusting the natural gas supply, thereby indirectly improving the stability of electricity production; During the peak electricity consumption period in summer, the higher-level power grid can meet additional electricity demand by adjusting the working status of natural gas power plants, while reducing the load pressure on other types of power plants. The higher-level gas network is not only responsible for ensuring the safe and reliable supply of natural gas, but also plays an important role in promoting effective connection between different forms of energy.

3.3.3 The Communication Methods of Different Proxy Modules

The communication methods of different proxy modules play a crucial role in the higher-level energy grid, ensuring efficient and orderly exchange of information and collaborative work between independent proxy modules. Firstly, in the combination of federated system communication and broadcast communication, each proxy module can dynamically join a virtual organization called a "federation", which provides a series of services for each member, including but not limited to registering new members, recording their abilities, tasks, and permission levels. In this way, the federation not only promotes mutual understanding between proxy modules, but also establishes a basic trust relationship. In addition, the communication mechanism in this mode also includes a broadcast function that allows the sender to publish messages to a specific group or all members, but in order to ensure information security and targeting, the sender needs to specify the unique address or conditions of the receiver. Only proxy modules that meet these conditions can receive and interpret this message. This design not only ensures the efficiency of information transmission, but also takes into account security.

On the other hand, the blackboard system provides a more open information sharing platform, which is actually a global database that acts as an intermediary for communication between proxy modules. In this system, the blackboard is carefully divided into multiple levels, each corresponding to a different problem area or work stage. The proxy module can write messages, publish results, or retrieve required information on the corresponding blackboard layer according to its own task requirements. It is worth noting that the hierarchical structure of the blackboard system enables proxy modules working at the same or adjacent levels to easily access relevant data, thereby achieving effective information flow and resource sharing. Meanwhile, any necessary information for the entire system can be posted on the blackboard for all authorized proxy modules to access, greatly improving the efficiency of collaboration between proxy modules and simplifying the management and execution process of complex tasks. Therefore, whether it is the combination of federal system communication with broadcast communication or the blackboard system, they all support effective communication between proxy modules at different levels, providing solid technical support for achieving the overall goals of the higher-level energy grid.

3.3.4 Development and Application of Proxy Module

The above example system includes a distributed energy grid proxy module and a higher-level energy grid leader module.

The distributed energy grid proxy module includes subunit components and all energy grid proxies. The subunit components are divided into PV units, FC units, GT units, and other components according to the actual situation and needs, and communicate with the proxy module in a request/response manner. Data exchange is achieved between energy grid agents through communication methods.

The control strategy is negotiated between the higher-level energy grid proxy module and the distributed energy grid proxy module through strategy negotiation, and the control strategy is transmitted to the distributed energy grid proxy module through response, and then the subunit components control the corresponding units themselves.

The following are application examples of the distributed energy grid proxy module:

(1) Subunit Components

(a) PV Unit

- The output characteristics of solar cells are nonlinear and vary with changes in light intensity and temperature. In order to fully utilize its photoelectric conversion capability, PV agents must have a maximum power point tracking function.
- Battery panel parameter detection and protection function: Monitor the basic parameters of solar panels. When the above parameters exceed the safe limit range, this component will activate protective measures, such as low voltage limit mode (LV), to ensure the safe operation of solar cells.

(b) GT Unit

- GT CHP supply: This function is an important part of the GT agency, where the energy generated by burning natural gas is converted into energy and supplied to distributed GTs. A part of the GT provides electrical energy for the distributed energy grid, while the generated waste heat is supplied to the heating system. When the heating system requires more energy, the system directly burns natural gas to provide energy for the heating system.
- Power variation restriction function: Due to the fact that in the case of CHP supply, various outputs of distributed GTs must meet the needs of heat, and their power variation is subject to certain limitations, the power variation restriction agent will ensure that the combined cooling, heating, and power supply of GTs remain within a limited range.
- Rectification/Inverter control function: Due to the speed issue of the GT, the output current of the GT cannot be directly utilized in the distributed energy grid. This agent will be responsible for effectively controlling the AC/DC rectifier and DC/AC inverter devices in the grid connection of the GT to input high-frequency power into the distributed energy grid.

(c) Battery Unit

- Monitoring function: Effective monitoring of the battery's status can enable effective control of the battery's energy before it exceeds the specified range. Including monitoring of battery voltage, current, and ES.
- Start-stop/limit function: When the battery reaches the limit value, it stops or starts the input/output of the battery (such as extreme charge discharge mode) to maintain its reliable operation.

(d) Wind Turbine Generator Set

- Anti interference function: Due to the instability of wind speed in WP generation, there are many harmonics generated. There should be filtering and power quality control parts in the wind turbine to ensure high-quality output of power quality.
- Wind speed related wind turbine blade control function: In WP generation, the link between wind speed and blade control is an important part, so agents need to be able to adjust the angle of the blades according to the wind speed and set the maximum wind speed adjustment. For example: when $V_{off} < V_{\max}$, the angle of the wind turbine blades should be adjusted to ensure the safe and stable operation of the fan.
- Reactive power function: The instability of power transmission caused by rapid changes in WP requires monitoring and control of reactive power facilities to achieve stable WP generation.

(e) Related Energy Flow Load

- Classify electric and thermal load, monitor load data, and cut off/grid load, while maintaining mutual communication with distributed energy grid agents.
- After initializing these functions, the corresponding control modules can be used to control the components they represent. At the same time, the basic information of the proxy components can be transmitted to the distributed energy grid proxy module through the communication module, and the data storage function can be used for users to query the information of this component. Finally, the grid module is used for grid connection/disconnection control with the distributed energy grid.

(2) Information and Communication Components

(a) Identity Recognition Function: Different distributed energy grid proxy modules have different identity IDs, which can be used to identify the identity of each proxy module and achieve correct communication.
(b) State Monitoring Function: Monitor various energy flows and unit states within the entire energy grid, such as the closed state of the switch in the connection line with the higher-level energy grid in the distributed energy grid system, the frequency and exchange of power between the distributed energy grid and the higher-level energy grid, etc.
(c) Power Control Function: Calculate active and reactive power reference values based on proxy status values, and adjust the power output and input of various units within the energy grid. When the proxy status value is updated, adjust the power output and input based on the new data.

References

Dezhdar A et al (2023) Transient optimization of a new solar-wind multi-generation system for hydrogen production, desalination, clean electricity, heating, cooling, and energy storage using TRNSYS. Renew Energy 208:512–537

Ding T et al (2022) Review of optimization methods for energy hub planning, operation, trading, and control. IEEE Trans Sustain Energy 13(3):1802–1818

Huang Y, Wang Y, Liu N (2022) A two-stage energy management for heat-electricity integrated energy system considering dynamic pricing of Stackelberg game and operation strategy optimization. Energy 244:122576

Hui H et al (2022) Coordination control of distributed generators and load resources for frequency restoration in isolated urban microgrids. Appl Energy 327:120116

Kim Y et al (2024) Development of a hybrid energy storage system for heat and electricity: application to green hydrogen production process integrated with a municipal solid waste incinerator. Energy Convers Manag 321:119009

Kumar K et al (2023) Intelligent controller design and fault prediction using machine learning model. Int Trans Electr Energy Syst 2023(1):1056387

Laghari AA et al (2021) A review and state of art of Internet of Things (IoT). Arch Comput Methods Eng 1–19

Li P et al (2021) The optimal decentralized coordinated control method based on the H∞ performance index for an AC/DC hybrid microgrid. Int J Electr Power Energy Syst 125:106442

Ma Y et al (2021) Modeling and optimization of combined heat and power with power-to-gas and carbon capture system in integrated energy system. Energy 236:121392

Malla SG et al (2022) Coordinated power management and control of renewable energy sources based smart grid. Int J Emerg Electr Power Syst 23(2):261–276

Mallier L et al (2021) A modelling framework for energy system planning: application to CHP plants participating in the electricity market. Energy 214:118976

Pinto G et al (2022) Enhancing energy management in grid-interactive buildings: a comparison among cooperative and coordinated architectures. Appl Energy 310:118497

Stennikov V et al (2022) Coordinated management of centralized and distributed generation in an integrated energy system using a multi-agent approach. Appl Energy 309:118487

Zhang Z et al (2021) Day-ahead optimal dispatch for integrated energy system considering power-to-gas and dynamic pipeline networks. IEEE Trans Ind Appl 57(4):3317–3328

Zhang Z et al (2022) Multitime scale co-optimized dispatch for integrated electricity and natural gas system considering bidirectional interactions and renewable uncertainties. IEEE Trans Ind Appl 58(4):5317–5327

Chapter 4
Day-Ahead Hierarchical Regulation and Control Technology for Distributed Energy Grid Clusters

The control of distributed energy grid clusters primarily aims to optimize the coordination of various energy resources, including DG sources, ES systems, thermal devices, gas supply equipment, and flexible loads. This is done by leveraging information within the grid to ensure the efficient use and coordinated operation of electricity, heat, and gas. Compared to traditional energy grids, the control of DEGs presents several key challenges. First, the control objects are diverse and dynamic, with various types and scales of distributed resources, each subject to frequent changes in their operating conditions. This makes it more complex to manage, as the boundaries between energy producers and consumers are becoming increasingly blurred, and the roles of energy production and consumption can switch at any time.

Additionally, DEGs typically have a large scale, with a wide variety of energy resources, each of which may require real-time adjustments in control strategies. Traditional centralized models struggle to meet real-time control demands, especially given the complexity and rapid response characteristics of distributed resources. The system's performance is further challenged by communication delays and the dynamic coupling between electricity, heat, and gas flows. With the high penetration of renewable energy sources, managing the interaction between these three types of energy becomes even more complex and flexible.

Moreover, DEGs must handle coordination across multiple time scales. Different resources have distinct response times: electrical storage can respond in milliseconds, thermal systems take minutes, and gas systems require hourly management. Thus, control strategies must manage everything from ultra-short-term responses to longer-term energy optimization, ensuring smooth coordination across these different time scales.

The control systems must also meet high standards in terms of reliability and flexibility. They need to address the intermittency and volatility of distributed energy resources while ensuring the stability and quality of electricity, heat, and gas supply. The grid must also be able to quickly respond to emergencies or faults. Furthermore,

T. Lu et al., *Operation Modeling and Collaborative Regulation of Distributed Energy Grid Clusters*, Smart Energy Systems, https://doi.org/10.1007/978-981-96-9942-1_4

it should ensure seamless integration with external grids in both islanded and grid-connected modes, adding another layer of complexity to the control system.

Overall, the control of distributed energy grid clusters requires balancing diverse and rapidly changing resources, managing large-scale optimization in real-time, and ensuring the stable and flexible operation of the system across multiple time scales and modes.

4.1 Architecture of Distributed Energy Grid Clusters

To address the challenges of complex processing and high computational demands faced by the central coordinator in a centralized control model for distributed energy grid clusters, a hierarchical control strategy has been proposed to achieve coordinated scheduling and control across different time scales. The hierarchical control divides the system into upper and lower levels. The lower-level system achieves local optimization of various energy units such as DG, ES, HPs, and gas supply through internal coordination and optimization, presenting stable characteristics to the upper-level system. The upper-level system, in turn, treats the lower-level system as a stable "black box" and focuses on optimizing the global energy grid. This strategy not only accelerates the control response time but also enhances the system's flexibility and adaptability to local changes and expanding scenarios.

Currently, hierarchical control has been widely applied in distributed energy grid clusters, consisting of both functional and physical layers. The functional layer focuses on coordinating functions across different time scales in the energy grid, including primary, secondary, and tertiary regulation of grid frequency in an electricity-heat-gas coupled system, as well as dynamic regulation of pressure and flow in the heat and gas grids. Based on functional layering, Pokharel (2021) proposed a secondary regulation method for voltage/frequency/pressure in multi-energy systems, while Bustos et al. (2023) analyzed a hierarchical optimization strategy for multi-energy system parameters. The physical layer refers to the coordination of management devices at different levels. To some extent, the functional and physical layers complement each other. In distributed energy grid clusters, various distributed energy units (such as DG sources, ES devices, HPs, and gas supply equipment) are located at the lower level of the physical layer and are primarily responsible for fast system regulation (in the millisecond range). The central coordination controller of the distributed energy grid cluster resides at the coordination level, mainly handling responses at the second to minute time scales.

4.1.1 Hierarchical Architecture Based on Multi-agent System

A MAS is a grid system composed of a group of intelligent agents that possess certain capabilities in sensing, computing, execution, and communication, which are linked through communication and other means. The concept of intelligent agents originated from early research in blackboard architectures, contract nets, and action-based systems, with the goal of replacing humans in handling complex or hazardous tasks as "proxy agents." The application potential of MAS has attracted significant attention in fields such as computer science, artificial intelligence, biology, communications, and systems and control science. In distributed energy grid clusters, MAS can serve as a flexible and efficient control strategy for coordinating the production, transmission, storage, and consumption of various forms of energy, such as electricity, heat, and gas.

From the perspective of distributed energy grid clusters, the MAS control system can meet the on-site demand for multiple types of energy, such as cooling, heating, and electricity, improving the reliability of multi-energy coordination, reducing energy transmission losses, and enhancing the overall energy efficiency on the user side. From an environmental perspective, MAS effectively reduces pollutant emissions and improves energy utilization efficiency by coordinating the use of renewable energy and low-carbon technologies, which is significant for mitigating climate change and protecting the environment. MAS has demonstrated its advantages in many issues within distributed energy grid clusters, and has been widely applied in areas such as distributed optimization and energy management. Amirkhani and Barshooi (2022) proposed a dynamic multi-agent coordination framework and designed a multi-energy scheduling algorithm considering safety domains; Wang et al. (2023) and Liu et al. (2024) focused on multi-energy consistency coordination strategies based on droop control.

Compared with traditional grid control systems, MAS in distributed energy grid clusters offers several advantages: (1) applicability to a wide range of tasks; (2) high efficiency and flexibility; (3) excellent system performance; (4) strong fault tolerance; (5) strong robustness; (6) low economic cost; (7) ease of development; (8) support for distributed sensing and action. The application of MAS provides strong support for achieving efficient coordination and low-carbon operation in distributed energy grid clusters.

In terms of spatial scale, the MAS in distributed energy grid clusters is composed of three functional layers: the central management layer, the intermediate coordination layer, and the bottom execution layer, as shown in Fig. 4.1. The central management layer is responsible for obtaining the target definitions and related constraints, including energy scheduling plans, functional evaluations, and optimization learning. The coordination layer activates task execution based on the tasks and steps defined by the management layer, and can respond quickly to energy events. The execution layer performs specific actions and provides real-time feedback on the execution results. Through top-down information instructions and bottom-up execution feedback, MAS enables the efficient coordination and interaction of the multi-energy system (electricity, heat, and gas), enhancing overall energy utilization efficiency.

(1) The Central Management Layer: This layer mainly consists of the distributed energy grid clusters management system. It is responsible for power flow monitoring and scheduling across the entire distributed energy grid clusters, market

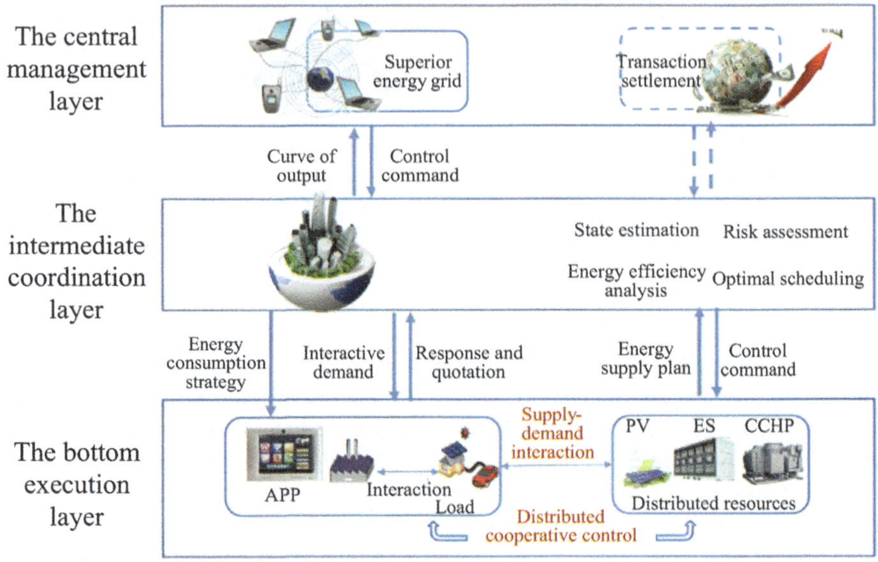

Fig. 4.1 Schematic diagram of the distributed energy grid clusters

trading settlements, planned maintenance, and other tasks. The central management layer coordinates the flow of electricity, heat, and gas to ensure smooth information exchange and energy transfer between the higher-level energy grid and the distributed energy grid, achieving the coordinated operation of electricity, heat, and gas. Through this system, the distributed energy grid clusters can optimize energy scheduling and achieve efficient collaboration between different energy carriers. The system dynamically adjusts the operating state of the distributed energy grid clusters based on real-time data such as the supply–demand balance of electricity, heat, and gas, operational indicators, etc. It sends load information and electricity shortages to the intermediate coordination layer and receives feedback on the execution of commands from the coordination layer, adjusting the scheduling strategy in real-time.

(2) The Intermediate Coordination Layer: This layer is responsible for coordinating the scheduling of multi-energy forms (electricity, heat and gas) in the distributed energy grid clusters, ensuring efficient interaction between information and energy flows. Its main tasks include real-time monitoring of operational status, responding to emergencies, optimizing resource allocation, and achieving energy balance and stable operation of the entire grid by coordinating lower-level execution units.

During operation, the intermediate coordination layer optimizes the energy distribution strategy based on real-time energy prices, distributed energy output, load forecasts, and other information. It adjusts the output and operation mode of distributed energy to ensure voltage stability, smooth heat and gas flow,

and overall system efficiency improvement. As the management center for each region, the intermediate coordination layer is responsible for coordinating energy and information flows between users and between users and the superior energy grid. By combining user energy preferences, production plans, and other factors, it creates recommended adjustment plans and interacts with users through an information-sharing platform to enhance user participation.
Additionally, the intermediate coordination layer comprehensively utilizes distributed resources such as ES devices, electric vehicles, and flexible loads to optimize operation strategies. It decomposes the macro goals of the central management layer into specific tasks and distributes them to lower-level units. By treating all participants fairly, the intermediate coordination layer aims to maximize the overall benefits, achieving efficient operation through the coordinated optimization of electricity, heat, and gas.

(3) The Bottom Execution Layer: This layer is mainly composed of commercial and industrial users, residential users, and dispatchable distributed resources (such as electric vehicles, PVs, ES devices, HPs, gas equipment, etc.). The bottom execution layer has an autonomous distributed management system and is responsible for energy management and production optimization of various energy forms (electricity, heat, gas) within its system. It also maintains bidirectional communication of information and energy with the intermediate coordination layer. The bottom execution layer receives operational instructions from the intermediate coordination layer and, based on its own equipment status, production processes, and adjustable capacities, responds quickly to maximize its own benefits. For example, it can meet control requirements by adjusting production plans, interrupting certain loads, optimizing the use of thermal and gas flows, or activating backup energy sources. If it is unable to fully respond to the instructions or encounters an emergency (such as equipment failure or supply interruption), it must promptly report back to the intermediate coordination layer for further guidance.

The main participation methods of the bottom execution layer include: interruptible load response, direct load control (DLC), production process optimization, multi-energy coordination and complementarity, dynamic charging and discharging of ES devices, optimized scheduling of electric vehicles, and efficient distribution and substitution of thermal and gas resources. Through flexible and efficient response capabilities, the execution layer achieves coordinated optimization of electricity, heat, and gas within the distributed energy grid clusters, supporting the stable operation and benefit enhancement of the entire grid.

The interactivity and overall encapsulation of the MAS in distributed energy grid clusters enable the achievement of system-wide optimization through interaction and coordination across the entire grid. The interactive MAS control has the following characteristics:

(1) High Reliability: The reliability of the control in distributed energy grid clusters is significantly improved. Many functions of traditional centralized controllers are delegated to the intermediate coordination layer and the execution layer,

avoiding the issue of system failure caused by a single point of failure in the centralized controller. At the same time, each layer of intelligent agents has strong autonomy and clear division of labor, with relatively low requirements for communication bandwidth, further enhancing the system's stability.

(2) Flexible Response: The bottom execution layer can quickly respond to fluctuations in demand for various forms of energy, such as DG, thermal loads, and gas loads, based on autonomous operation. The intermediate coordination layer optimizes the scheduling of the lower-level execution units according to the system's state, making the system operation more flexible and meeting the operational requirements of the distributed energy grid clusters across different time scales.

(3) Strong Scalability: It supports "plug-and-play" for distributed energy resources and can adapt to the dynamic integration and withdrawal of various distributed resources like electricity, heat, and gas, ensuring the normal operation of the distributed energy grid clusters.

(4) Lower Cost: There is no need for centralized processing of large-scale data or the construction of complex communication grids, reducing the construction cost of secondary systems and significantly improving cost-effectiveness. Moreover, the distributed optimization of intelligent agents at each layer reduces computational complexity.

(5) Good Information Privacy: The agents only need to exchange minimal external characteristic information, without sharing internal model parameters or operational data.

The dynamic multi-agent-based intelligent hierarchical control framework adopts a three-level hierarchical control strategy, consisting of the optimization scheduling layer, secondary control layer, and primary control layer, as shown in Fig. 4.2. In terms of spatial scale, the framework design combines the coordinated characteristics of electricity, heat, and gas in distributed energy, clearly defining the functional modules at each layer. The optimization target curve is transmitted step by step from top to bottom, while control capacity feedback is transmitted from bottom to top to achieve the coordinated optimization control of multiple energy forms across the entire grid, addressing the volatility of distributed energy output and the complexity of multi-energy coordination, thereby enhancing system stability and efficiency.

(1) The Primary Control Layer: This layer includes various distributed energy agents as well as multi-load agents for commercial, industrial, and residential users. The distributed energy agents have functions such as SA and control execution, enabling them to perform energy interactions across multiple energy forms based on instructions from the system control center and their own operational constraints. They sense external environmental changes and achieve automatic switching of control modes and execution of commands through real-time information exchange among agents. The load agents have functions for judgment, prediction, and communication, allowing them to upload real-time load information for electricity, heat, gas, etc., provide feedback on load status, and execute operations such as load shedding and energy conversion optimization.

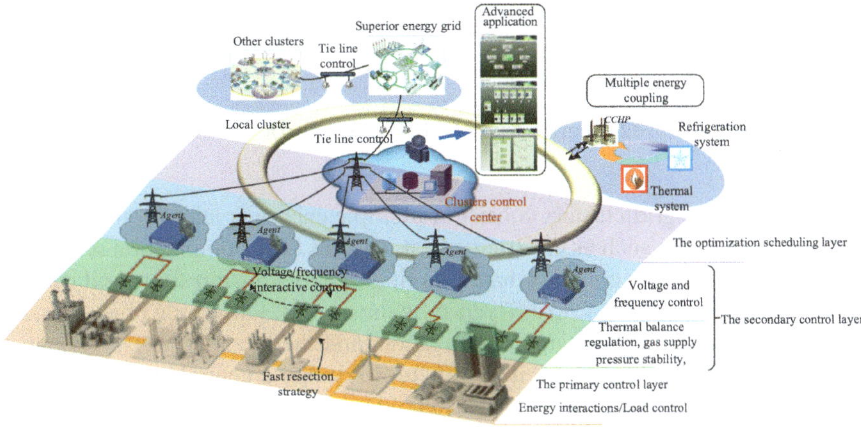

Fig. 4.2 Hierarchical control structure based on MAS

(2) The Secondary Control Layer: Composed of the distributed energy grid cluster system agents for each region, this layer includes functional modules for voltage and frequency control, thermal balance regulation, gas supply pressure stability, regional coordination control, and real-time communication. The secondary control layer receives optimized scheduling and intelligent control commands from the upper layer and issues commands based on load and distributed energy priorities. Through multi-agent consensus control algorithms, the secondary control layer ensures "centralized management, decentralized control" of the primary sensing layer agents within the region, ensuring the stability of voltage, frequency, thermal load, and gas pressure in the region. At the same time, the secondary control layer regulates multi-energy coordination between regions, considering energy efficiency constraints, to meet the dynamic adjustment requirements of the distributed energy grid clusters.

(3) The Optimization Scheduling Layer: Composed of the distributed energy grid clusters control center agents, this layer includes a scheduling decision-making functional module. The optimization decision-making layer integrates operational information from lower-level agents and multi-agent status data, performing complementary coordination and optimization scheduling for multi-energy types and various users. By decomposing hierarchical optimization scheduling into three-time scales—day-ahead planning, intra-day rolling, and real-time correction—the optimization decision layer coordinates controllable resources with different control response speeds, progressively eliminating forecast errors and disturbances. This layer proposes a hierarchical optimization scheduling strategy for electricity, heat, gas, and diverse users, achieving overall optimization of the energy grid cluster and user interaction, thereby enhancing system efficiency and flexibility.

4.1.2 Interactive Architecture of Distributed Energy Grid Clusters

(1) Game-theory-based Energy Management Model for Distributed Energy Grid Clusters

Distributed energy grid clusters consist of multiple energy organizations, involving various energy forms such as electricity, heat, and gas. The optimization scheduling of such a system is more complex than for a single energy organization, as it needs to coordinate more factors. Due to differences in the owners, scheduling objectives, and task requirements of each energy organization, and the fact that all parties pursue their own benefit maximization, it is difficult to design a centralized optimization scheduling solution that satisfies all stakeholders. Therefore, a distributed energy management model based on a master–slave game is proposed, suitable for multi-energy coordination in distributed energy grid clusters. In this model, coordination and cooperation between energy cells maximize energy supply benefits; whereas within the entire regional energy grid cluster, cooperation among energy organizations minimizes operational costs. This distributed energy management model can be integrated with a Model Predictive Control (MPC) architecture to effectively address the intermittency and volatility of renewable energy output and the uncertainty of load demand, thereby improving the overall operational efficiency and stability of the system.

In the EMS of distributed energy grid clusters, there are multiple decision-making entities, each with different roles and functions in the system. Some strategies need to be determined in advance as known conditions for other decision-making problems; this type of game is called a master–slave game. The game-theory-based decision model of a distributed energy grid cluster mainly consists of three sub-decision models: the internal game model on the user side, the internal game model on the electricity/gas/heat supply side, and the game model between the electricity/gas/heat supply side and the user side. Since the game optimization research for energy stations (such as large power plants, gas stations, and coal-fired plants) has already been quite mature, the internal game of these stations is not considered here. Instead, the focus is on the joint operation issues between the energy stations and ES systems. Therefore, the game-theory-based decision model of the distributed energy grid cluster includes two main sub-decision models: the internal game model on the user side and the game model between the electricity/gas/heat supply side and the user side. The following describes these two game models.

1. Internal Game Model on the User Side

Participants: As shown in Fig. 4.3, the participants are mainly composed of N energy organizations on the user side.

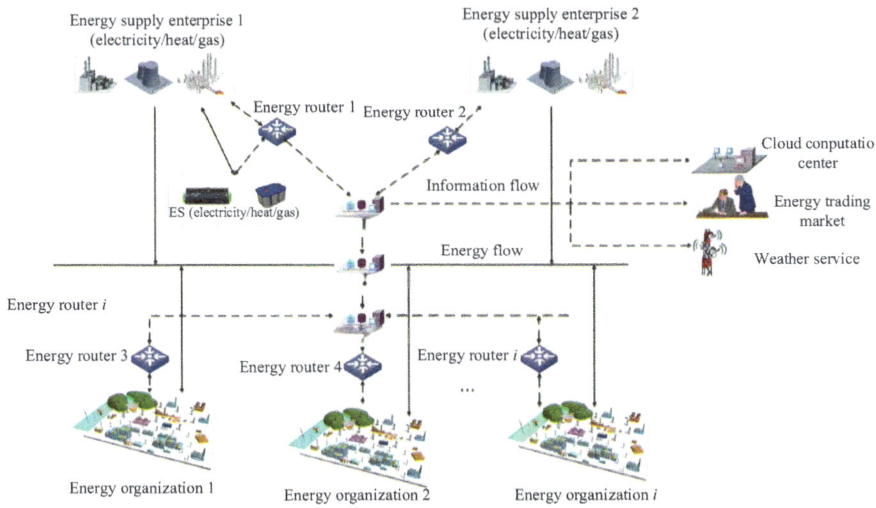

Fig. 4.3 Master–slave game framework of distributed energy grid clusters

Strategy: The energy organizations' strategy involves the amount of energy they buy or sell during each time period t. The buying/selling behavior and its power are influenced by all controllable energy cells within each energy organization (such as PVs, ES systems, electric vehicles, etc.).

Profit: The profit of energy organization i during the t period in the k-th system game iteration can be expressed by its operational cost. The lower the cost, the higher the profit, meaning the energy organization performs better in the game.

2. Game Model Between the Electricity/Gas/Heat Supply Side and the User Side

Participants: As shown in Fig. 4.3, the participants mainly include two entities: the energy management center on the electricity/gas/heat supply side and the energy management center on the demand side.

Strategy: The strategy of the energy management center on the electricity/gas/heat supply side is the amount of energy supplied by the energy stations, which includes the power output from the electricity/gas/heat stations as well as the operation power of the ES systems (including charging and discharging power). The strategy of the demand-side energy management center is the total amount of purchased energy, which is the sum of the energy bought and sold by each energy organization.

Supply–Demand Balance: According to the system's supply–demand balance requirements, the supply of energy from the electricity/gas/heat stations at any given time should be equal to the total purchased energy on the demand side. The coordination and optimization of electricity, gas, and heat (multi-energy forms) are achieved through a game between the two sides.

3. Energy Management Model of Energy Organizations

Since the energy management game models in the distributed energy grid clusters involve energy organizations both in the internal game model on the user side and in the game model between the electricity/gas/heat supply side and the user side, it is necessary to first establish the energy management model for each energy organization. To make the energy management model of energy organizations more universal, a typical energy organization i is selected for modeling. This energy organization includes various types of energy cells such as loads, ES devices, and renewable energy generation equipment, and the energy purchase and sale prices in the local energy grid may differ at the same time.

4. Load Cell Model

The load cells within the energy organization mainly include the following four types: critical load, power-adjustable load, operating time-adjustable load, and load with both operating time and power adjustable.

5. Energy Organization and External Electricity/Gas/Heat Grid Power Interaction Model

In the open energy market, to incentivize demand-side users to utilize local energy more efficiently, differentiated purchase and sale prices for energy at the same time are applied. The energy purchase and sale prices between energy organization users will align with the system prices. Therefore, to fully describe the changes in energy organization i's purchase and sale of energy during period t in the k-th system game iteration under the differentiated energy pricing model, it is necessary to establish a separate power interaction model between the energy organization and the external electricity/gas/heat grid.

6. ES Cell System Model

The ES model needs to be suitable both for the ES system within energy organization i and for the ES systems of electricity/gas/heat supply enterprises.

7. Renewable Energy Cell Output Constraint Model

In the optimization scheduling model, it is required that the PV output and wind turbine output of energy organization i during period t should be within their rated operational range.

8. Energy Organization Supply and Demand Balance Constraint Model

The premise for the optimal operation of energy organization i is that its internal supply and demand balance is satisfied at all times.

9. Energy Management Model of Electricity/Gas/Heat Supply Enterprises

As the main power source of energy organizations, the output capacity of power plants, natural gas stations, and CHP plants needs to meet the energy demand of the supply-side users during period t.

10. System Supply and Demand Balance Constraints

In the regional energy grid optimization scheduling model, the supply and demand balance between the demand side and the electricity/gas/heat supply side must be satisfied at any period t.

11. System Game Revenue and Equilibrium

When designing the game model, the main focus is on the revenue of each decision-maker and whether the game model can reach equilibrium. This includes two aspects: the revenue of the demand-side energy organization user i during period t in the k-th system game iteration and the revenue of the electricity/gas/heat supply enterprises. According to the definition of game equilibrium, the game model reaches equilibrium when no participant can independently change their strategy to achieve higher revenue.

(2) Distributed Solution Method

Since in the regional energy grid energy management model, each participant aims to maximize their profit, it is difficult to use centralized optimization methods to calculate the equilibrium point of the game model. Therefore, this project proposes a parallel distributed game model solution method, which can maximize the reduction in communication bandwidth and data transmission requirements, and can quickly reach the game equilibrium point. The specific solution process is shown in Fig. 4.4.

The specific solution process is as follows:

a. Input the raw data and parameters required for the distributed energy grid clusters game decision model. This includes parameters such as the types and quantities of renewable energy, ES, loads, and other energy cells within each energy organization, as well as the operational parameters of the units on the electric/gas/heat station side, electricity prices, gas prices, and heating price adjustment factors.
b. Based on the system configuration and operational objectives of each participant, construct the game strategy set, decision optimization functions, etc. for each participant.
c. Provide the initial state values of the energy organization users on the demand side at the start of the game and upload them to the distributed energy grid clusters operation and scheduling center.
d. The management center of the energy supply enterprises, such as large power plants, gas stations, and thermal power plants, solves their respective profit objective functions based on the demand values sent by the distributed energy grid clusters operation and scheduling center. This is done to determine the unit output plan, ES charging/discharging plan, and basic energy price, while storing the relevant data and uploading the sales energy price information.

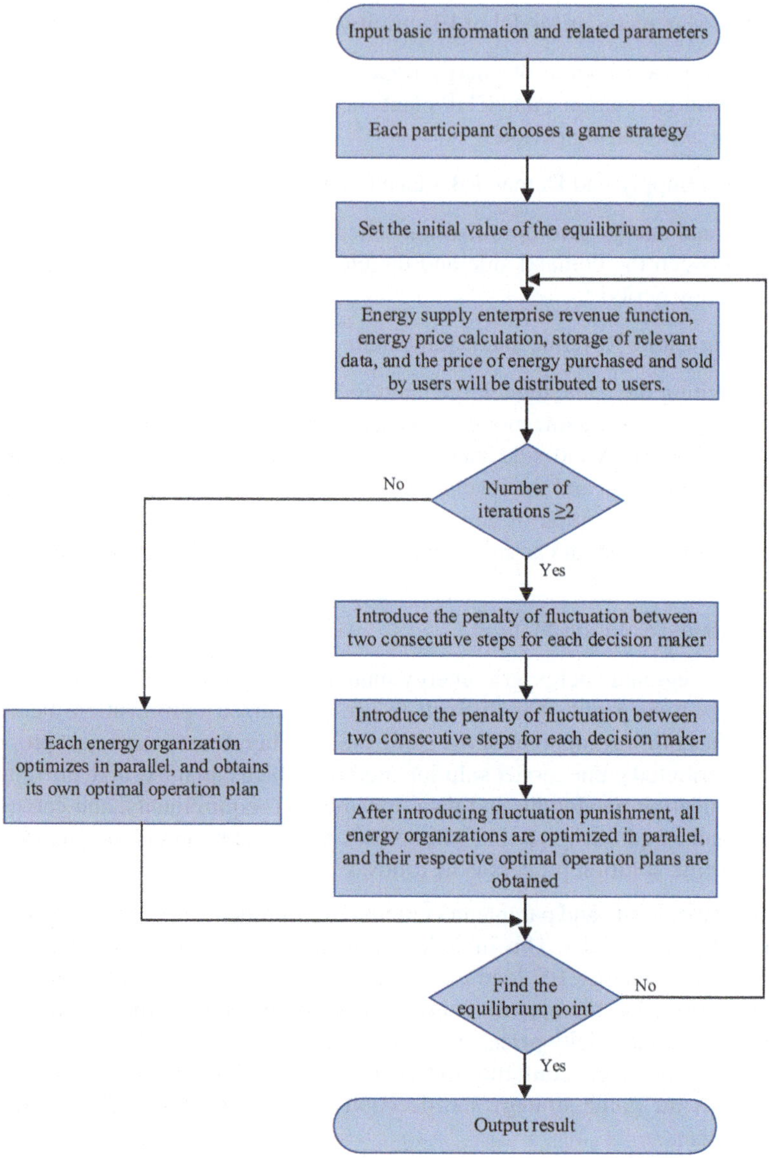

Fig. 4.4 Solution process for the game model of distributed energy grid clusters based on parallel distribution

e. Determine whether the system game iteration count is greater than or equal to 3. If it is greater than 3, execute step f; otherwise, execute step g.
f. Calculate the operational fluctuation penalty function for each user between two consecutive steps.
g. Each demand-side energy organization user, based on the energy price information sent by the distributed energy grid clusters operation and scheduling center, independently and in parallel determines their own energy purchase/sale plans, load operation plans, ES charging/discharging plans, etc., and stores the relevant data. Meanwhile, the demand-side management center uploads the total energy purchase plan for the demand side.
h. Determine whether the system game has reached the game equilibrium. If equilibrium has been achieved, stop the iterative solution and output the results. Otherwise, return to step d.

4.2 Day-Ahead Hierarchical Operation Model of Distributed Energy Grid Clusters

4.2.1 Operation Optimization Model Between Clusters and Superior Energy Grid

Figure 4.5 shows the system structure of the distributed energy grid clusters studied in this section. The DEGs in different regions manage their respective distributed energy sources (such as wind turbines (WT), PVs, GTs, FCs, P2G, ES, etc.) and connect to the superior energy grid through various public connection points. The energy trading between the distributed energy grid group and the superior energy grid is bidirectional.

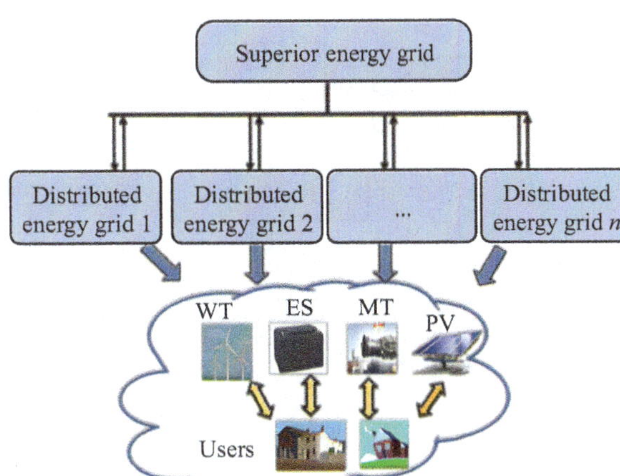

Fig. 4.5 System structure of distributed energy grid clusters

A two-layer decision-making model is proposed for the distributed energy grid clusters. From the perspective of the distributed energy grid, the goal of each grid point is to minimize its own operating costs while meeting the overall scheduling requirements. In the optimization framework, the upper-layer superior energy grid operation optimization model minimizes system losses, temperature, and pressure deviations by controlling the input power at each energy grid interface, ensuring the stable and coordinated operation of electricity, heat, and gas. Since the distributed energy grid clusters connected to the superior energy grid affect their dispatchable capacity and operational strategies, it is necessary to obtain the input power limits for all PCCs based on the output prediction data of each energy grid before performing upper-layer optimization. The lower-layer multi-energy grid operation optimization model ensures the maximization of economic benefits under the coordinated operation of electricity, heat, and gas by deciding the output of distributed energy sources, ES systems, and electrical and thermal loads at each grid point (Fig. 4.6).

The objective function of the superior energy grid optimization model consists of multiple objective components and requires consideration of the coordinated optimization of electricity, heat, and gas.

$$\min Z = \sum_{i \in \mathcal{N}} \left(\alpha_1 \cdot L_{\text{elec},i} + \alpha_2 \cdot L_{\text{heat},i} + \alpha_3 \cdot L_{\text{gas},i} \right) + \alpha_4 \cdot \Delta V + \alpha_5 \cdot \Delta T \quad (4.1)$$

where $\mathcal{N}$ represents the set of distributed energy grid clusters, including the nodes of the electricity grid, thermal grid, and gas grid. $L_{\text{elec},i}$ denotes the electrical energy loss at node i, measured in kilowatt-hours (kWh). $L_{\text{heat},i}$ represents the thermal energy loss at node i, measured in kilowatt-thermal (kWth). $L_{\text{gas},i}$ denotes the gas power loss at node i, measured in kilowatt-gas (kWh). ΔV represents the voltage deviation

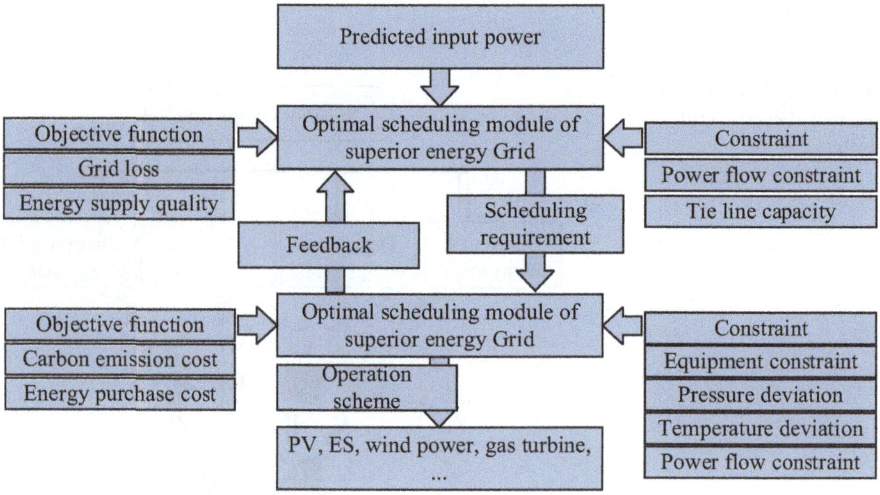

Fig. 4.6 Bi-level optimization model

of the entire grid, indicating the stability of the electricity grid, measured in volts. ΔT represents the temperature deviation of the entire grid, indicating the stability of the thermal grid, measured in degrees Celsius. α_i denotes the weight coefficients, reflecting the influence of different factors on the objective function.

Transmission line capacity and distributed energy generation output constraints:

$$\begin{cases} P_{Gi,\min} \le P_{Gi}^{t} \le P_{Gi,\max} \\ Q_{Gi,\min} \le Q_{Gi}^{t} \le Q_{Gi,\max} \\ P_{Gi}^{i,t} - P_{Gi}^{i,t-1} \le R_{up,Gi} \\ P_{Gi}^{i,t-1} - P_{Gi}^{i,t} \le R_{down,Gi} \\ P_{\min} \le P_{PCC}^{t} \le P_{\max} \end{cases} \tag{4.2}$$

where $R_{up,Gi}$ and $R_{down,Gi}$ represent the lower and upper limits of the ramping rate of the distributed energy sources. P_{PCC}^{t} denotes the transmission line power, satisfying $P_{PCC}^{t} = \sum_{i=1}^{I} P_{\mu Gi}^{t}$. P_{Gi}^{t} represents the output of the distributed energy sources connected to energy grid i. The lower and upper limits of the transmission line power are denoted as $P_{Gi,\min}$ and $P_{Gi,\max}$

The decision variables of the optimization operation model at the distributed energy grid cluster level are the output power of distributed energy sources and ES systems. The lower-level optimization model aims to minimize the daily integrated operational cost, considering a 24 h time horizon.

$$\min C_{cost} = C_{fuel} + C_{O\&M} + C_{DEP} + C_{ENV} + C_{PE} + I_{SE} + C_{PVWT} \tag{4.3}$$

The cost function consists of the following six components:

1. Fuel Cost:

$$C_{fuel} = \sum_{i=1}^{I} \sum_{t=1}^{24} \sum_{m=1}^{M^i} K_{fuelm} P_m^{i,t} \tag{4.4}$$

   where K_{fuelm} represents the unit fuel cost of distributed energy source m; $P_m^{i,t}$ denotes the output of distributed energy source m in energy grid i at time t; $K_{fuelm} P_m^{i,t}$ represents the fuel cost of distributed energy source m at time t; $\sum_{m=1}^{M^i} K_{fuelm} P_m^{i,t}$ denotes the total fuel cost of energy grid i at time t; $\sum_{t=1}^{24} \sum_{m=1}^{M^i} K_{fuelm} P_m^{i,t}$ represents the 24 h total fuel cost of energy grid i.

2. Operation and Maintenance Cost:

$$C_{O\&M} = \sum_{i=1}^{I} \sum_{t=1}^{24} \sum_{m=1}^{M^i} K_{OMm} P_m^{i,t} \tag{4.5}$$

   where K_{OMm} represents the unit operation and maintenance cost of distributed energy source m.

3. Investment Depreciation Cost:

$$C_{DEP} = \sum_{i=1}^{I}\sum_{t=1}^{24}\sum_{m=1}^{M^i}\left(\frac{r_m(F + r_m)^{n_m}}{(F + r_m)^{n_m} - 1}\right)\left(\frac{C_{invm}}{8760l}\right)P_m^{i,t} \tag{4.6}$$

In the equation, C_{invm} represents the annual average installation cost of distributed energy source m; F is the average capacity factor; r_m represents the depreciation rate of distributed energy source m; and n_m denotes the service life of distributed energy source m.

4. Environmental Management Cost:

$$C_{ENV} = \sum_{i=1}^{I}\sum_{t=1}^{24}\sum_{m=1}^{M^i}\sum_{k=1}^{K^m} R_k K_{ENVmk} P_m^{i,t} \tag{4.7}$$

where M represents the quantity of pollutants; K_{ENVmk} denotes the amount of pollutant k emitted per unit of electricity generated by distributed energy source m; R_k represents the treatment cost of pollutant k.

5. Interaction Cost:

$$\begin{aligned} C_{PE} &= \begin{cases} 0, \ P_{PCC}^t < 0 \\ \sum_{t=1}^{24} c_{pe}^t P_{PCC}^t, \ P_{PCC}^t \geq 0 \end{cases} \\ I_{SE} &= \begin{cases} \sum_{t=1}^{24} i_{se}^t P_{PCC}^t, \ P_{PCC}^t < 0 \\ 0, \ P_{PCC}^t \geq 0 \end{cases} \end{aligned} \tag{4.8}$$

where c_{pe}^t represents the price at which the distributed energy grid purchases energy from the superior energy grid; i_{se}^t denotes the price at which the distributed energy grid sells energy to the superior energy grid; P_{PCC}^t represents the interaction power between the distributed energy grid and the superior energy grid; C_{PE} is the total cost of energy purchased by the distributed energy grid; and I_{SE} represents the revenue from energy sales.

6. Curtailed Wind and Solar Energy Cost
Due to the integration of high-penetration distributed energy sources into the distributed energy grid clusters, and the power output restrictions imposed by the superior energy grid, each distributed energy grid must implement wind and solar curtailment measures to ensure power balance in situations where wind and solar output increases and load demand decreases, in order to meet the PCC connection power constraints. The curtailed wind and solar energy cost mainly arises from carbon emission trading, and can be expressed as:

$$C_{PVWT} = \sum_{t=1}^{24}\sum_{i=1}^{I}(c_{\text{wc}} p_{\text{wc}}^{i,t} + c_{\text{pvc}} p_{\text{pvc}}^{i,t}) \tag{4.9}$$

The lower-level optimization model's constraints mainly include the following four components:

1. Power Balance Constraint:

$$\sum_{i=1}^{I}\left(\sum_{m=1}^{M^i} P_m^{i,t} - P_{Load}^{i,t}\right) = P_{PCC}^t \tag{4.10}$$

   where $P_{Load}^{i,t}$ represents the load of distributed energy grid i. Since the structure of DEGs is often relatively simple, this section deliberately neglects the grid losses within the distributed energy grid.
2. Distributed Energy Output Constraint:

$$P_{m,\min} \leq P_m^{i,t} \leq P_{m,\max} \tag{4.11}$$

   where $P_{m,\max}$ and $P_{m,\min}$ represent the lower and upper output limits of distributed energy source m, respectively.
3. Micro-turbine (MT) Ramping Constraint:

$$P_{MT}^{i,t} - P_{MT}^{i,t-1} \leq R_{up,MT} \tag{4.12}$$

$$P_{MT}^{i,t-1} - P_{MT}^{i,t} \leq R_{down,MT} \tag{4.13}$$

   where $R_{up,MT}$ and $R_{down,MT}$ represent the ramping down and ramping up coefficients of the MT, respectively; $P_{MT}^{i,t}$ denotes the output of the MT at time t.
4. ES Performance Constraint
   The energy storage system (ESS) in this section consists of batteries, and its operational constraints include the following aspects:

   a. Adjacent Time-step ES State Constraint:

$$E_{bat}^{i,t} = E_{bat}^{i,t-1}(1-\eta_{bat,l}) + (P_{batc}^{i,t} - P_{batd}^{i,t})\Delta t \quad (1 \leq t \leq 24) \tag{4.14}$$

   where $E_{bat}^{i,t}$ and $E_{bat}^{i,t-1}$ represent the ES state of the battery in distributed energy grid i at time t and $t+1$, respectively; $\eta_{bat,l}$ denotes the battery's self-loss rate; $P_{batc}^{i,t}$ and $P_{batd}^{i,t}$ represent the charging and discharging power of the battery in distributed energy grid i at time t; Δt denotes the time interval for each time step.

   b. ES State at the Beginning and End of the Operation Period Constraint
   Since the optimization is performed over a cycle (in this section, one day), the ES state at the beginning and end of the operation period should be equal, i.e.:

$$E_{bat}^{i,0} = E_{bat}^{i,24} \tag{4.15}$$

c. Charging and Discharging Power Constraint
The charging and discharging constraints of the battery at time t are as follows:

$$0 \leq P_{batc}^{i,t} \leq X_{bat}^{i,t} \gamma_{batc,\max} C_{bat} X_{bat}^{i,t} = \{0, 1\} \tag{4.16}$$

$$0 \leq P_{batd}^{i,t} \leq Y_{bat}^{i,t} \gamma_{batd,\max} C_{bat} Y_{bat}^{i,t} = \{0, 1\} \tag{4.17}$$

where C_{bat} represents the ES capacity
Since charging and discharging cannot occur simultaneously, the charging and discharging state constraint is:

$$X_{bat}^{i,t} + Y_{bat}^{i,t} \leq 1 \tag{4.18}$$

where $\gamma_{batc,\max}$ and $\gamma_{batd,\max}$ represent the maximum charging and discharging rates, respectively; $X_{bat}^{i,t}$ and $Y_{bat}^{i,t}$ represent the charging and discharging status indicators.

d. Battery Energy State Constraint:

$$SOC_{\min} \leq E_{bat}^{i,t} \leq SOC_{\max} \tag{4.19}$$

where the battery's state of charge is limited by $SOC_{\min}$ and $SOC_{\max}$, representing the minimum and maximum state of charge, respectively.

4.2.2 Operational Optimization Model Between Distributed Energy Grids

The operational optimization model between distributed energy grid clusters consists of an upper-level integrated optimization decision module and a lower-level distributed energy grid management module. The upper-level integrated optimization module aims for the coordinated scheduling of electricity, heat, and gas across the entire grid. It integrates the forecast data, energy flow information, and grid status transmitted from each energy grid, constructing an optimization model across grid clusters and formulating global operational strategies. The lower-level management module is responsible for the local operational optimization and real-time regulation of each distributed energy grid, coordinating the operation of internal distributed energy sources, ES systems, and thermal-gas sources to meet the scheduling requirements from the upper-level and achieve local supply–demand balance. This model enhances the overall operational efficiency and economic performance of the energy grid clusters through dynamic interaction and distributed collaborative optimization between the upper and lower levels.

The operational optimization model is based on a two-level decision structure, designed to explain the coordinated interactions of electricity, heat, and gas within the distributed energy grid clusters. The upper-level is a multi-objective optimization model, aiming to reduce power fluctuations and losses while improving the distribution of voltage, electricity, thermal energy, and gas sources within an acceptable cost range. The lower-level focuses on improving energy utilization efficiency and ensuring the economic and efficient operation of each energy system.

(1) Superior Energy Grid Model

When multiple DEGs with high penetration of renewable energy sources are integrated into the superior energy grid, they can significantly impact system operations. For example, they may cause fluctuations in power at the common connection point, increase grid losses, and lead to fluctuations in voltage, electrical, thermal, and gas supply. Therefore, the main objective of the distribution grid operation is to minimize the negative impacts of multi-energy flow coordination while ensuring the reliability of the system and maintaining high quality in electricity, thermal energy, and gas supply. The objective components in the upper-level decision model include: reducing power fluctuations, minimizing grid losses, optimizing voltage, enhancing the stability of electricity, heat, and gas supply, and ensuring the economic and reliable operation of the overall system.

$$\min F_1 = \sum_{t=1}^{T} \sqrt{\sum_{f \in F} (U^{f,t} - 1)^2 / F} \tag{4.20}$$

$$\min F_2 = \sqrt{\sum_{t=1}^{T} \left(\sum_{i \in I} P_{\mathrm{TL}}^{i,t} - k \right)^2 / T} \tag{4.21}$$

$$\min F_3 = \sum_{t=1}^{T} P_{\mathrm{loss}}^{t} \tag{4.22}$$

$$\min F_4 = \sum_{t=1}^{T} \left(\sum_{j=1}^{H} \left| Q_j(t) - Q_j^{ref}(t) \right|^2 + \sum_{k=1}^{G} \left| G_k(t) - G_k^{ref}(t) \right|^2 \right) \tag{4.23}$$

where F_1 represents the standard deviation of the voltage at each node over the entire operating period, used to measure the degree of voltage deviation during the operation period; F_2 represents the fluctuation of the total exchange power (the exchange power of multiple DEGs) with the superior energy grid, around a constant k, k is a constant greater than zero and less than the superior energy grid's adjustable capacity. The purpose of setting this component is to limit the total exchange power of the multiple DEGs to a specific value and smooth the fluctuation around this value, reducing the impact of high-penetration distributed energy sources on the overall superior energy grid when integrated; F_3 represents the sum of grid losses over the entire operating period; T is the total number of system operation periods; t is the t-th time period; F is the total number of grid nodes; f is the f-th grid node; I is the total number of DEGs; i is the i-th distributed energy grid; $U^{f,t}$ is the voltage at node f at time t (per unit value); $P_{\mathrm{TL}}^{i,t}$ is the exchange power of distributed energy grid i at time t; P_{loss}^{t} is the grid loss at time t; $Q_j(t)$ is the thermal energy output at the thermal energy supply point j at time t; $Q_j^{ref}(t)$ is the reference thermal energy demand; $G_k(t)$ is the gas supply output at the gas supply point k at time t; and $G_k^{ref}(t)$ is the reference gas demand.

Since the superior energy grid layer needs to coordinate multiple energy flows of electricity, heat, and gas, the following constraints must be considered.

a. Power Flow Constraint

Ensure that the power flow at each node in the grid complies with physical laws, while maintaining the stability of voltage and power flow.

$$P_i(t) = \sum_{j \in \mathcal{N}_i} B_{ij}(V_i(t), V_j(t), \theta_{ij}) + \epsilon_P, \quad \forall i \tag{4.24}$$

where $P_i(t)$ is the power output at node i at time t, $V_i(t)$ and $V_j(t)$ are the voltages at nodes i and j, θ_{ij} is the phase difference between nodes i and j, and B_{ij} is the admittance matrix element of the grid.

b. Thermal Flow Constraint

The flow of thermal energy must also satisfy specific transmission and distribution rules, similar to the transmission of electric power.

$$Q_{heat,i}(t) = \sum_{j \in \mathcal{N}_i} C_{ij}(T_i(t), T_j(t)) + \epsilon_Q, \quad \forall i \tag{4.25}$$

where $Q_{heat,i}(t)$ is the thermal energy output at node i, at time t, $T_i(t)$ and $T_j(t)$ are the temperatures at nodes i and j, and C_{ij} is the thermal flow transmission matrix.

c. Gas Flow Constraint

The gas flow constraint primarily ensures that the flow of natural gas or other gas sources can meet the demand and avoid gas source overload.

$$G_k(t) = \sum_{j \in \mathcal{N}_k} D_{kj}(P_{gas,i}(t), T_{gas,j}(t)) + \epsilon_G, \quad \forall k \tag{4.26}$$

where $G_k(t)$ is the gas output at gas source node k at time t, $P_{gas,i}(t)$ is the gas output power, $T_{gas,j}(t)$ is the gas flow temperature, and D_{kj} is the gas flow transmission matrix.

The voltage at the nodes connected to the distribution grid must meet system safety and equipment operation requirements.

$$V_{\min} \leq V_i(t) \leq V_{\max}, \quad \forall i \tag{4.27}$$

where $V_i(t)$ is the voltage at node i at time t, $V_{\min}$ and $V_{\max}$ are the minimum and maximum voltage limits, respectively.

For the thermal energy system, the thermal energy output at each node must be constrained within the operational range of its equipment to prevent overload or failure to meet demand.

$$Q_{\min}^{heat,i} \leq Q_{heat,i}(t) \leq Q_{\max}^{heat,i}, \quad \forall i \tag{4.28}$$

where $Q_{\min}^{heat,i}$ and $Q_{\max}^{heat,i}$ are the minimum and maximum thermal energy outputs at node i at time t, and $Q_{heat,i}(t)$ is the thermal energy output at node i at time t.

Similarly, the output of the gas source system is also subject to certain limits to ensure that the system does not exceed the maximum supply capacity of the gas source equipment.

$$G_{\min}^{gas,i} \leq G_{gas,i}(t) \leq G_{\max}^{gas,i}, \quad \forall i \tag{4.29}$$

where $G_{\min}^{gas,i}$ and $G_{\max}^{gas,i}$ are the minimum and maximum gas source outputs at node f at time t, and $G_{gas,i}(t)$ is the gas source output at node i at time t.

(2) Distributed Energy Grid Layer Model

The core objective of this model is to ensure the reliability, efficiency, and environmental benefits of the operation, while considering the collaborative characteristics of multiple energy flows. The multiple energy systems (including electricity, heat, and gas) integrated into the distributed energy grid form a cooperative relationship. For example, when a distributed energy grid experiences issues such as power deficits, insufficient heat, or gas source fluctuations, other DEGs or energy systems connected through energy channels can provide free electricity, heat, or gas source support. This support is based on the dispatchable capacity of each distributed energy grid and energy system. Additionally, a mutual support relationship is established between the distribution grid, distributed energy grid clusters, and heating and gas systems in the distributed energy grid. The specific objectives of the lower-level decision model include:

$$\begin{aligned} f_1 = &\sum_i \left(c_e \cdot P_{\text{trade},i}^e + c_h \cdot Q_{\text{trade},i}^h + c_g \cdot G_{\text{trade},i}^g\right) \\ &+ \sum_i \left(C_{\text{el},i} \cdot P_{\text{el},i} + C_{\text{th},i} \cdot Q_{\text{th},i} + C_{\text{gas},i} \cdot G_{\text{gas},i}\right) \end{aligned} \tag{4.30}$$

$$f_2 = \sum_{t=1}^{T} \sum_{i \in I} \sum_{j \in J^i} (p_{CO_2} \cdot N_{CO_2}^j + p_{NO_x} \cdot N_{NO_x}^j + p_{SO_2} \cdot N_{SO_2}^j) \cdot P_{DG}^{j,i,t} \tag{4.31}$$

where f_1 is the basic cost required to ensure the reliable operation of the entire distributed energy grid layer. $P_{\text{trade},i}^e$, $Q_{\text{trade},i}^h$, and $G_{\text{trade},i}^g$ are the power transactions (electricity, heat, and gas) between the i-th distributed energy grid and the superior energy grid. c_e, c_h, and c_g are the unit prices for electricity, heat, and gas, respectively. $C_{\text{el},i}$, $C_{\text{th},i}$, and $C_{\text{gas},i}$ represent the production costs for electricity, heat, and gas in the i-th distributed energy grid. $P_{\text{el},i}$, $Q_{\text{th},i}$, and $G_{\text{gas},i}$ are the production power for electricity, heat, and gas in the i distributed energy grid. Since the distributed energy grid structure is simple, grid losses are not considered. f_2 represents the total penalty cost for pollutant gas emissions to measure environmental indicators. p_{CO_2}, p_{NO_x}, and p_{SO_2} are the emission penalty prices for CO_2, NO_x and SO_2, respectively. $N_{CO_2}^j$, $N_{NO_x}^j$, and $N_{SO_2}^j$ are the emissions per 1 kW h of CO_2, NO_x and SO_2, produced by the j-th type of distributed energy source.

The following are the constraint conditions for the distributed energy grid layer optimization model.

The power balance constraint is:

$$\begin{cases} P_{\text{el},i} + P_{\text{trade},i}^e = L_{\text{el},i} & \forall i \\ Q_{\text{th},i} + Q_{\text{trade},i}^h = L_{\text{th},i} & \forall i \\ G_{\text{gas},i} + G_{\text{trade},i}^g = L_{\text{gas},i} & \forall i \end{cases} \tag{4.32}$$

where $P_{\text{el},i}$ is the electricity production power of the i distributed energy grid. $L_{\text{el},i}$ is the electricity load demand of the i-th distributed energy grid. $P_{\text{trade},i}^e$ is the electricity transaction power between the i-th distributed energy grid and the superior energy grid. The same applies to heat and gas power, so they are not repeated here.

The DG unit constraint is:

$$\begin{cases} P_{\text{MT}}^{\min} \le P_{\text{DG}}^{1,i^1,t} + x^{1,i^1,t} P_{\text{DGEX}}^{1,i^1,t} \le P_{\text{MT}}^{\max} \\ P_{\text{FC}}^{\min} \le P_{\text{DG}}^{2,i^2,t} + x^{2,i^2,t} P_{\text{DGEX}}^{2,i^2,t} \le P_{\text{FC}}^{\max} \\ -P_{\text{SB}}^{\max} \le P_{\text{DG}}^{3,i,t} + y^{i,t} P_{\text{SBEX}}^{i,t} \le P_{\text{SB}}^{\max} \\ Q_{\text{MT}}^{\min} \le Q_{\text{DG}}^{1,i^1,t} \le Q_{\text{MT}}^{\max} \\ Q_{\text{FC}}^{\min} \le Q_{\text{DG}}^{2,i^2,t} \le Q_{\text{FC}}^{\max} \\ -Q_{\text{SB}}^{\max} \le Q_{\text{DG}}^{3,i,t} \le Q_{\text{SB}}^{\max} \end{cases} \tag{4.33}$$

where i^1 is the distributed energy grid containing the microturbine, and i^2 is the distributed energy grid containing the FC. This chapter assumes that each distributed energy grid contains a battery. $P_{\text{MT}}^{\min}$, $P_{\text{MT}}^{\max}$, $P_{\text{FC}}^{\min}$, $P_{\text{FC}}^{\max}$, $-P_{\text{SB}}^{\max}$, $P_{\text{SB}}^{\max}$, $Q_{\text{MT}}^{\min}$, $Q_{\text{MT}}^{\max}$,

Q_{FC}^{min}, Q_{FC}^{max} and $-Q_{SB}^{max}$ are the lower and upper limits of the output power for the microturbine, FC, and battery, respectively.

The ramping constraints for the microturbine and FC are:

$$\begin{cases} P_{DG}^{1,i^1,t} + x^{1,i^1,t} P_{DGEX}^{1,i^1,t} - P_{DG}^{1,i^1,t-1} \leq R_{MT}^{up} \\ P_{DG}^{1,i^1,t-1} - \left(P_{DG}^{1,i^1,t} + x^{1,i^1,t} P_{DGEX}^{1,i^1,t} \right) \leq R_{MT}^{down} \\ P_{DG}^{2,i^2,t} + x^{2,i^2,t} P_{DGEX}^{2,i^2,t} - P_{DG}^{2,i^2,t-1} \leq R_{FC}^{up} \\ P_{DG}^{2,i^2,t-1} - \left(P_{DG}^{2,i^2,t} + x^{2,i^2,t} P_{DGEX}^{2,i^2,t} \right) \leq R_{FC}^{down} \end{cases} \tag{4.34}$$

where R_{MT}^{up}, R_{MT}^{down}, R_{FC}^{up} and R_{FC}^{down} are the limits for the increase and decrease of active power for the microturbine and FC, respectively.

The battery capacity constraint is:

$$\begin{cases} E_{T}^{i,t} = E_{T}^{i,t-1} - P_{DG}^{3,i,t} \\ E^{i,t} = E_{T}^{i,t} - y^{i,t} P_{SBEX}^{i,t} \\ E^{min} \leq E^{i,t} \leq E^{max} \end{cases} \tag{4.35}$$

where $E^{i,t}$ is the remaining capacity of the battery in distributed energy grid i at time t, and E^{min}, E^{max} are the lower and upper limits of the battery's capacity, respectively.

The interactive reserve capacity constraint is:

$$\sum_{i \in I} R_{ex}^{i,t} \leq \varepsilon R_{OMG}^{t} \tag{4.36}$$

$$R_{MG}^{t} = (1 - \varepsilon) R_{OMG}^{t} \tag{4.37}$$

where R_{OMG}^{t} represents the total capacity at time t. ε is defined as the support game coefficient, with $0 < \varepsilon < 1$, which reflects the allocation of reserve capacity between the distributed energy grid and the superior energy grid, while also indicating the level of support provided by the distributed energy grid layer to the superior energy grid layer.

(3) Cooperative Operation Behavior Analysis Between DEGs

When multiple DEGs are interconnected through energy channels and operate in coordination, they actually provide more options for the scheduling department but also make scheduling decisions more complex. Therefore, the cooperative operation behavior of DEGs can be analyzed and modeled based on alliance game theory and information flow game theory.

In alliance game theory, different DEGs can form alliances. The formation of alliances is based on benefit distribution strategies, aiming to maximize the total system benefits. When the distribution strategy within the alliance is optimal, the individuals within the alliance can obtain more benefits than they would without forming an alliance. In this process, DEGs interact cooperatively through multi-energy flow channels, including electricity, thermal energy, and gas, creating a mutually supportive cooperative relationship. Through energy sharing, power scheduling, and other means, DEGs can optimize the overall system's operation, improve energy utilization efficiency.

$$\begin{cases} \boldsymbol{x} = (x1, \ldots, xi) \\ \sum_{i=1}^{n} xi = v(N) \\ \sum_{i \in S} xi \geq v(S), \forall S \in N \end{cases} \tag{4.38}$$

where x_i and $v(\cdot)$ represent the profit function and characteristic function, respectively. In a cooperative game, $(N, v), N = \{1, 2, \ldots, n\}$ will be allocated to each participant i $(i \in N)$, and it satisfies the condition $\boldsymbol{x} = (x1, \ldots, xi)$, $\sum_{i=1}^{n} x_i = v(N)$, which means that $x_i \geq v(\{i\})$, is the allocation strategy for the coalition S. $v(S)$ represents the maximum utility of the game between S and $N - S = \{i | i \in N, i \notin S\}$.

In a coalition S of multiple DEGs, at least one energy demander i and one energy supplier j exist. Define $N_i = P_{loadi} - P_i$ as the energy required by i, where P_{loadi} is the load of distributed energy grid i, and P_i is the output of the distributed energy sources in grid i. The supplier j provides the remaining dispatchable interactive backup capacity. The utility parameter λ ($/kW) is defined as follows:

$$v(S) = \lambda \sum_{k \in S} P_k + \lambda \sum_{i \in S_D, j \in S_P} P_{ij} \tag{4.39}$$

where P_{ij} represents the power transmitted from distributed energy grid i to distributed energy grid j. Therefore, the profit function for distributed energy grid i can be expressed as:

$$x_i = \lambda \pi_i + \lambda \pi_i^{\mathrm{ex}} \tag{4.40}$$

where π_i represents the output of distributed energy grid i in the alliance, and π_i^{ex} represents the total energy support from other DEGs in the alliance. Let us assume that each alliance S satisfies $\sum_{i \in S} N_i = \sum_{i \in S_D, j \in S_P} P_{ij}$, then we can derive the following:

$$
\begin{aligned}
x(N) &= \lambda \sum_{i \in N} (\pi_i + \pi_i^{ex}) = \lambda\left(\sum_{i \in N} P_i + \sum_{m \in S'_D, l \in S'_P} P_{ml}\right) \\
&= v(N), S'_D \cup S'_P = N \\
x(S) &= \lambda \sum_{i \in S} (\pi_i + \pi_i^{ex}) = \lambda\left(\sum_{i \in S} P_i + \sum_{i \in S} N_i\right) \\
&= \lambda\left(\sum_{i \in S} P_i + \sum_{i \in S_D, j \in S_P} P_{ij}\right) \\
&= v(S), S_D \cup S_P = S, \forall S \in N
\end{aligned}
\tag{4.41}
$$

According to the condition, the allocation strategy for interactive backup capacity reaches its optimal point. π_i and λ are constants. Therefore, the distributed energy grid with a larger π_i^{ex} will gain more benefits, meaning the interactive backup capacity needs to be utilized as much as possible. However, from the perspective of dynamic game theory, the supply and demand balance within the distributed energy grid range will not always be satisfied within an alliance, and λ will not always remain constant. Thus, it is necessary to design rules to guide the specific allocation of interactive backup capacity and the interactive operation of DEGs.

When $N \neq 0$, distributed energy grid i will have a pending game set $\{a_1, a_2, a_3, a_4\}$, which contains four decision variables: a_1 for the renewable energy generation of i, a_2 for the output of other DG units of i (including battery charging and discharging), a_3 for the part of the backup capacity from the adjacent distributed energy grid that belongs to microturbines and FCs, and a_4 for the part of the backup capacity from the adjacent distributed energy grid that belongs to batteries. The following table provides the game strategies using electricity as an example.

In the first round of the game, to fully utilize renewable energy and absorb its generation fluctuations, the decision a1 should be made first. a_1 is the first decision in the game and will always be executed, so the 0–1 decision variable x_1 is always set to 1. After executing a_1, in the second round of the game, there are two possible scenarios: if the energy from a_1 equals the deficiency, the game ends and no further decisions need to be executed; if the energy from a_1 is greater than or less than the deficiency, the battery in a_2 will need to store or release energy. After executing this, in the third round of the game, two scenarios can also occur: if the energy from a_2 equals the deficiency, the game ends; if the energy from a_2 is less than the deficiency, besides a_2, a secondary decision variable needs to be set to decide the generation order of different neighboring DEGs, similar to the previous step. The game continues in this manner until the last round. When the decision variable is set to 1, it indicates the decision is executed under specific conditions; when set to 0, the decision is not executed. The final game set is formed after all the rounds (Table 4.1).

(4) Interactive Reserve Game Matrix

As discussed earlier, the game involving a_1 and a_2, since the operational total cost of WT and PV power generation is lower than that of other distributed generators in the lower-level objective function, these two decision variables will change according

Table 4.1 Game strategies

Game order	Decision variable	0–1 Decision variable	Analysis (conditions for 0–1 decision variable value)	Post-game behavior
1	a_1	$x_1 = 1$	a_1 should be fully utilized and always executed first	If $a_1 = P_{\text{load}}$, the game ends (remaining 0–1 variables are set to 0); if not, proceed to Order 2
2	a_2	$x_2 = 1$	a_2 needs to store excess energy or compensate for P_{load}	If $a_1 + a_2 = P_{\text{load}}$, the game ends; if not, proceed to Order 3
		$x_2 = 0$	Achieve energy balance	The game ends
3	a_3	$x_3 = 1$	If $a_1 + a_2 < \text{P}_{\text{load}}$, interaction support is required to compensate for P_{load}	If $a_1 + a_2 + a_3 = \text{P}_{\text{load}}$, the game ends; if not, proceed to Order 3
		$x_3 = 0$	Achieve energy balance or require other DEGs' batteries to store excess energy	If $a_1 + a_2 = P_{\text{load}}$, the game ends; if $a_1 + a_2 > P_{\text{load}}$, proceed to Order 4
4	a_4	$x_4 = 1$	If $a_1 + a_2 + a_3 < P_{\text{load}}$, controllable distributed energy sources in interaction support should compensate first, leaving flexible batteries for emergency situations; if $a_1 + a_2 > P_{\text{load}}$, these batteries store excess energy	After a_4 is executed, the game ends
		$x_4 = 0$	Achieve energy balance	The game ends

to the direction of the objective function during optimization, prioritizing the parts with lower costs. Therefore, the distributed generators in the distributed energy grid itself do not require 0–1 decision variables. However, $C^{i,t}$ changes with $R_{\text{ex}}^{i,t}$, so 0–1 decision variables are needed to guide the operational behavior of the interactive reserve. Below is an introduction to the interactive reserve game matrix:

$$R_{\text{ex}}^{i,t} = a^{i,t} X^{i,t} P_{\text{DGEX}}^{i,t} + b^{i,t} Y^{i,t} P_{\text{SBEX}}^{i,t} \tag{4.42}$$

$$X^{i,t} = \left[x^{1,h,t} \ldots x^{M^h,h,t} \ldots x^{M^{H^i},H^i,t} \right] \tag{4.43}$$

$$Y^{i,t} = \left[y^{h,t} \ldots y^{H^i,t} \right] \tag{4.44}$$

$$P_{\text{DGEX}}^{i,t} = \left[P_{\text{DGEX}}^{1,h,t} \cdots P_{\text{DGEX}}^{M^h,h,t} \cdots P_{\text{DGEX}}^{M^{H^i},H^i,t} \right]^{\text{T}} \tag{4.45}$$

$$P_{\mathrm{SBEX}}^{i,t} = \left[P_{\mathrm{SBEX}}^{h,t} \cdots P_{\mathrm{SBEX}}^{H^i,t}\right]^{\mathrm{T}} \tag{4.46}$$

$$a^{i,t} = \begin{cases} 1 \ P_{\mathrm{ML}}^{i,t} - \sum\limits_{j \in J^i} P_{\mathrm{DG}}^{j,i,t} > 0 \\ 0 \ P_{\mathrm{ML}}^{i,t} - \sum\limits_{j \in J^i} P_{\mathrm{DG}}^{j,i,t} \leq 0 \end{cases} \tag{4.47}$$

$$x^{m,h,t} = \begin{cases} 1 & \begin{aligned} &\alpha^{m,h,t} = \alpha(P_{\mathrm{DG\,max}}^{m,h,t} - P_{\mathrm{DG}}^{m,h,t}) = \\ &\max\{\alpha^{m',h',t} \big| m' \in M^{h'}, h' \in H^i\} \end{aligned} \\ or & \begin{aligned} &\alpha^{m',h',t} = P_{\mathrm{DGEX}}^{m',h',t}, \alpha^{m',h',t} \in \{\alpha^{m',h',t} \big| \alpha^{m',h',t} > \\ &\alpha^{m,h,t}, m' \in M^{h'}, h' \in H^i\} \end{aligned} \\ 0 & others \end{cases} \tag{4.48}$$

$$b^{i,t} = \begin{cases} 1 & \begin{aligned} &P_{\mathrm{ML}}^{i,t} - \sum_{j \in J^i} P_{\mathrm{DG}}^{j,i,t} < 0 \ or \ \alpha^{m,h,t} = \\ &P_{\mathrm{DGEX}}^{m,h,t}, \forall \alpha^{m,h,t} \in \{\alpha^{m',h',t} \big| m' \in M^{h'}, \ h' \in H^i\} \end{aligned} \\ 0 & others \end{cases} \tag{4.49}$$

$$y^{h,t} = \begin{cases} & \beta^{h,t} = \beta E_{\mathrm{T}}^{h,t} = \max\{\beta^{h',t} \big| h' \in H^i\} \\ 1 & or \ \beta^{h',t} = P_{\mathrm{SBEX}}^{h',t}, \beta^{h',t} \in \{\beta^{h'',t} \big| \beta^{h'',t} > \beta^{h,t}, h'' \in H^i\} \\ & or \ P_{\mathrm{SBEX}}^{h,t} < 0 \\ 0 & others \end{cases} \tag{4.50}$$

$$C^{i,t} = \frac{\sum\limits_{h \in H^i} \left[\sum\limits_{m \in M^h} (K_{\mathrm{fuel}}^m + K_{\mathrm{om}}^m) P_{\mathrm{DGEX}}^{m,h,t} + K_{\mathrm{om}}^3 P_{\mathrm{SBEX}}^{h,t}\right]}{\sum\limits_{h \in H^i} \left[\sum\limits_{m \in M^h} P_{\mathrm{DGEX}}^{m,h,t} + P_{\mathrm{SBEX}}^{h,t}\right]} \tag{4.51}$$

where $R_{\mathrm{ex}}^{i,t}$ represents the variable integrated into the interactive reserve game matrix; i is the distributed energy grid receiving cooperative support; H^i is the set of all DEGs connected, and its largest index is the element; h is the distributed energy grid belonging to set H^i; M^h are the set of types of machines contained in distributed energy grid h; m is the type of machine within set; $P_{\mathrm{DGEX}}^{i,t}$ is a matrix representing the interactive reserve capacity provided by all DEGs h for MT and FC, existing between the unit output and the maximum allowable output at that time; $P_{\mathrm{DGEX}}^{m,h,t}$ represents the output of machine type m in distributed energy grid h used to support distributed energy grid i at time t; $a^{i,t}$ is the primary 0–1 decision variable matrix, indicating whether to execute the decision when distributed energy grid i's own generation is insufficient to compensate for the power shortfall; $X^{i,t}$ is the secondary 0–1 decision variable matrix of $P_{\mathrm{DGEX}}^{i,t}$, determining the order of execution of parallel elements in the set, with a game strategy that prioritizes those with larger interactive reserve capacities to maintain balance in the overall distributed energy grid cluster's reserve capacities; and α is the extreme coefficient of $P_{\text{DGEX}}^{m,h,t}$ to prevent distributed energy grid h from keeping the machine in full operation to support other grids, thus ensuring the machine's reliable operation. When this limit is reached, the second-largest reserve capacity from another grid will provide support, and so on. $P_{\text{SBEX}}^{i,t}$ represents the interactive reserve capacity provided by the battery storage system, and $P_{\text{SBEX}}^{h,t}$ indicates the battery's use of interactive reserve capacity at time t in distributed energy grid h. When this value is in a transitional state, the battery's capacity lies between the transitional battery capacity and the maximum capacity. The primary 0–1 decision variable is $P_{\text{SBEX}}^{i,t}$, which can be executed in two cases: one where MT and FC reserves cannot compensate for the power shortfall, and thus, the battery storage system's reserve is reserved for emergencies; and the second when the supported grid has surplus power after consumption within the grid (including selling excess energy to the higher-level energy grid), in which case the battery system will be charged. The secondary 0–1 decision variable matrix $Y^{i,t}$ executes the battery reserve with the larger transitional capacity first, similar to the principle of the MT and FC reserve. β represents the extreme coefficient of $P_{\text{SBEX}}^{h,t}$. The detailed flowchart is shown in Fig. 4.7.

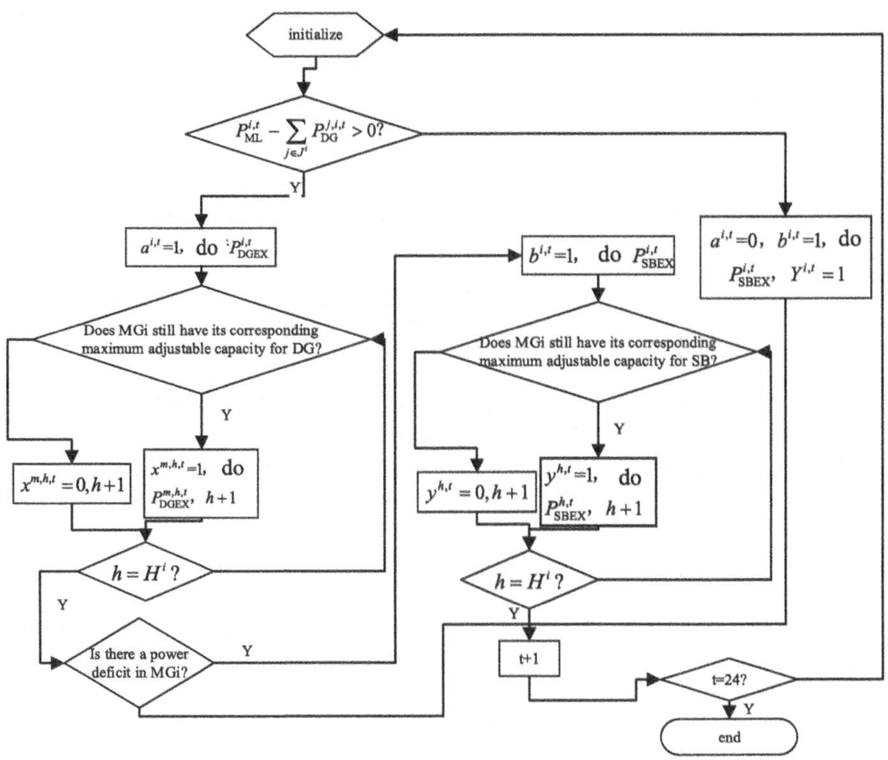

Fig. 4.7 Flowchart of the execution of the interactive alternate game matrix

4.3 Case Study and Simulation

4.3.1 Case Study 1

This section adopts a DEG Cluster system consisting of the IEEE 33-node distribution grid and the 32-node thermal grid of Bali, Indonesia. DEG 1 is connected to both distribution grid node 13 and thermal grid node 17, DEG 2 is connected to distribution grid node 18 and thermal grid node 24, while DEG 3 is solely connected to distribution grid node 30. The grid configuration is illustrated in Fig. 4.8.

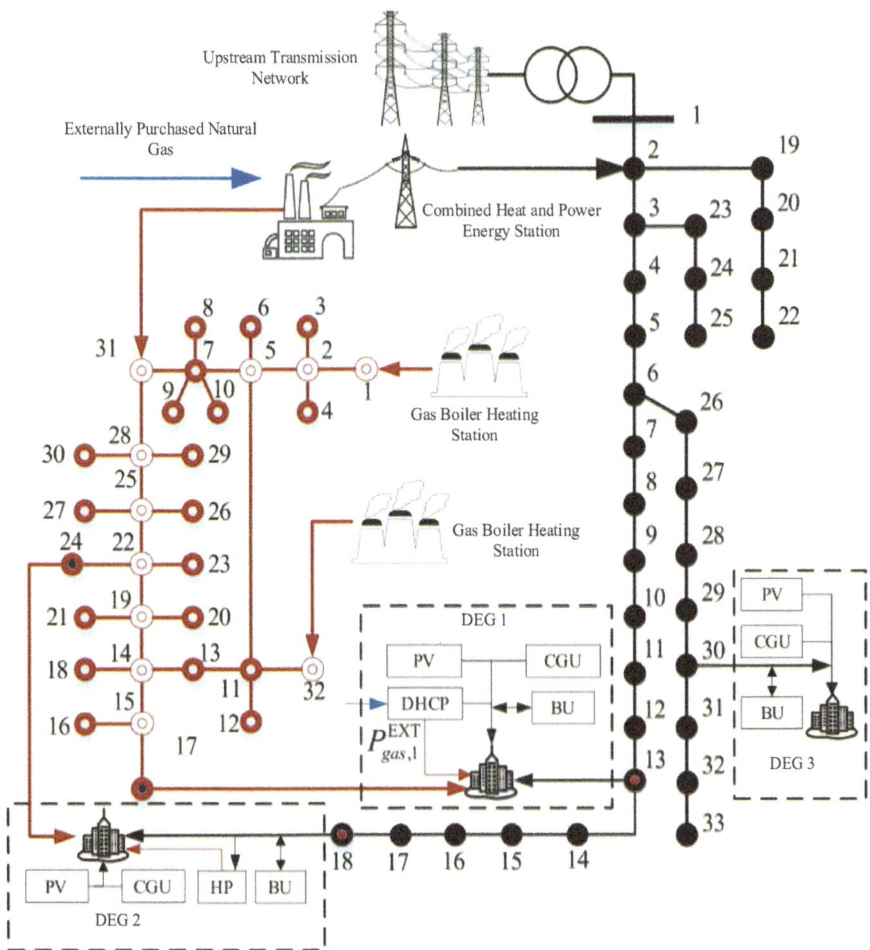

Fig. 4.8 DEG cluster system

The parameters of diesel generators and ES within the DEGs are detailed in Tables 4.2 and 4.3, respectively. A scheduling period of 15 min is adopted, aiming to achieve the lowest total cost over the entire day through day-ahead dynamic economic dispatch. Real-time transaction electricity prices are obtained from Li et al. (2019), and equipment operation and maintenance costs are derived from Jiang et al. (2020). The PV output is shown in Fig. 4.9.

The penalty function multipliers λ_m and ω_m are initialized at 0.5, the multiplier update coefficient κ is set to 1.5, and the convergence precisions ε_1 and ε_2 are 0.01. The scaling factor s values are provided in Table 4.4.

Table 4.2 Small diesel generator parameters

DEGs	Output upper limit (kW)	Output lower limit (kW)	a ($/(MW)2h)	b ($/MWh)	c ($/h)
1	50	300	0.5	10	5
2	100	500	1.2	15	3
3	100	500	0.8	12	4

Table 4.3 ES system parameters

DEGs	Minimum energy (%)	Maximum energy (%)	Output upper limit (kW)	Output lower limit (kW)	Charging efficiency	Discharge efficiency	Self-loss coefficient
1	10	90	6	8	0.95	0.95	0.995
2	10	90	6	8	0.95	0.95	0.995
3	10	90	6	8	0.95	0.95	0.995

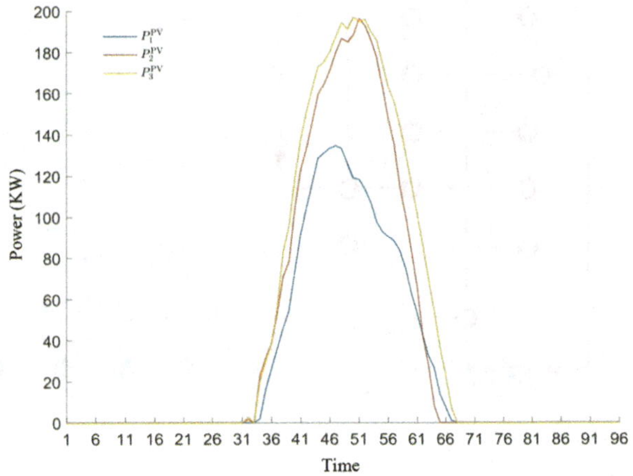

Fig. 4.9 Profiles of forecast distributed PV power

Table 4.4 Value of scaling factor

$s^{(1)}$	0.003	$s_1^{(1)}$	0.002	$s_2^{(1)}$	0.005	$s_3^{(1)}$	0.003
$s^{(2)}$	0.003	$s_1^{(2)}$	0.002	$s_2^{(2)}$	0.005	$s_3^{(2)}$	0.003
$s^{(3)}$	0.006	$s_1^{(3)}$	0.002	$s_2^{(3)}$	1		

4.3.1.1 Hierarchical Optimization Dispatch Results

Figure 4.10 presents the results of the hierarchical optimization dispatch. Due to higher daytime temperatures and lower nighttime temperatures, the heating load peaks at night, while the electricity power curve remains at a low point during the night. Therefore, there is a significant difference between the peaks and valleys of the electricity and heat load curves. Considering the heat loss during the thermal transmission process, the total heat supply is greater than the total heat demand. From the figure, it can be observed that the system's thermal output does not balance in real-time with the heat load, primarily due to different transmission delays caused by varying distances from the heat source to the DEGs. Additionally, the thermal output curve is not a simple translation of the heat load curve. For example, during scheduling periods 61–65, from 15:00 to 16:15, the system's thermal output is lower than the heat load, causing the room temperature in DEG 1 to gradually decrease to the minimum allowable temperature. The temperature at each node of the thermal grid changes over time, as shown in Fig. 4.10. During this period, the temperature of the supply-return water pipes in the thermal grid significantly decreases, indicating that part of the thermal energy in the grid is used to compensate for the insufficient thermal output, leading to a reduction in the grid's thermal storage level (Fig. 4.11). The terminal thermal energy can be supplied either by the thermal grid or local heat sources, based on cost considerations. Through electricity-heat coordination and thermal storage in the heat grid responding to peak-valley electricity prices, the system's energy costs can be reduced.

Table 4.5 compares the dispatch costs under three different scenarios. When the GBs power limit is 1,000 kW and the heat grid's thermal storage capacity and thermal inertia are not considered, the thermal load peak exceeds the system's maximum thermal output, resulting in an inability to maintain thermal balance. When the GBs power limit is increased to 1,200 kW, and the dispatch mode accounts for the thermal storage and thermal inertia of the grid, the economic benefits of the scheduling mode are significantly improved. This also demonstrates that the dynamic economic model established in this section, which considers the dynamic characteristics of the heat grid, fully utilizes the flexibility of the thermal grid. It also takes into account the inherent connections between the system's operational periods, making it more reflective of real-world conditions.

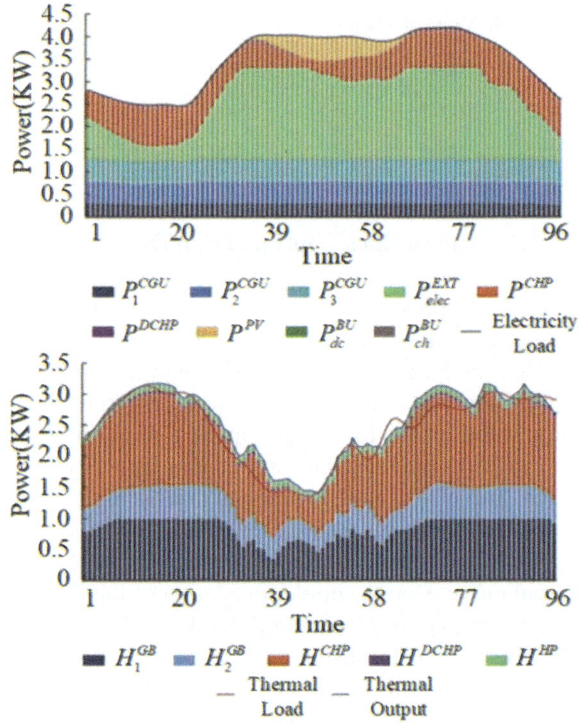

Fig. 4.10 Optimized dispatch results of the DEG cluster

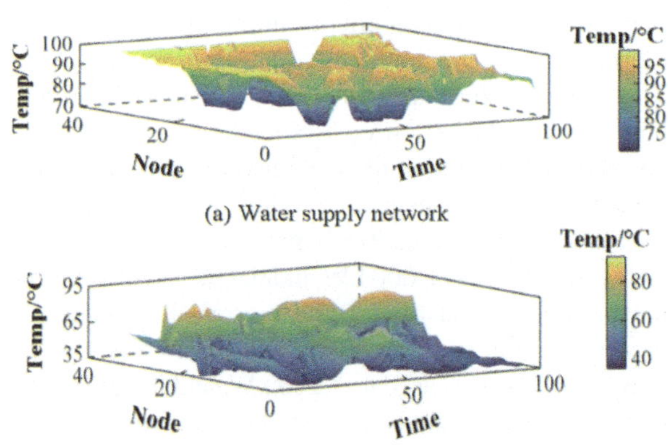

(a) Water supply network

(b) Return Water Network

Fig. 4.11 Temperature at each node of the thermal grid

4.3.1.2 Comparison of Hierarchical Autonomous Algorithm and Centralized Algorithm

As an example where PV forecasting uncertainty is not considered, the comparison between the hierarchical autonomous algorithm and the centralized algorithm in

Table 4.5 Comparison of dispatch economics under three different scenarios

Mode	Whether to consider the thermal storage of the grid	Whether to consider thermal inertia	Total cost (USD)	
			The upper limit of thermal output is 1000 kW	The upper limit of thermal output is 1200 kW
1	N	N	\	26,806.67
2	N	Y	\	26,724.99
3	Y	Y	26,168.6	26,083.95

terms of computation results and solution time is shown in Table 4.6. Both algorithms solve the active power exchange between the DEGs and the upstream grid. The active power exchange of DEG 1 is shown in Fig. 4.12. Positive values indicate that active power flows from the DEG to the upstream distribution grid. The centralized heating power of DEG 1 is shown in Fig. 4.13.

From the Fig. 4.12, it can be seen that the trends of the interaction power solved by the two methods are generally consistent, with the real-time power differences primarily caused by the scaling factor. The cost errors calculated by both methods are no higher than 0.3%, indicating that the hierarchical optimization method proposed in this paper ensures the global optimality of the interests of each entity and the total system production cost with high accuracy. The hierarchical autonomous algorithm converges after 13 iterations. Since this paper uses a single computer to perform serial solutions for multiple DEGs in both the upper and lower layers, the total duration is

Table 4.6 Comparison of different algorithms in dispatch economics

Algorithm	Total cost (USD)	Deviation from centralized (%)	Serial solution time (s)	Number of iterations	Maximum power deviation (kW)
Hierarchical autonomous algorithm	26 168.60	0.276	13.696 0	13	0.006
Centralized algorithm	26 096.61	\	1.321 1	1	\

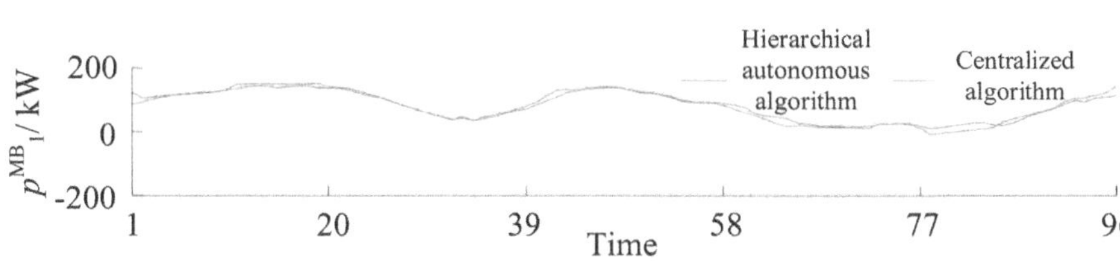

Fig. 4.12 Active power flow curves from DEGs to the distribution grid under the hierarchical autonomous algorithm and centralized algorithm

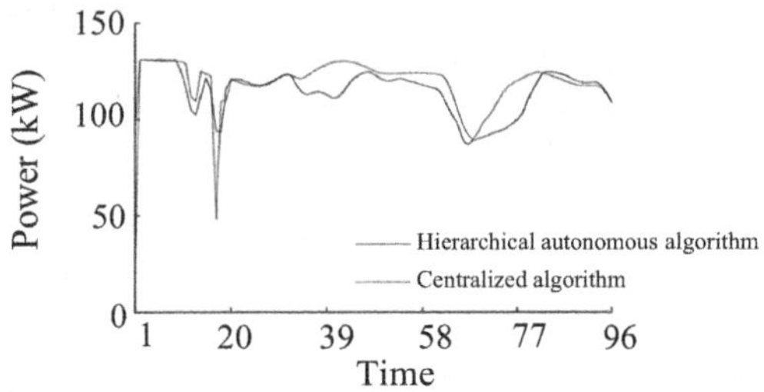

Fig. 4.13 Centralized heating power curves under the hierarchical autonomous algorithm and centralized algorithm

longer than that of the centralized algorithm. However, in reality, each lower-layer DEG is optimized independently and in parallel, so the actual solution time would be further reduced.

4.3.1.3 Algorithm Convergence Analysis

The convergence process of the interaction power between the three DEGs and the upper layer in the 31st scheduling period is shown in Fig. 4.14a. By adding a penalty function, the interaction power gradually converges within the permissible error range. The changes in the penalty function value, convergence criterion value, and total production cost during the iteration process are shown in Fig. 4.14b–d, respectively.

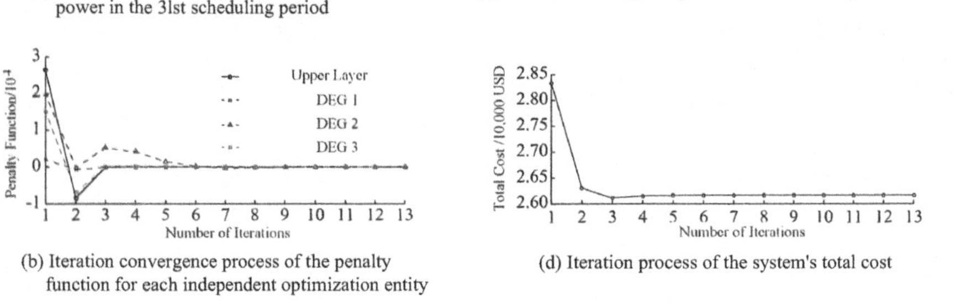

Fig. 4.14 Curves of boundary power, penalty function, convergence criterion, and total cost variation during the iteration process

Table 4.7 Device parameter information

Equipment name	Efficiency	Operating cost (CNY)	Wind and solar curtailment penalty cost (CNY)
Wind turbine	–	2450	3500–5000
PV	–	3575	3500–5000
Electric refrigeration	0.7	2400	–
CCHP	0.6	4800	–
ES	0.9	8760	–
Gas storage	0.9	3200	–
Thermal storage	0.9	4300	–
Ice storage	0.9	4000	–

Since the first iteration is an independent optimization without considering the penalty function, the initial penalty term and interaction power deviation are large. After a limited number of coordination calculations, the interaction power deviation gradually decreases. At the same time, the penalty multiplier increases, ensuring that the penalty function effectively reduces the power deviation during the iteration process. Finally, after 13 iterations, both the maximum power deviation and the total cost change rate are less than 0.01, and the total cost converges to 26,168.6 yuan.

4.3.2 Case Study 2

Actual data from a specific region is selected, considering the coordinated conversion and utilization of various flexibility resources in the DEG. Among these, clean energy sources include wind, solar, and geothermal energy. Under the framework of a DEG that integrates multiple clean energy sources and their interconversion, uncertainties in wind and solar output as well as load are taken into account. A probabilistic model describing the uncertainty of random variables is established.

At the same time, controllable load and ES are considered as flexibility resources. Using a hierarchical coordination optimization method, the optimal solution is obtained. The equipment parameter information used in the case study is shown in Table 4.7.

4.3.2.1 Scenario Generation Results Analysis

In the hierarchical coordination optimization of the DEG, it is crucial to consider the correlation of different uncertain energy sources and forecast errors. If various uncertainties are not taken into account, the results of wind and solar power forecasting will deviate from reality. During the winter, when the weather is cold and solar irradiance is low, wind speeds tend to be higher, and heating load is the most frequently used type of load. In contrast, during the summer, when the weather is hot, solar irradiance is stronger, and wind speeds are lower, the types and amounts of load used can differ significantly from those in winter. The uncertainty probability distributions of electricity and heating loads in different typical seasons are distinct. If uncertainty factors are not considered, there will be a significant deviation from the actual situation.

Considering the correlation between multiple uncertain factors in scenario generation, 1,000 samples are taken for summer solar radiation, wind speed, and electricity load based on the distribution function characteristics of each random variable. The comparison between two directly generated results, one considering correlation and the other ignoring correlation, is shown in Fig. 4.15.

As shown in Fig. 4.14, Fig. 4.14a illustrates the case where the correlation between uncertainties in wind speed, PV output, and load is not considered, with

Fig. 4.15 Comparison of sampling results

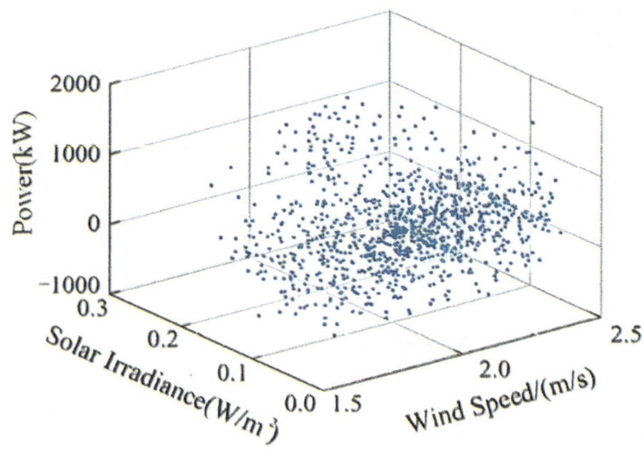

(a) Exclusion of correlation constraints

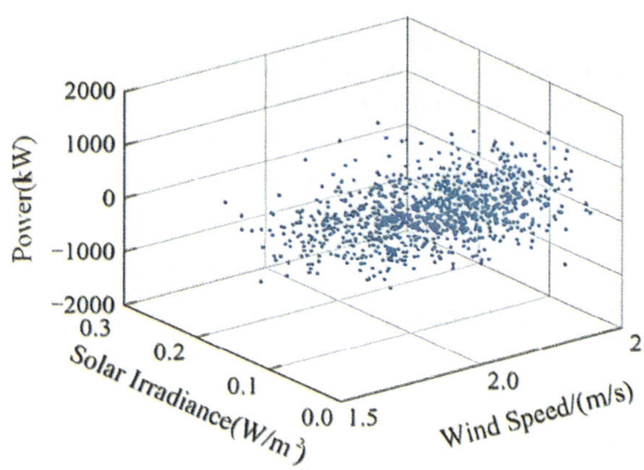

(b) Consideration of correlation constraints

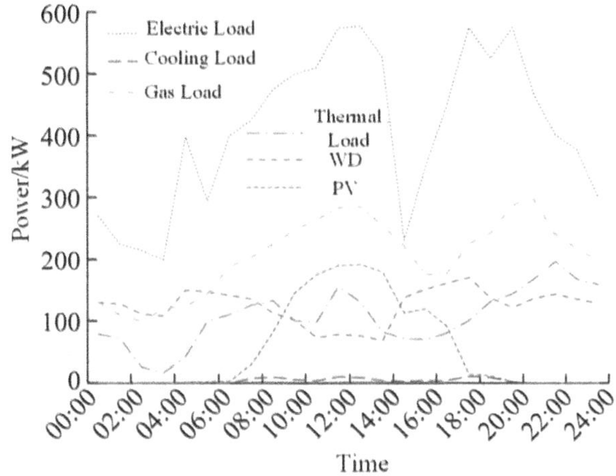

Fig. 4.16 Typical winter scene

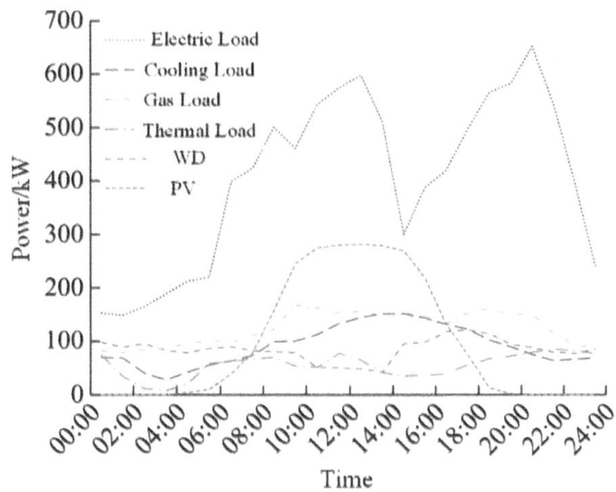

Fig. 4.17 Typical summer scene

a more dispersed data distribution. Figure 4.14b shows the case where the correlation between uncertainties in wind speed, PV output, and load is considered, and the correlation is visually evident through the clustering degree. After applying scenario reduction techniques and incorporating forecasting errors, the typical winter and summer scenarios are obtained, as shown in Figs. 4.16 and 4.17.

4.3.2.2 Analysis of Hierarchical Optimization Results

To more intuitively analyze the impact of the scenario generation method that considers forecast errors on the system's economy, two hierarchical scheduling schemes are set for comparative analysis:

Table 4.8 Cost comparison of different scheme

Scheme	Upper layer cost (10,000 CNY)	Lower layer cost (10,000 CNY)	Total cost (10,000 CNY)	Wind and solar curtailment penalty (10,000 CNY)
1	4.281	7.145	11.426	0.415
2	3.171	6.262	9.433	0.372

Scheme 1: Hierarchical coordination optimization based on the wind, solar, and load random scenario generation method that does not consider forecast errors

Scheme 2: Hierarchical coordination optimization based on the wind and solar random scenario generation method that considers forecast errors

Based on the parameter settings, the hierarchical scheduling method is used to calculate the results, and the economic comparison under different scenarios is shown in Table 4.8. The wind and solar curtailment amounts are shown in Fig. 4.18.

As clearly shown in Table 4.8, the total cost and the wind and solar curtailment penalty cost for Scheme 1 are both higher than those for Scheme 2. From Fig. 4.18, it can be seen that the wind and solar curtailment amount for Scheme 2 is significantly lower than that for Scheme 1, leading to a much lower wind and solar curtailment penalty cost in Scheme 2. Therefore, considering the correlation between uncertainties in wind speed, PV output, and load, as well as forecasting errors, can effectively reduce the wind and solar curtailment amount and economic costs of the DEG.

In conclusion, based on the typical summer and winter scenarios of wind and solar output, it is essential to consider forecasting errors and perform hierarchical optimization scheduling for the DEG cluster.

4.3.2.3 Comparative Analysis of Hierarchical Coordination Optimization Method and Centralized Method

A hierarchical optimization approach is adopted to reduce the curtailment of wind and solar power in DEGs while more accurately minimizing economic costs. The iterative curves are shown in Figs. 4.19 and 4.20.

A comparison between the hierarchical coordination optimization method and the centralized optimization method is shown in Table 4.9. When considering hierarchical coordination optimization, the cost difference from the centralized optimization algorithm does not exceed 0.3%. This indicates that the hierarchical coordination optimization method enhances the economic performance of the distributed energy grid while maintaining accuracy.

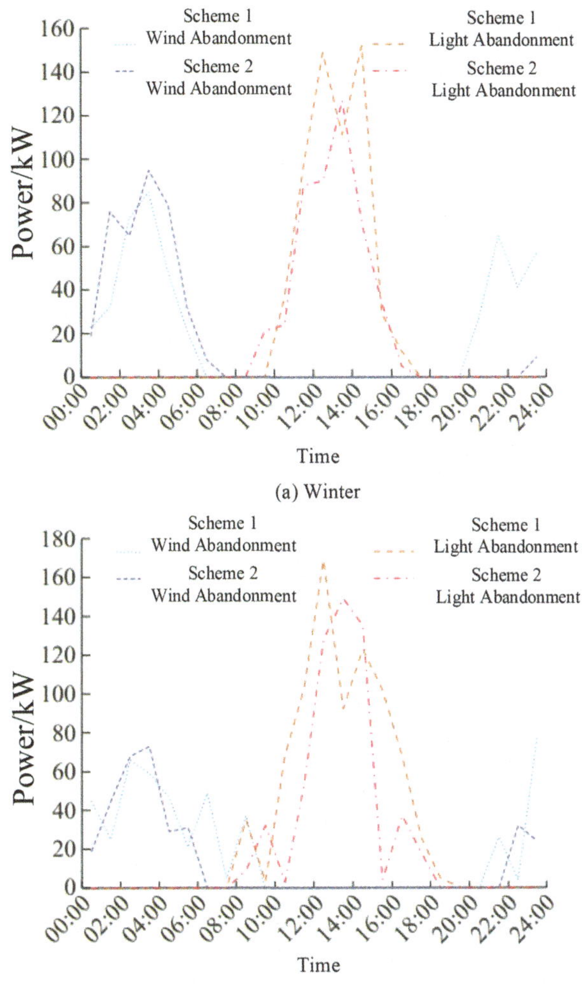

Fig. 4.18 Comparison of wind and solar abandonment

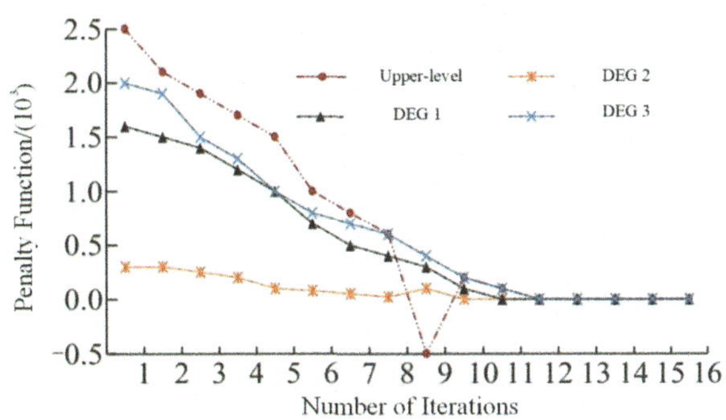

Fig. 4.19 Penalty function iterative convergence procedure

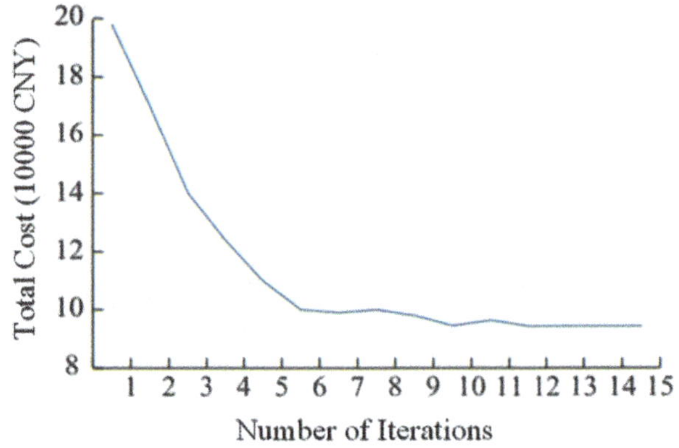

Fig. 4.20 Total cost iterative convergence process

Table 4.9 Cost comparison of different algorithms

Algorithm	Total cost (10,000 CNY)		Deviation from centralized method (%)	Number of iterations
	Total cost	Total cost		
Hierarchical coordination optimization method	3.171	6.262	0.279	13
Centralized method	6.262		0	1

References

Amirkhani A, Barshooi AH (2022) Consensus in multi-agent systems: a review. Artif Intell Rev 55(5):3897–3935

Bustos R et al (2023) Hierarchical energy management system for multi-microgrid coordination with demand-side management. Appl Energy 342:121145

Jiang Y, Wan C, Chen C, Shahidehpour M, Song Y (2020) A hybrid stochastic-interval operation strategy for multi-energy microgrids. IEEE Trans Smart Grid 11(1):440–456. https://doi.org/10.1109/TSG.2019.2923984

Li R, Wei W, Mei S, Hu Q, Wu Q (2019) Participation of an energy hub in electricity and heat distribution markets: an MPEC approach. IEEE Trans Smart Grid 10(4):3641–3653. https://doi.org/10.1109/TSG.2018.2833279

Liu C et al (2024) Enhanced power recovery strategy in PV-ESS-VSG systems using coordinated virtual impedance and virtual damping control. J Energy Storage 104:114664

Pokharel B (2021) A voltage based optimal relay coordination in distributed generation penetrated distribution network using particle swarm optimization. Dissertation, Pulchowk Campus

Wang Q, Guo Y, Wang P (2023) Coordinated control strategy of grid-connected converters based on maximum power optimization during grid voltage sags. Int J Electr Power Energy Syst 144:108472

Chapter 5
Adaptive Adjustment Technology for Distributed Energy Grid Clusters

5.1 Rolling Optimization Modeling

5.1.1 Voltage/Frequency Modeling Based on Multi-agent Consistency

This section proposes an adaptive droop control strategy based on multi-agent consensus, utilizing distributed communication among intelligent agents, to address the issues of frequency and voltage deviation, system stability, and power allocation accuracy in traditional droop control. Based on traditional droop control, a second-order dynamic model for active power allocation, active frequency, and reactive voltage is established. Based on considering communication delay, applying leaderless consistency control to the active output of distributed power sources meets the requirements of traditional droop control. Apply consistency correction with virtual leaders to address frequency and voltage deviations in traditional droop control. At the same time, the asymptotic stability of the system is verified through the direct Lyapunov method, thereby solving the impact of the active power droop coefficient on system stability in traditional control. On this basis, the impact of communication disturbances on control results and stability is analyzed.

(1) Introduction

Affected by primary energy, most distributed power sources are connected to the grid through power converters. Typical distributed power sources include power control, voltage control, current control, pulse width modulation (PWM), primary energy and ESSs, inverter circuits, and filtering circuits. The typical distributed power sources and their constituent units are shown in Fig. 5.1.

The power flow and vector diagram between two nodes in distributed energy grid clusters are shown in Fig. 5.2. The trend of the line can be expressed as:

T. Lu et al., *Operation Modeling and Collaborative Regulation of Distributed Energy Grid Clusters*, Smart Energy Systems, https://doi.org/10.1007/978-981-96-9942-1_5

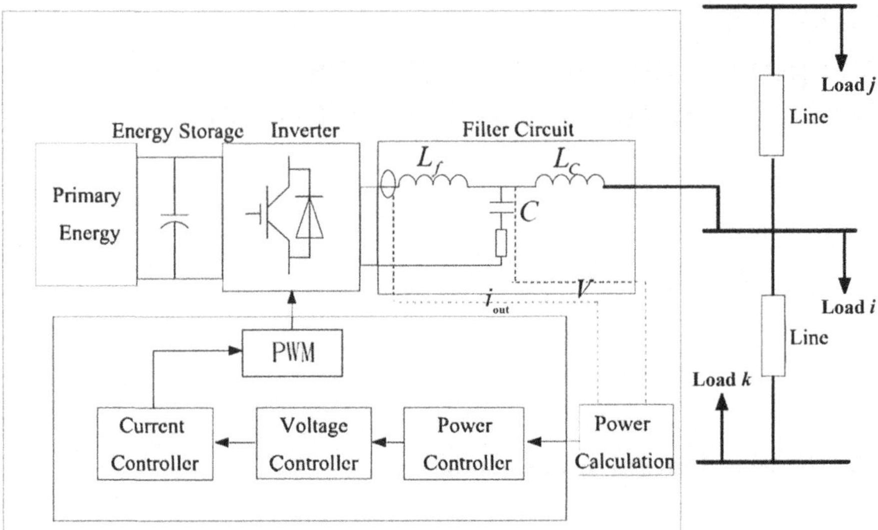

Fig. 5.1 Typical distributed power structure

$$P = \frac{V_1^2}{z}\cos\theta - \frac{V_1V_2}{z}\cos(\theta + (\sigma_1 - \sigma_2)) \tag{5.1}$$

$$Q = \frac{V_1^2}{z}\sin\theta - \frac{V_1V_2}{z}\sin(\theta + (\sigma_1 - \sigma_2)) \tag{5.2}$$

where P and Q are the active and reactive power on the line, v_1, δ_1 and v_2, δ_2 respectively represent the voltage and phase angle of two nodes, Z and θ represent line impedance and impedance angle, respectively. Multiplying Eq. (5.1) by reactance X minus Eq. (5.2) by resistance R, and multiplying Eq. (5.1) by resistance R plus Eq. (5.2) by reactance X, yields:

Fig. 5.2 Line current between two nodes

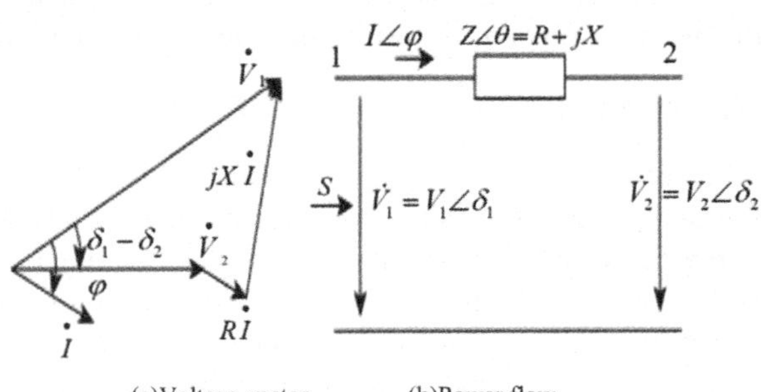

$$\begin{cases} XP - RQ = U_1U_2\sin(\sigma_1 - \sigma_2) \\ RP + XQ = U_1^2 - U_1U_2\cos(\sigma_1 - \sigma_2) \end{cases} \tag{5.3}$$

$$\begin{cases} \left|U_1 \sin\sigma = \frac{X}{U_2}P\right| \\ \left|U_1 - U_2\cos\sigma = \frac{X}{U_1}Q\right| \end{cases} \tag{5.4}$$

Without considering the line resistance, traditional droop control can be expressed as:

$$\omega_i = \omega_{0i} - m_iP_i \tag{5.5}$$

$$V_{di}^* = V_{0i} - n_iQ_i \tag{5.6}$$

$$V_{qi}^* = 0 \tag{5.7}$$

where ω is the system frequency, V_{di}^* and V_{qi}^* are the components of the output voltage of the distributed power source on the vertical and horizontal axes, P and Q are the active and reactive power of the distributed power source, ω_{0i} and V_{0i} are the frequency and voltage when the output of the distributed power source is 0, respectively. m and n represent the droop coefficients of active and reactive power, respectively.

Figure 5.3 shows the control block diagram of droop control, where LPF is a low-pass filter with a cut-off frequency ω_c.

In distributed energy grid clusters, small distributed power sources are mainly connected to the grid through voltage-type converters, and the system inertia is approximately zero. By simulating the power regulation characteristics of traditional generators, the droop control in Eqs. (5.5)–(5.7) is proposed to solve the problem of parallel operation of multiple distributed power sources, to achieve coordinated control of active and reactive power of distributed power sources. In traditional droop control, distributed power sources only need to collect port voltage and frequency, and do not need to communicate with each other, but cannot coordinate with other

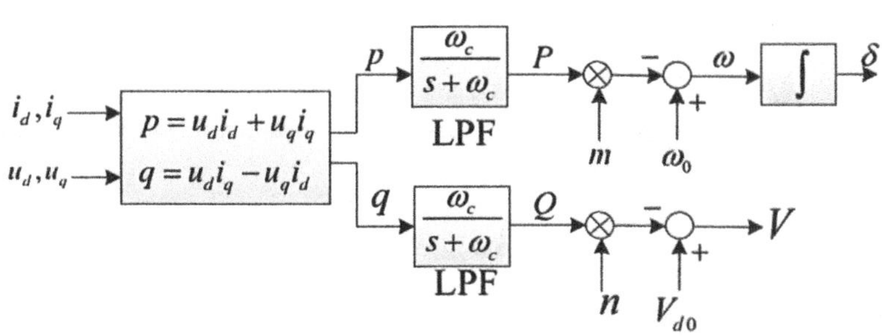

Fig. 5.3 Droop control diagram

distributed power sources to achieve system-level goals. In the energy system, communication networks are inevitably present due to the need for EMS. Based on the local communication between distributed power sources, this chapter proposes the application of consistency control to enable distributed power sources to coordinate and cooperate and achieve system-level goals such as frequency regulation. Based on the active power allocation and voltage/frequency regulation characteristics in traditional droop control, a droop parameter adaptive control strategy considering communication network delay and disturbance is proposed, which enables distributed power sources to maintain system frequency and voltage at target values while meeting the requirements of droop control characteristics. Meanwhile, by applying the Lyapunov direct method to analyze the stability of the system, the results show that under the control of the strategy proposed in this chapter, the system is asymptotically stable.

(2) Voltage/Frequency Control Model Based on Feedback Linearization

The basic idea of the feedback linearization method is to simplify a nonlinear system into a linear or partially linear system under reasonable assumptions, and then design a controller using the linear system design method in classical control theory, to transform many complex nonlinear problems into relatively simple linear problems. According to different forms of feedback, feedback linearization can be divided into two types: state and output feedback linearization. The former is a complete linearization of the state equation, while the latter emphasizes linearization between input and output. This chapter applies the state feedback linearization method to model the active frequency, voltage reactive power, and active power distribution of distributed power sources.

1. Active frequency control

In traditional droop control, ω_0 and V_0 are used to control the initial values of frequency and voltage, which are constants and are often overlooked in small signal stability analysis. On this basis, the improved droop strategy mainly improves system performance by controlling other electrical quantities or increasing auxiliary control quantities. This chapter mainly adjusts these two control variables through adaptive control strategies to meet the requirements of the system.

By taking the derivative of Eq. (5.5), we can obtain:

$$\dot{\omega}_i = \dot{\omega}_{0i} - m_i \dot{P}_i \equiv x_{\omega i} \tag{5.8}$$

where $x_{\omega i}$ represents the first-order differentiation of frequency and voltage. Further differentiation yields:

$$\dot{x}_{\omega i} = u_{c\omega i} \tag{5.9}$$

where $u_{c\omega i}$ represents the second-order differentiation of frequency and voltage. Equations (5.8) and (5.9) together constitute a second-order dynamic regulation model for frequency regulation of distributed energy grid clusters.

2. Reactive power voltage control

By taking the derivative of Eq. (5.9), it can be obtained:

$$\dot{V}_{di} = \dot{V}_{0i} - n_i\dot{Q}_i \equiv x_{vi} \tag{5.10}$$

where x_{vi} represents the first-order differentiation of frequency and voltage. By taking the derivative of Eq. (5.10) again, we can obtain:

$$\dot{x}_{vi} = u_{cvi} \tag{5.11}$$

where u_{cvi} represents the second-order differentiation of frequency and voltage. Equations (5.10) and (5.11) together form the second-order dynamic regulation model for group voltage control in distributed energy grid clusters.

3. Active power control

For the droop control of Eq. (5.5), the active power distribution among distributed power sources needs to meet the distribution characteristics of Eq. (5.12).

$$m_1P_1 = m_2P_2 = \cdots = m_nP_n \tag{5.12}$$

Similar to the modeling model of frequency and voltage, the control model for active power allocation in distributed power sources can be obtained by taking the derivative of Eq. (5.12) twice in a row:

$$\begin{cases} |m_i\dot{P}_i = x_{Pi} \\ |\dot{x}_{Pi} = u_{cPi}| \end{cases} \tag{5.13}$$

From the above model, it can be seen that adjusting $u_{c\omega i}$, u_{cvi}, u_{cpi} in Eqs. (5.9), (5.11), and (5.13) reasonably can control $x_{\omega i}$, x_{vi}, x_{pi} to be 0, simultaneously adjust ω_0 and V_0 to ensure that the system operates at rated frequency and voltage.

(3) Consistency Based Voltage/Frequency Control

1. Active power control

Since the active power of distributed power sources varies with changes in load, there is no fixed target value. Assuming t_d is the delay in the communication system, this chapter proposes the leaderless consensus implementation of active power control for distributed power sources as shown in Eqs. (5.14)–(5.18).

$$u_{cPi} = u_{cP\alpha i} + u_{cP\beta i} \tag{5.14}$$

$$u_{cP\alpha i} = -k_{P1}\sum_{j=1,j\neq i}^{n} a_{ij}(\Delta mP_{ij}) \tag{5.15}$$

$$u_{cP\beta i} = \sum_{i=1,i\neq j}^{n} a_{ij} sign(x_{Pji})|x_{Pji}|^{k_{P2}} \tag{5.16}$$

$$\Delta mP_{ij} = mP_i(t) - mP_j(t - t_d) \tag{5.17}$$

$$x_{Pji} = x_{Pj}(t - t_d) - x_{Pi}(t) \tag{5.18}$$

where $u_{cp\alpha i}$ and $u_{cp\beta i}$ represent two independent parts of u_{cpi}, $u_{cp\alpha i}$ is mainly used to control the product of the active droop coefficient and active power of all distributed power sources (i.e., *mp* is equal). $u_{cp\beta i}$ is mainly used to control its micro-components ($m\dot{P}$ is equal). k_{p1} and k_{p2} are control parameters for collaborative control, and $k_{p1} > 0$, $sign(x)$ represents a sign function concerning x, as follows:

$$sign(x) = \begin{cases} |1, x > 0 \\ | - 1, |x < 0| \end{cases} \tag{5.19}$$

By combining the active power regulation model of Eq. (5.13), the equilibrium point for active power regulation can be obtained. The following two equations:

$$m_i P_i(t) = m_j P_j(t - t_d) \tag{5.20}$$

$$m_i \dot{P}_i(t) = m_j \dot{P}_j(t - t_d) \tag{5.21}$$

When the system runs stably, both the first and second derivatives of active power are 0, indicating that the output power of distributed power sources is constant. In traditional first-order consistency control, only Eq. (5.20) holds. The equilibrium point of the system may have periodic components with t_d. As the time, it may even lead to system instability, and therefore communication delay has a significant impact on the control results and stability of the system. According to the control principle, the stability of the system needs further analysis. The following section will apply the Lyapunov direct method to analyze the asymptotic stability of the system. According to the principle of the Lyapunov direct method. When the energy function $V(x)$ satisfies the following four requirements, the system is asymptotically stable.

(a) $V(0) = 0$.
(b) When $x \neq 0$ exists, there exists $V(x) > 0$.
(c) When $\|x\| \to \infty$ exists, there exists $V(x) \to \infty$.
(d) For all $x \neq 0, \dot{V}(x) < 0$.

For the previously constructed system, choose the energy function shown in Eq. (5.22).

$$V(x) = \frac{1}{2}\sum_{i=1}^{n} x_{Pi}^2 + \frac{1}{2}\sum_{i=1}^{n}\sum_{j=1,j\neq i}^{n} k_{P1}a_{ij}(\Delta mP_{ij})^2 \tag{5.22}$$

By analyzing the characteristics of the sub-functions, it can be seen that Eq. (5.22) satisfies the first three requirements of Lyapunov, and its derivative still needs further analysis.

$$\dot{V}(x) = \sum_{i=1}^{n} x_{Pi}u_{cPi} + \sum_{i=1}^{n}\sum_{j=1,j\neq i}^{n} k_{P1}a_{ij}(\Delta mP_{ij})(m\dot{P})_{ij} \tag{5.23}$$

Substituting the governing equations of Eqs. (5.14)–(5.17) into Eq. (5.23) yields.

$$\begin{aligned}\dot{V}(x) = &\sum_{i=1}^{n} a_{ij}x_{Pi}\sum_{j=1,j\neq i}^{n}\left(-k_{P1}(\Delta mP_{ij}) + sign(x_{Pji})|x_{Pji}|^{k_{P2}}\right)\\ &+\sum_{i=1}^{n}\sum_{j=1,j\neq i}^{n} k_{P1}a_{ij}(\Delta mP_{ij})x_{Pij}\end{aligned} \tag{5.24}$$

The first and third parts of Eq. (5.24) satisfy:

$$\sum_{i=1}^{n} x_{Pi}\sum_{j=1,j\neq i}^{n} k_{P1}a_{ij}(\Delta mP_{ij}) = \sum_{i=1}^{n}\sum_{j=1,j\neq i}^{n} k_{P1}a_{ij}(\Delta mP_{ij})x_{Pij} \tag{5.25}$$

Therefore, $\dot{V}(x)$ can be simplified as:

$$\dot{V}(x) = \sum_{i=1}^{n}\sum_{j=1,j\neq i}^{n} a_{ij}(x_{Pij})sign(x_{Pji})|x_{Pji}|^{k_{P2}} \tag{5.26}$$

For all $x \neq 0$, $\dot{V}(x) < 0$ constants, it holds that under the consistency protocol of Eqs. (5.14)–(5.18), the system is asymptotically stable. In traditional first-order consistency protocols, communication delay t_d has a significant impact on the stability of the system. Especially when t_d exceeds the absolute value of the second-largest eigenvalue of the Laplacian matrix, the system is unstable.

2. Active frequency droop control

This section mainly analyzes the adaptive adjustment correction of system frequency deviation through ω_{0i}. In the energy system, the target of system frequency is the pre-set rated value, which means that all distributed power sources operate at the rated frequency under stable operating conditions. Apply consistency with virtual leadership to achieve adaptive frequency control, and the control equations are shown in Eqs. (5.27)–(5.34).

$$u_{coi} = u_{co\alpha i} + u_{co\beta i} + u_{co\gamma i} \tag{5.27}$$

$$u_{c\omega\beta i} = -k_{\omega 1} \sum_{j=1, j \neq i}^{n} a_{ij}\omega_{ij} \tag{5.28}$$

$$u_{c\omega\beta i} = \sum_{i=1, i \neq j}^{n} a_{ij} sign(x_{\omega ji})|x_{\omega ji}|^{k_{\omega 2}} \tag{5.29}$$

$$u_{c\omega\gamma i} = k_{\omega 3}\left[-k_{\omega 1}\omega_{iL} + sign(x_{\omega Li})|x_{\omega Li}|^{k_{\omega 2}}\right] \tag{5.30}$$

$$\omega_{ij} = \omega_i(t) - \omega_j(t - t_d) \tag{5.31}$$

$$\omega_{iL} = \omega_i - \omega_L \tag{5.32}$$

$$x_{\omega ji} = x_{\omega j}(t - t_d) - x_{\omega i}(t) \tag{5.33}$$

$$x_{\omega Li} = x_{\omega L} - x_{\omega i} \tag{5.34}$$

where L represents a virtual leader with control objectives, which can be a central controller. $u_{c\omega\alpha i}$, $u_{c\omega\beta i}$, $u_{c\omega\gamma i}$ represent the three sub-control components in consistency control respectively. The first two control parts have similar functions to Eq. (5.14), while the third part is mainly used to control the frequency and virtual leadership of distributed power sources. $k_{\omega 1}$, $k_{\omega 2}$ and $k_{\omega 3}$ are control parameters in the consistency protocol, and $k_{\omega 1}$, $k_{\omega 3} > 0$. In frequency control, since the rated frequency in the virtual leader is pre-set, the communication delay between the virtual leader and the distributed power source can be ignored.

$$\omega_i(t) = \omega_j(t - t_d) = \omega_N \tag{5.35}$$

$$\dot{\omega}_i(t) = \dot{\omega}_j(t - t_d) = 0 \tag{5.36}$$

Equation (5.35) indicates that in a stable state, the frequency of the output voltage of each distributed power source is the rated frequency of the system, which is the target of system frequency control. Similarly, select the energy function of Eq. (5.37) to analyze the stability of the system.

$$V(x) = \frac{1}{2}\sum_{i=1}^{n}(x_{oiL})^2 + \frac{1}{2}\sum_{i=1}^{n}\sum_{j=1,j\neq i}^{n} k_{o1}a_{ij}\omega_{ij}^2 + \sum_{i=1}^{n} k_{o1}k_{o3}\omega_{iL}^2 \tag{5.37}$$

By taking the derivative of Eq. (5.37) and applying the analytical methods of Eqs. (5.23)–(5.25), we can simplify it as follows:

$$\dot{V}(x) = \sum_{i=1}^{n}\hat{x}_{oi}\sum_{j=1,j\neq i}^{n} a_{ij}sign(\hat{x}_{oji})|\hat{x}_{oji}|^{k_{oj}} + k_{oj}\sum_{i=1}^{n}\hat{x}_{oi}sign(-\hat{x}_{oi})| - \hat{x}_{oi}|^{k_{oj}} \tag{5.38}$$

where $\hat{x}_{\omega i} = x_{\omega i} - x_{\omega L}$. The energy function of Eq. (5.37) satisfies all the requirements of the Lyapunov direct method. Therefore, under the consistency protocol of Eqs. (5.27)–(5.34), the system is asymptotically stable. Therefore, by combining Eqs. (5.8), the parameter adjustment models of Eqs. (5.9) and (5.27)–(5.34) can calculate ω_{0i} as:

$$\omega_{oi} = \int (m_i\dot{P}_i + \int u_{coi}dt)dt + \omega_{oi0} \tag{5.39}$$

where ω_{oi0} is the initial value of ω_{oi}, to improve convergence speed, it can be set to the rated value of the system frequency. Combined with the active power control in the previous section. The $m\dot{P}$ in Eq. (5.39) is obtained from Eqs. (5.13)–(5.18), that is:

$$m_i\dot{P}_i = \int u_{cP_i}dt + m_i\dot{P}_{i0} \tag{5.40}$$

where $\dot{P}_{i0}$ is the initial value of the first derivative of active power $\dot{P}_i$. Since the goal of control is to keep P constant, $\dot{P}_i$ can be selected as 0. The joint Eqs. (5.5) and (5.40) can obtain adaptive droop control of frequency, and the control block diagram is shown in Fig. 5.4.

3. Reactive power voltage droop control

In traditional droop control, the reactive power of distributed power sources does not meet specific relationships due to differences in line resistance and node voltage. This chapter mainly uses adaptive droop control parameters to maintain the node voltage of distributed power sources at the rated value, thereby providing high-quality power quality for the load. The adaptive droop control of voltage is similar to frequency when the rated voltage is known. u_{cv} can be obtained using the consistency protocol of Eqs. (5.27)–(5.33).

$$V_i = \int \left(\int u_{cvi}dt\right)dt + V_{oi0} - n_iQ_i \tag{5.41}$$

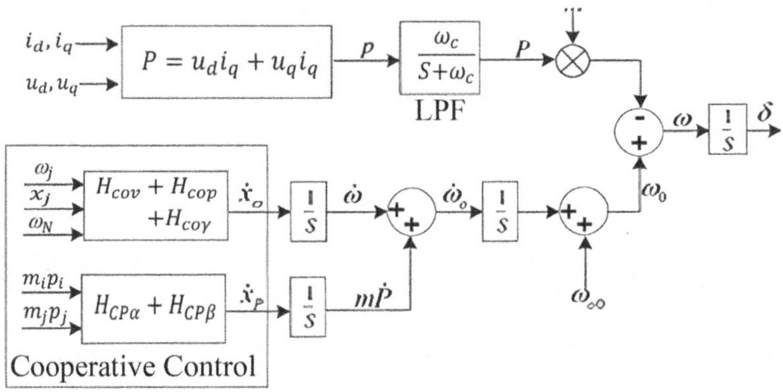

Fig. 5.4 Frequency adaptive control block diagram

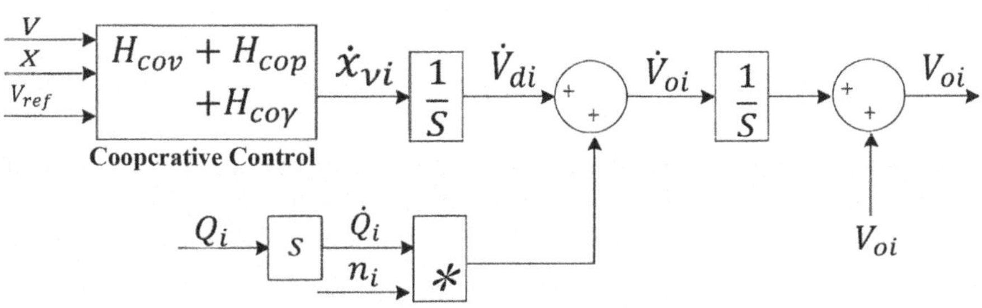

Fig. 5.5 Voltage control block diagram

where V_{0i0} is the initial value of V_{0i}, to accelerate the stabilization speed, it can be set to the rated voltage (Fig. 5.5).

4. Control characteristics under disturbance

In actual distributed energy grid clusters, disturbances such as data collection, transmission accuracy, and communication errors in the controller may affect the control accuracy and even stability of distributed power sources. Take frequency regulation as an example to analyze the control characteristics with disturbances. Assuming $\delta_{\omega j}$ and $\delta_{x\omega j}$ respectively represent the disturbances in the frequency and derivative received by agent i from agent j. In frequency adjustment, Eqs. (5.31) and (5.33) need to be corrected to:

$$\hat{\omega}_{ij} = \omega_i(t) - (\omega_j(t - t_d) + \delta_{\omega j}) \tag{5.42}$$

$$\hat{x}_{\omega ji} = (x_{\omega j}(t - t_d) + \delta_{x\omega j}) - x_{\omega i}(t) \tag{5.43}$$

Similarly, the system equilibrium points of Eqs. (5.35) and (5.36) are:

$$\omega_i(t) = \omega_j(t - t_d) + \delta_{\omega j} = \omega_N \tag{5.44}$$

$$x_{\omega i}(t) = x_{\omega j}(t - t_d) + \delta_{x\omega j} = 0 \tag{5.45}$$

According to the analysis results of Eqs. (5.44) and (5.45), it can be seen that in a stable state, the i-th distributed power source operates at the frequency of virtual leadership, which means that the control objective of the i-th intelligent agent is basically not affected by disturbances in the transmitted information and has strong robustness.

In stability analysis, replacing ω_{ij} and $x_{\omega ij}$ with $\underset{ij}{\widehat{\omega}}$ and $\underset{\omega ij}{\widehat{x}}$ containing perturbation amounts can also yield the result of Eq. (5.38), which states that the system remains asymptotically stable even under communication disturbances.

(4) Simulation analysis

This section adopts the collaborative control strategy proposed in the distributed energy network group test as shown in Fig. 5.6, and its distributed communication network is shown in Fig. 5.7.

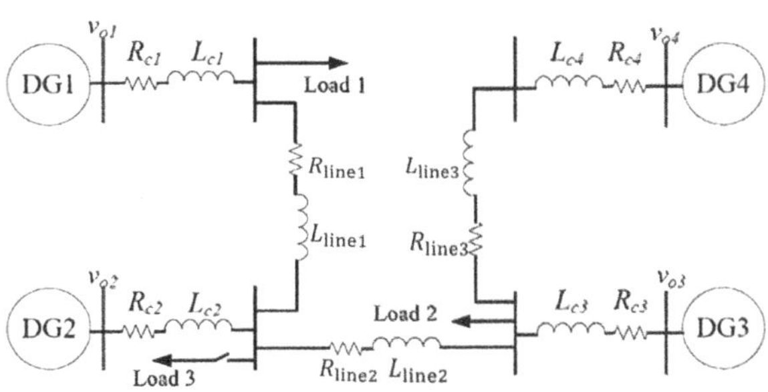

Fig. 5.6 Test simulation system

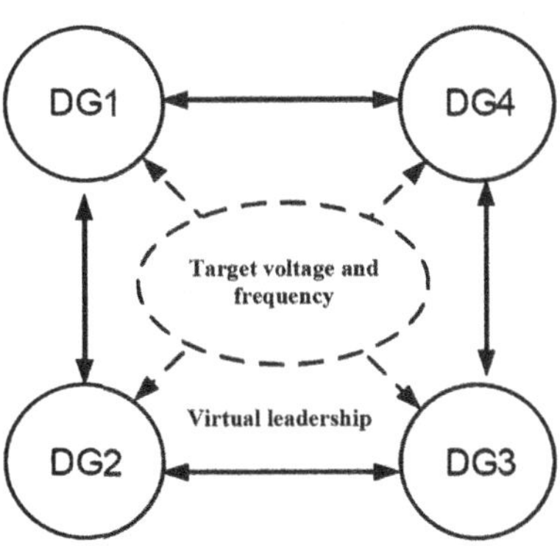

Fig. 5.7 Communication system with virtual leadership

According to graph theory analysis, the Laplacian matrix of the distributed system shown in Fig. 5.7 is:

$$L = \begin{bmatrix} 0 & -1 & 0 & -1 \\ -1 & 0 & -1 & 0 \\ 0 & -1 & 0 & -1 \\ -1 & 0 & -1 & 0 \end{bmatrix} \tag{5.46}$$

The line and load parameters used in the distributed energy grid clusters are listed in Table 5.1. The parameters for droop control and consistency control are shown in Table 5.2. This section mainly verifies the effectiveness of the proposed method through three simulation cases.

Example 1: Simulation without delay and disturbance

The simulation results ignoring communication delay and disturbance are shown in Fig. 5.8. To demonstrate the effectiveness of the strategy proposed in this chapter,

Table 5.1 Distributed energy grid clusters

DGs and lines	$R_{c1} = R_{c2} = R_{c3} = R_{c4}$		$L_{c1} = L_{c2} = L_{c3} = L_{c4}$	
	0.2Ω		1 mH	
	$R_{\text{line1}} = R_{\text{line3}}$	$L_{\text{line1}} = L_{\text{line3}}$	R_{line2}	L_{line2}
	0.23Ω	0.318 mH	0.35Ω	1.847mH
Loads	Load_1		Load_3	
	P_1 (kW)	Q_1 (kVar)	P_3 (kW)	Q_3 (kVar)
	36	36	45.9	22.8
	Load_2			
	P_2 (kW)	Q_2(kVar)		
	36	36		

Table 5.2 Control parameters

Droop parameters	$m_{p1} = m_{p2}$	$n_{Q1} = n_{Q2}$	$m_{p3} = m_{p4}$	$n_{Q3} = n_{Q4}$
	0.000094	0.0013	0.000125	0.0015
	V_{oi}		ω_{oi}	
	330 V		325.3 rad/s	
Consistency control parameters	Frequency	$k_{\omega 1}$	$k_{\omega 2}$	$k_{\omega 3}$
		50	0.8	800
	Voltage	K_v	K_v	K_v
		150	0.8	3
	Active power	K_p	K_p	$k_{\omega 3}$
		500	0.8	

Fig. 5.15 shows the simulation results of traditional droop control under the same parameter conditions. During the simulation process, loads 1 and 3 were continuously connected, and load 2 was connected to the system at $t = 2$ s. As shown in Fig. 5.8a, b, the control strategy proposed in this section can adaptively adjust the droop parameters under load changes, thereby maintaining the system frequency at the rated value. In Fig. 5.15a, although the initial frequency of the system is set near the rated value through the initial value, there is a significant deviation in the system frequency under load changes. The simulation results of voltage control are shown in Fig. 5.8c, d. In adaptive control, corresponding adjustments are made as the load changes to maintain the terminal voltage of the distributed power source at the rated value. In Fig. 5.9b, the terminal voltage of each distributed power source varies with the change of load. The above results indicate that the adaptive droop control proposed in this section can effectively correct frequency and voltage deviations in traditional droop control, and improve power supply quality. As shown in Fig. 5.8e, while adjusting the frequency and voltage, the active power distribution of distributed power sources still meets the requirements of Eq. (5.12). Compared to the active power in Fig. 5.15c, the active power output of the distributed power source has increased because the terminal voltage of the nodes is maintained at the rated value, and the load in the simulation is a constant impedance load. In adaptive droop control, as the node voltage is maintained at the rated value, the reactive load of load 2 at $t = 2$ s is mainly borne by distributed power source 2, achieving the local balance of reactive power. The result is shown in Fig. 5.9f.

Example 2: Simulation considering delay

In MASs, agents generally exchange information with each other through serial communication, including RS485, CAN, and Ethernet. Among the above communication methods, RS485 has the slowest communication speed. Considering communication stability, a baud rate of 9600 bps (bits/second) is generally chosen. In this chapter, the data that needs to be exchanged between intelligent agents includes seven analog quantities: the product of active power and active droop coefficient of distributed power sources (mP) and its derivative, reactive power Q, output frequency and its derivative, output voltage and its derivative, etc. To improve the accuracy of data transmission, the above-mentioned transmission quantities are all transmitted using 32-bit signed shaping. Considering the control bits (usually 16 bits) and checksum (usually using CRC checksum, 16 bits) for serial communication. Based on the above analysis, the effective data for a single communication between intelligent agents is 256 bits. Considering other data for serial communication, the unidirectional transmission of data by intelligent agents should not exceed 512 bits. Therefore, the maximum time for data exchange is once $t_{\max} = 512 \times 2 \div 9600 = 0.107$ s. In the communication system of Fig. 5.10, each agent needs to exchange data with two adjacent agents. Therefore, the total time for a single intelligent agent and adjacent intelligent agents to complete data exchange is 0.214 s. Considering the influence of other factors, such as the computing speed of distributed power sources, it can be assumed that the delay in the distributed communication system is 0.5 s.

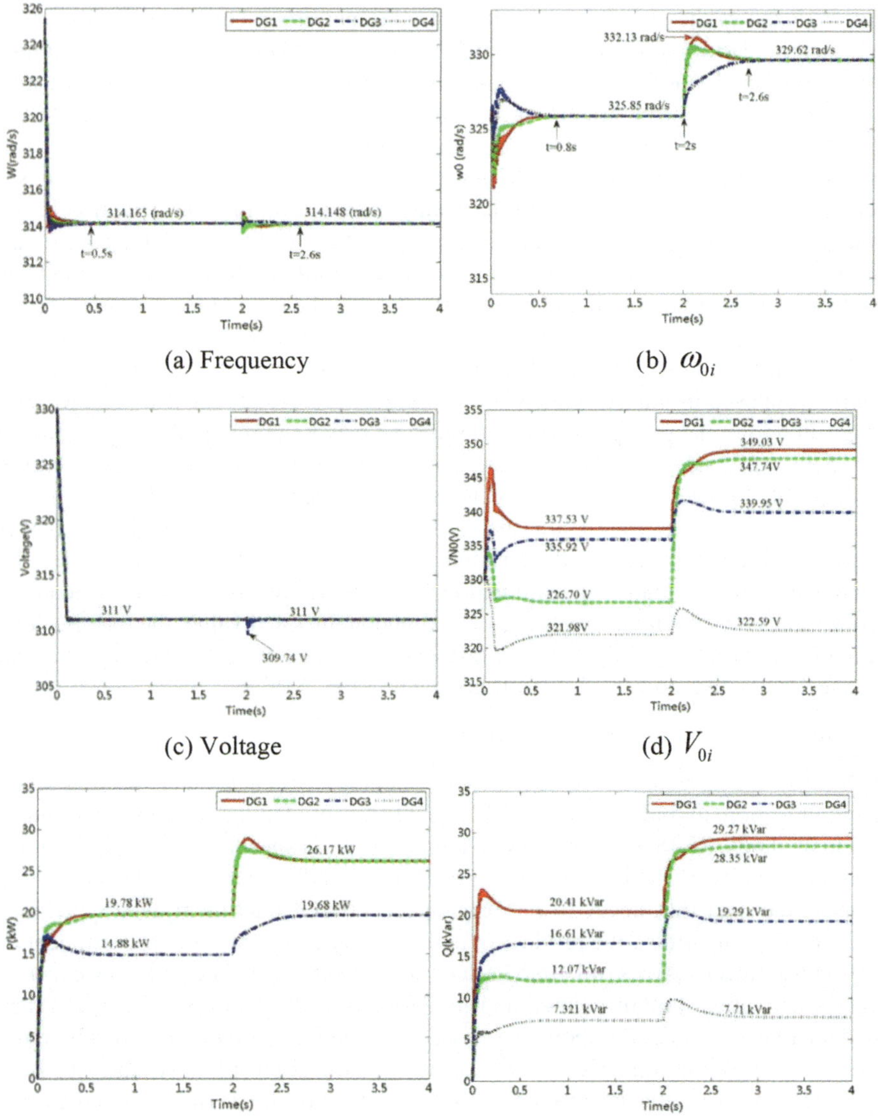

(a) Frequency (b) ω_{0i}

(c) Voltage (d) V_{0i}

(e) Active power (f) Reactive power

Fig. 5.8 Simulation results of adaptive droop control

Simultaneously set load 2 to be connected to the system at $t = 10$ s. The simulation results in this situation are shown in Fig. 5.11.

According to the simulation results, within $t = 10$–10.5 s, due to the influence of communication delay, the information received by the distributed power source from adjacent intelligent agents cannot be updated. As shown in Eqs. (5.27)–(5.33),

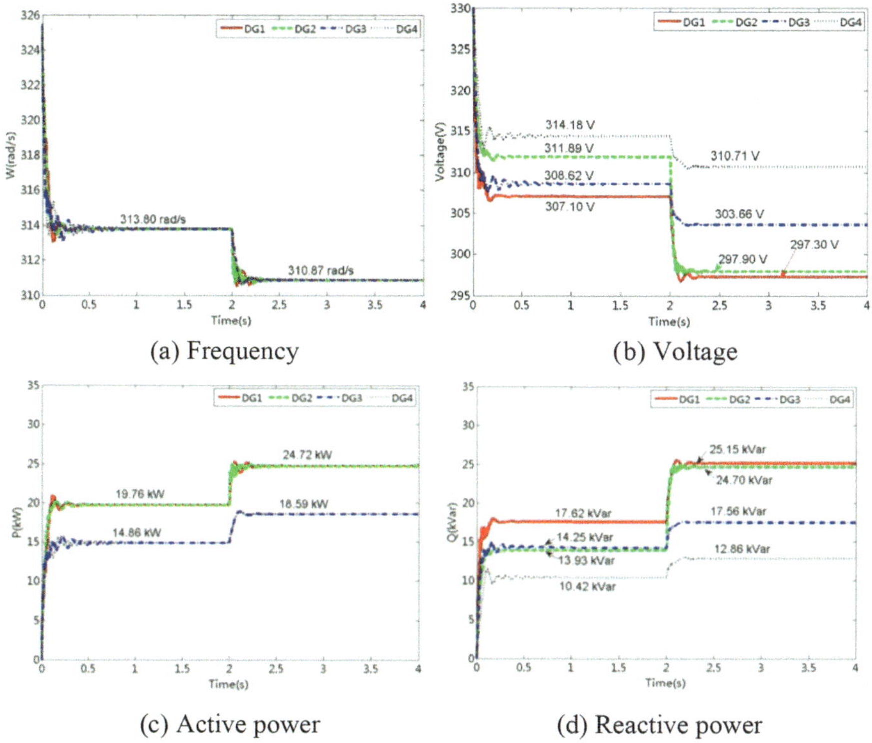

(a) Frequency (b) Voltage

(c) Active power (d) Reactive power

Fig. 5.9 Simulation results of traditional droop control

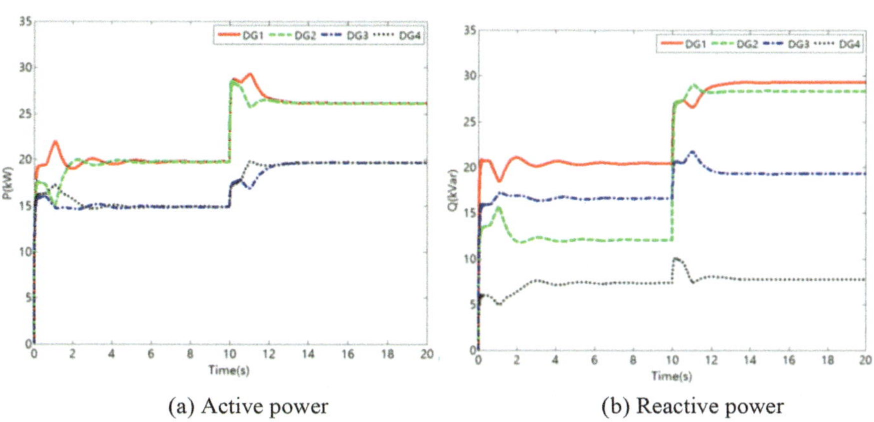

(a) Active power (b) Reactive power

Fig. 5.10 Simulation results considering delay

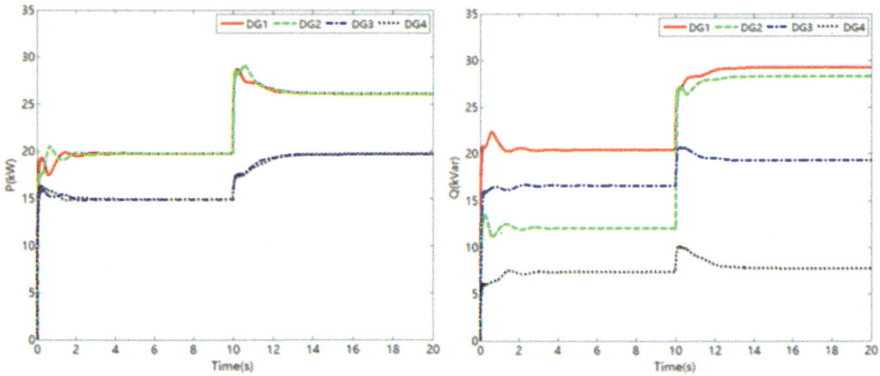

(a) Active power (b) Reactive power

Fig. 5.11 Simulation results considering delay and disturbance

the auxiliary control variable in the droop control still maintains the state value before the load change, therefore, ω_{0i} and V_{0i} in Eqs. (5.40) and (5.41) remain constant, distributed power generation is controlled according to the traditional droop control law. When $t > 10.5$ s, the distributed power source adaptively changes the droop parameters upon receiving updates on power, frequency, and voltage from neighboring distributed power sources. According to the stability analysis results, the adaptive droop control strategy proposed in this chapter is asymptotically stable, that is, although communication is delayed, the state of the distributed power source is still adjusted according to Eqs. (5.14)–(5.18) and (5.27)–(5.33), and operates in a stable state. According to the analysis results of Eqs. (5.44) and (5.45), it can be concluded that after a period of adjustment, the system will operate stably at the rated frequency and voltage. Under the control of the parameters in Table 5.2, the system will be adjusted to the stable operation state after 2.5 s. The results of Fig. 5.10 indicate that the adaptive coordinated control strategy proposed in this section still has excellent control effects in systems with communication delays.

Example 3: Simulation considering delay and disturbance conditions

Assuming a delay of 0.5 s in a distributed communication system, the disturbances in frequency and active power are $\delta_{\omega j} = 1.25\sin(2\pi * t)$ and $\delta_{pj} = 50\sin(2\pi * 500t)$. The simulation results considering both communication delay and disturbance are shown in Fig. 5.11.

In the case of communication delay and fixed disturbance, the system operates stably under constant load (t = 0–10 s), that is, the stable state of the system is not affected by communication delay and disturbance. When the load increases at t = 10 s, the distributed power source operates with traditional droop control during the communication delay period, achieving rapid response to the load. At the same time, after receiving updated information from other distributed power sources, the droop coefficient is adaptively adjusted to maintain the system's rated frequency and voltage operation. However, due to the influence of disturbances, the adjustment time of the system has also increased. Under the simulation parameters selected in this section, the adjustment time is about 3 s.

The distributed control of distributed energy grid clusters can avoid the use of a central controller and reduce the requirements of the control system on the communication network, resulting in high system reliability. Taking into account communication delay and disturbance, this section proposes an adaptive droop parameter control strategy based on the unified consistency of MASs. Regarding the allocation law of active power in droop control, leaderless consistency is applied to coordinate the active power of distributed power sources, enabling them to quickly respond to load changes. Furthermore, this section proposes a consensus-based adaptive droop parameter with virtual leaders, real-time frequency, and voltage dynamic deviations, to ensure power supply quality. Meanwhile, under the control of the strategy proposed in this section, the system is asymptotically stable, thus avoiding the influence of the high-gain active droop coefficient on stability in traditional droop control. The simulation results show that the strategy proposed in this chapter has a good effect on simultaneously improving system stability and improving system voltage and frequency.

5.1.2 Multi-regional Intraday Rolling Optimization Scheduling Model Based on Autonomous Decentralized Control

(1) Introduction

With the increasing scale of distributed energy, the operation control of a single distributed power source and the scheduling control of multiple distributed power sources between different regions have encountered some new problems and challenges. To better and more flexibly utilize distributed energy in multiple regions and improve the utilization rate of distributed energy, further research is needed on the optimization scheduling problem between distributed power sources in multiple regions. Based on existing research, this section adopts the method of autonomous decentralized control to study the optimization scheduling problem of multi-regional distributed power sources. Firstly, a consistency algorithm is used in autonomous decentralized systems to obtain global system information. On this basis, a basic strategy for autonomous decentralized control of multi-regional distributed power sources is proposed, taking into account the economic efficiency of system operation and the solvability of the model, and conducting intraday optimization in layers. During the day, considering the randomness of the output of uncontrollable units in distributed power sources, a self-regulating decentralized control strategy was adopted to adjust the output of each distributed power source. Finally, a numerical example is constructed to validate the effectiveness of the model and strategy.

(2) Design of Autonomous Distributed System Based on Consistency Algorithm

1. Autonomous decentralized system

Autonomous decentralized systems are composed of subsystem units that are controllable and coordinated, and each subsystem unit contains data nodes and data domains. In an autonomous decentralized system, the non-operational state of any individual subsystem unit, such as faults, maintenance, etc., is a normal state for the entire autonomous decentralized system, that is, the structure of the autonomous decentralized system is not fixed and unchanging, but determined by the state of each subsystem unit.

In autonomous decentralized systems, each subsystem unit has the characteristics of equality, locality, and homogeneity. Firstly, the ability of each subsystem to manage itself and interact with other subsystems is self-controllable. Secondly, each subsystem unit is equal and can complete its own control and management without being managed by other subsystem units, and there is no master–slave relationship between them. In addition, each subsystem unit can manage itself and coordinate with other subsystem units by relying solely on local information. In autonomous decentralized systems, each subsystem unit only exchanges information with adjacent subsystem units, and the loose coupling structure between subsystem units is maintained only through data exchange.

Due to the independence and equality of each subsystem unit in the autonomous decentralized system, and the absence of any membership relationship between each subsystem unit, to ensure the coordinated and effective operation of the entire system, each subsystem unit needs to jointly select and produce a leader. If any subsystem unit fails to function, other subsystem units can still complete their assigned tasks normally, and each subsystem unit can coordinate and redistribute the tasks that the system should complete, thereby achieving the overall operational control objectives of the system. When the leader's function fails, other subsystem units will reelect a new leader to ensure the continuous coordinated operation of the system.

Therefore, autonomous decentralized systems are composed of independent subsystem units and are driven by information generated by external and other subsystem units. If each subsystem unit is counted as S_i, it can be defined using function f_i. For any input u_i, the output can be expressed as:

$$y_{i,k+1} = f_i(u_{i,k+1}, y_{1,k}, y_{2,k}, \ldots y_{i,k}, \ldots y_{n,k}) \tag{5.47}$$

where $y_{i,k}$ refers to the k-th output of subsystem unit i. Refers to the k-th input of subsystem unit i. n refers to the number of subsystems in an autonomous decentralized system.

When subsystem unit S_i is driven by input u_i, its output will drive other subsystem units normally, and each subsystem unit in the system will generate a series of effective outputs, which are expressed as follows:

$$Y_S = H_S \cdot (y_1, y_2, \ldots y_i, \ldots y_n)^T \tag{5.48}$$

When the control objective $\tilde{Y}$ of the autonomous decentralized system is given, feedback control $u_{i,k+1} = k_i(Y_{S,i})$ can be designed to make the actual output of the system as close as possible to the set target output. In a self-regulating decentralized system, when a subsystem unit S_i fails, the overall topology of the system will also change accordingly, and the system structure will become S'. The output results of other subsystem units will no longer drive this subsystem unit, and other subsystem units will continue to achieve the system goals under the structure of system S'.

2. Consistency Algorithm

In the process of optimizing distributed systems based on autonomous decentralized systems, information acquisition and communication among various distributed units are the key and difficult points for the entire system to complete the optimization process. The consistency algorithm is applied in autonomous decentralized systems, and its function is to enable each subsystem unit to obtain global information about the system only through communication with adjacent subsystem units, which is the basis for achieving autonomous decentralized control. The basic idea is to converge the information of each subsystem unit to the same value through multiple iterations, and the specific algorithm formula is as follows:

$$x_{i,k+1} = x_{i,k} + \sum_{j=1}^{n} a_{ij} \cdot (x_{j,k} - x_{i,k}) \tag{5.49}$$

where $x_{i,k}$ refers to the information obtained by iteration subsystem unit i for the k-th time. a_{ij} refers to the coefficient of information exchange between subsystem i and j. n refers to the number of subsystem units involved in information exchange. From the perspective of the entire system, it can be written in matrix form as follows:

$$\begin{aligned} \mathbf{X}_{k+1} &= A \cdot \mathbf{X}_k \\ &= \begin{bmatrix} 1-\sum a_{1j} & a_{12} & \dots & a_{1n} \\ a_{21} & 1-\sum a_{2j} & \dots & a_{2n} \\ \dots & \dots & & \dots \\ a_{n1} & a_{n2} & \dots & 1-\sum a_{nj} \end{bmatrix} \cdot \begin{bmatrix} x_{1,k} \\ x_{2,k} \\ \dots \\ x_{n,k} \end{bmatrix} \end{aligned} \tag{5.50}$$

where A refers to the matrix of information exchange coefficients. Due to the interconnected nature of distributed systems, when the communication between subsystem units in a distributed system is bidirectional and equivalent, the elements in the information exchange coefficient matrix A are:

$$a_{ij} = \begin{cases} \frac{1}{d_i} & j \in N_i \\ 0 & j \notin N_i \end{cases} \tag{5.51}$$

where N_i refers to the collection of subsystem units adjacent to subsystem unit i. d_i refers to the number of subsystem units adjacent to subsystem i. After multiple iterations, the information obtained by each subsystem unit finally reached consistency, and the matrix X converged to X_{AVE}, which is expressed as follows:

$$X_{AVE} = \lim_{k \to +\infty} X_k = \frac{1}{n} \cdot (I^T \cdot X_0) \cdot \mathbf{I} \tag{5.52}$$

where X_0 refers to the column vector composed of the initial information obtained by each subsystem unit. **I** refers to all elements being n-dimensional column vectors. From Eq. (5.68), it can be seen that when the system reaches convergence, the elements in each subsystem unit are equal and only related to the initial information they each obtain.

(3) Autonomous Decentralized Control of Multi-regional Distributed Power Sources

1. Multi-regional distributed power structure

Distributed power sources can aggregate different types, capacities, and even regions of distributed energy through communication and control technologies, enabling them to participate in energy market competition and energy system operation as a whole. This not only reduces the energy grid connection risk of various distributed energy sources, but also provides additional profits compared to operating distributed energy sources separately. This section sets up distributed power sources in each region that contain controllable traditional power generation units (such as thermal power), uncontrollable power generation units (such as wind or PV power), ES equipment, interruptible loads, and fixed loads. In the multi-regional distributed power network framework, each region is interconnected through energy and information networks, and distributed power sources in different regions belong to individuals with independent interests. In the process of participating in the operation of power systems with distributed power sources in multiple regions, when energy saturation or shortage occurs in a certain region, distributed power sources in other regions can help absorb or supply energy to it, as shown in Fig. 5.12.

In the system model constructed in this section, multi-regional distributed power sources form an autonomous decentralized control system as a whole. Each regional distributed power source is a subsystem unit in the autonomous decentralized control system. Each distributed power source obtains information through communication with adjacent regional distributed power sources, thereby achieving coordinated interaction among multi-regional distributed power sources and ensuring the overall power balance of the system.

2. Autonomous and decentralized control strategy for distributed resources

Each DG aggregates the distributed energy within its region as a whole and forms a multi-region distributed power interconnection network structure with the distributed power in other regions that have independent interests but communicate with each

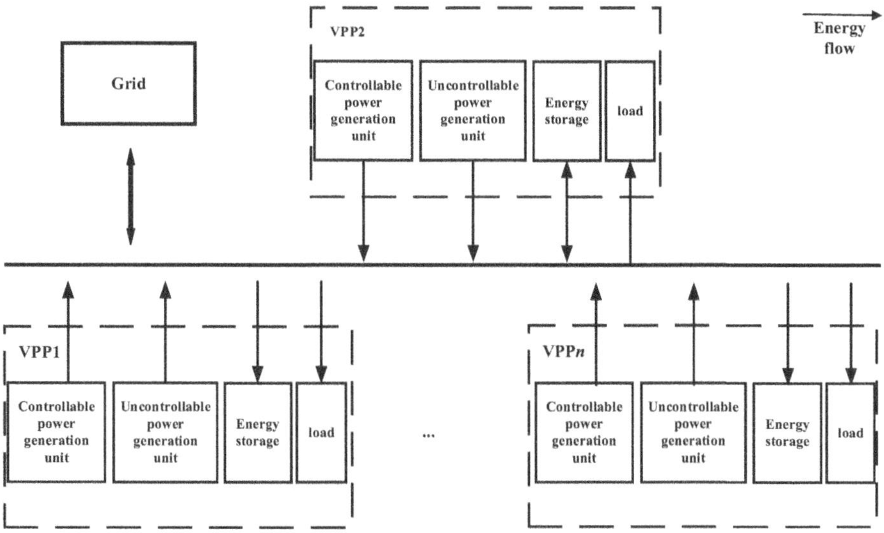

Fig. 5.12 Structure diagram of regional distributed power generation

other. Because distributed power sources in each region are independent entities with their interests, they are unable to obtain network structure information from other regions and all power generation information from other distributed power sources. Therefore, in the pre-optimization stage of distributed power sources, each distributed power source only optimizes to maximize its interests, obtains the overall pre-output plan of distributed sources (i.e. the pre-planned electricity quantity and price of distributed sources), and reports it. After receiving the daily output plan information reported by various distributed sources, the power grid trading center integrates network information to maximize overall social benefits for market clearing, thereby determining the overall bidding results of the market and issuing them to each distributed power source.

However, considering that there are uncontrollable units such as wind and light aggregated within distributed power sources, their output has a certain degree of randomness and fluctuation. Only using models to predict the output of uncontrollable units in advance will result in certain errors. Therefore, distributed power sources need to make secondary adjustments to the output of each distributed energy source within the day. To ensure the stability and economy of the overall operation of multi-regional distributed power sources, when the output of uncontrollable power generation units fluctuates within a certain region, the distributed power source with the lowest average regulation cost is used to bear more output, thereby ensuring that the overall regulation cost of multi-regional distributed power sources is lower. Therefore, in this chapter, the self-regulating decentralized control method is adopted to adjust the output fluctuations of various distributed power sources within a day, and the power regulation cost serves as the consistency variable for daily adjustment to maintain the overall balance of the operation of multi-regional distributed power sources. The self-regulating decentralized control strategy for multi-regional distributed power sources within a day is as follows:

(a) When the actual output of uncontrollable units within a multi-regional distributed power source is less than the planned output, the output fluctuation is jointly borne by the multi-regional distributed power sources. The consistency algorithm is used to solve the output fluctuation values that each distributed power source needs to bear. If there is a distributed power source whose adjustable cost or adjustable capacity reaches its limit, it will be removed from the network topology. The adjacent distributed power sources will update the network status after receiving the information so that the remaining distributed power sources can coordinate with each other to continue to achieve the power regulation goal. This ensures the overall stability of multi-regional distributed power sources while maximizing their economic efficiency.
(b) When the actual output of uncontrollable units within a multi-regional distributed power source exceeds the planned output, targeted consumption will be carried out based on the type of distributed energy aggregated by each regional distributed power source. If there are ES devices in the aggregation of distributed power sources, the ES can be charged according to the operating status of the ES devices. If distributed power sources aggregate electric vehicles, price incentives can be used to guide electric vehicle charging. If the distributed power source contains small hydropower, it can be charged through pumped storage.
(c) When multi-regional distributed power sources are unable to absorb the power fluctuations caused by internal distributed energy, the overall power balance is achieved by balancing market purchases and sales, thereby ensuring the safety and stability of the overall operation of multi-regional distributed power sources. At this point, the purchasing cost of distributed power sources in the balanced market is borne by the distributed power sources in each region with power shortages according to the proportion of the shortfall, while the sales revenue of distributed power sources in the balanced market is obtained by the corresponding distributed power sources in each region according to their power generation.

Due to the use of autonomous decentralized control to schedule the intraday fluctuations of multi-regional distributed power sources, during the scheduling process, when the leader distributed power source reaches its own adjustable cost or adjustable capacity limit, it needs to be removed from the system network according to the control rules, and a new leader needs to be reelected by the remaining distributed power sources. Therefore, to simplify the consistency algorithm solving process, the distributed power source with the highest adjustable cost is selected as the initial leader of the algorithm, ensuring that there is no need to replace the leader.

3. Intraday optimization scheduling model for distributed power generation

Due to the randomness of the output of uncontrollable units such as wind and solar power, each distributed power source needs to optimize the internally distributed members daily, and also adjust and control the output of each distributed member twice within the day based on the error between their actual output and planned output. In this chapter, the self-regulating decentralized control method is used to make intraday adjustments to multi-regional distributed power sources. The power regulation cost is selected as the consistency variable, and the power regulation cost of the i-th distributed power source at time t is:

$$r_{i,t} = \Delta C_{i,t}^{con} + \Delta C_{i,t}^{es} + \Delta C_{i,t}^{il} \tag{5.53}$$

where $\Delta C_{i,t}^{con}$ refers to the regulation cost of the controllable unit of the i-th distributed power source at time t. $\Delta C_{i,t}^{es}$ refers to the regulation cost of ES for i distributed power sources at all times. $\Delta C_{i,t}^{il}$ refers to the cost of regulating the interruptible load of the i-th distributed power source at time t. To ensure the stability and economy of daily scheduling of multi-regional distributed power sources, an objective function is established within each distributed power source to minimize the average regulation cost of each distributed power source:

$$\min \tilde{r}_{i,t} = \frac{r_{i,t}}{\Delta P_{i,t}^{con} + \Delta P_{i,t}^{es} + \Delta P_{i,t}^{il}} \tag{5.54}$$

During the optimization process, the following constraints must be met: power balance constraint:

$$\sum_{i=1}^{n} \Delta P_{i,t}^{con} + \Delta P_{i,t}^{es} + \Delta P_{i,t}^{il} = \sum_{i=1}^{n} (P_{i,t}^{unc} - P_{i,t}^{unc,pre}) + P_{t}^{rt} \tag{5.55}$$

where $\Delta P_{i,t}^{con}$ refers to the regulated power of the controllable unit in the i-th distributed power source at time t. $\Delta P_{i,t}^{es}$ refers to the regulated power of the energy stored in the i-th distributed power source at time t. $\Delta P_{i,t}^{il}$ refers to the regulated power of interruptible loads in the i-th distributed power source at time t. $P_{i,t}^{unc}$ refers to the actual output of the uncontrollable unit in the i-th distributed power source at time t. $P_{i,t}^{unc,pre}$ refers to the planned output of the uncontrollable unit in the i-th distributed power source at time t. P_{t}^{rt} refers to the total trading volume between i-th distributed power sources and the real-time market at time t. The optimized revenue and cost of each distributed power source in a balanced market are:

$$I_{vpp,i}^{rt} = \sum_{t=1}^{24} I_{t}^{rt} \cdot \frac{\Delta P_{i,t}^{rt}}{\Delta P_{t}^{rt}} \tag{5.56}$$

$$C_{vpp,i}^{rt} = \sum_{t=1}^{24} r_{i,t} + \sum_{j=1,j\neq i}^{n} \sum_{t=1}^{24} r_{j,t} \cdot \left(\frac{\Delta P_{i,j,t}}{\Delta P_{j,t}} - \frac{\Delta P_{j,i,t}}{\Delta P_{j,t}} \right) \tag{5.57}$$

$$I_t^{rt} = \begin{cases} \lambda_t^{rt+} \cdot P_t^{rt}, & P_t^{rt} \geq 0 \\ \lambda_t^{rt-} \cdot P_t^{rt}, & P_t^{rt} \leq 0 \end{cases} \tag{5.58}$$

where I_t^{rt} refers to the overall purchase cost or sales revenue of distributed power sources in the real-time market at time t. ΔP_t^{rt} refers to the total amount of electricity purchased or sold by distributed power sources in the real-time market at time t. $\Delta P_{i,t}^{rt}$ refers to the real-time market purchase and sale electricity provided by the i-th distributed power source at time t. λ_t^{rt+} refers to the real-time market electricity price at time t. λ_t^{rt-} refers to the real-time market purchase price of electricity at time t. $\Delta P_{j,t}$ refers to the power regulated by the N-th distributed power source at time t. $\Delta P_{j,i,t}$ refers to the power that the i-th distributed power source helps regulate the i-th distributed power source at time t. The overall benefits of multi-regional distributed power sources participating in the energy market are:

$$I = \sum_{i=1}^{n} (I_{vpp,i}^{da} + I_{vpp,i}^{rt} - C_{vpp,i}^{rt}) = \sum_{i=1}^{n} I_{vpp,i}^{da} + \sum_{t=1}^{24} I_t^{rt} - \sum_{i=1}^{n} \sum_{t=1}^{24} r_{i,t} \tag{5.59}$$

(4) Example Analysis

To verify the above strategy, this chapter constructs a multi-area distributed power system as shown in Fig. 5.13.

The controllable units G1, G2, and G3 have a capacity of 8 MW, and their operating cost coefficients are shown in Table 5.3.

The predicted output and load curves of wind and PV power are shown in Fig. 5.14. The rated power for charging and discharging ES is 1.5 MW, with an upper limit of 6 MW and a lower limit of 0.1 MW for stored electricity. The charging efficiency is 0.85 and the discharging efficiency is 0.9, with an operating cost coefficient of 0.056

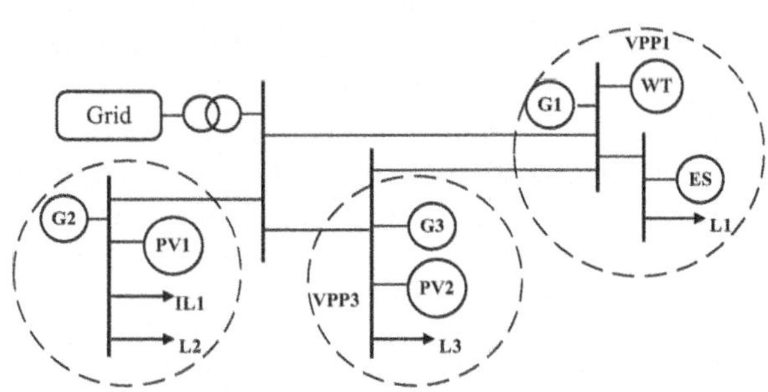

Fig. 5.13 System architecture diagram

Table 5.3 Controllable unit operating cost coefficient

Parameters	a/(yuan/kW^{2}h)	b/(yuan/kWh)	c/(yuan/h)
G1	$0.96 * 10^{-6}$	0.21	0.16
G2	$1.43 * 10^{-6}$	0.17	0.34
G3	$1.21 * 10^{-6}$	0.18	0.23

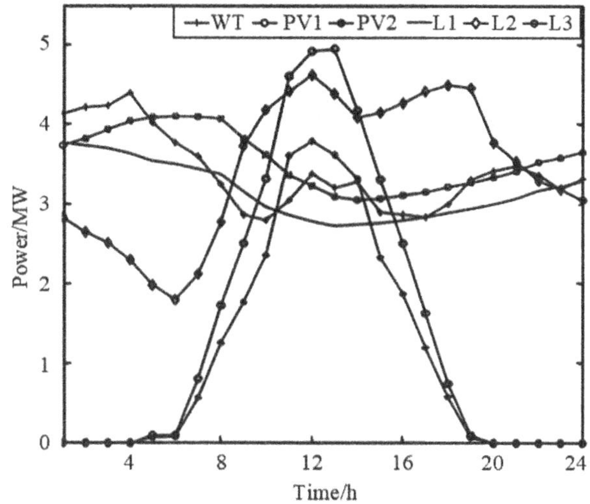

Fig. 5.14 Wind, solar prediction curve and load curve

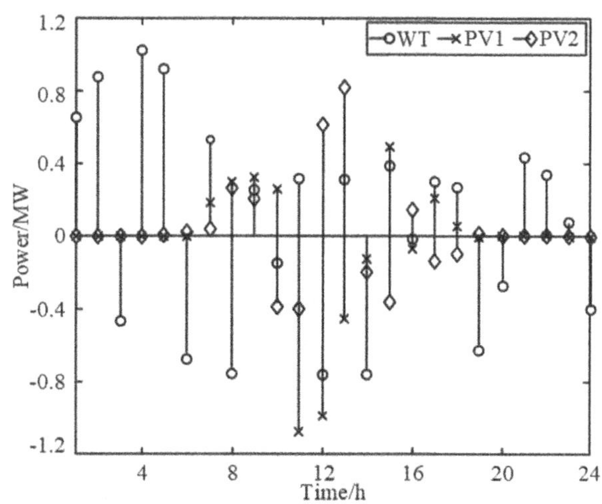

Fig. 5.15 Wind and solar power output fluctuation

yuan/(kWh). The maximum interruptible power of the interruptible load is 5 MW, with a cost coefficient of 0.41 yuan/(kWh).

The current market and real-time market electricity prices are shown in Table 5.4.

Due to the uncertainty of the output of uncontrollable units such as wind and solar power, it is necessary to adjust their output fluctuations within the day. The fluctuation of wind and solar power output is shown in Fig. 5.15.

This section adopts autonomous decentralized control, with power regulation cost as the consistency variable, and uses a consistency algorithm to calculate the regulation power cost of distributed power sources in each region, which is calculated

Table 5.4 Daily market and real-time market electricity prices (yuan/KWh)

t	λ_t^{da}	λ_t^{rt+}	λ_t^{rt-}	t	λ_t^{da}	λ_t^{rt+}	λ_t^{rt-}
1	0.32	0.29	0.34	13	0.39	0.36	0.41
2	0.31	0.28	0.31	14	0.37	0.35	0.40
3	0.31	0.31	0.33	15	0.36	0.36	0.39
4	0.31	0.30	0.32	16	0.49	0.49	0.52
5	0.38	0.35	0.42	17	0.74	0.74	0.76
6	0.56	0.51	0.57	18	0.61	0.56	0.62
7	0.58	0.55	0.64	19	0.53	0.51	0.56
8	0.57	0.56	0.62	20	0.50	0.50	0.55
9	0.53	0.48	0.55	21	0.41	0.39	0.43
10	0.47	0.43	0.51	22	0.36	0.33	0.39
11	0.45	0.45	0.48	23	0.33	0.31	0.33
12	0.41	0.37	0.44	24	0.27	0.25	0.28

as Scheme 1. At the same time, the centralized dispatching strategy and the separate dispatching strategy of each distributed power generation are used for intraday optimization, which is calculated as schemes 2 and 3 respectively. The optimized dispatching results are shown in Table 5.5.

It can be seen from the above table that compared with the centralized dispatching strategy, the regulation cost of each distributed power supply under the autonomous decentralized control strategy is higher, but the autonomous decentralized control strategy ensures the independence of each distributed power supply, making the control of the entire network system simpler and the system operation simpler. Compared to individual scheduling strategies, the cost of regulating each distributed power source is lower under the autonomous decentralized control strategy. This is because through the coordination and mutual assistance between distributed power sources, flexible scheduling of distributed energy sources in multiple regions is achieved, thereby improving the overall economic efficiency of the system operation. Among all the scheme strategies, the power regulation cost of distributed power source 2 is the highest under the autonomous decentralized control strategy. This is because distributed power source 2 has the highest adjustable cost. During the intraday power adjustment phase, distributed power source 2 helps distributed power

Table 5.5 Optimization results of each plan

Scheme		DG 1/¥	DG 2/¥	DG 3/¥	Total
1	r_i	450.48	620.67	356.18	1427.33
	$C_{vpp,i}^{rt}$	475.64	583.42	368.27	1427.33
2		432.17	567.26	347.21	1346.64
3		512.86	596.75	422.46	1532.07

sources 1 and 3 regulate their power fluctuations. Therefore, the autonomous decentralized control strategy reduces the difficulty of overall system operation control and model solving, while ensuring the economic efficiency of system operation.

5.2 Real-Time Adjustment Optimization Model

5.2.1 A Hierarchical Real-Time Optimization Model Based on Multi-agent Consistency

Economic operation, frequency and voltage control and system stability are the main components of real-time control of distributed energy grid clusters. Based on the multi-agent consistency protocol, a hierarchical real-time optimization strategy is proposed in this chapter. Under the condition of only considering the DG output constraints, the first-order consistency is applied to achieve the control of the active output of the DG and optimize the operation cost; secondly, a new second-order consistency protocol is proposed for the accurate control of the DG power with respect to the droop control characteristics; finally, the consistency protocol is applied to coordinate the droop parameters of the DG in real time to achieve the optimization of the frequency and voltage of the system. Meanwhile, the stability analysis of the proposed strategy is carried out by Lyapunov direct method, which proves that the system is asymptotically stable.

(1) Introduction

The economic operation of a distributed energy grid cluster refers to minimizing the total cost of the system by coordinating the output between distributed generation while satisfying the system and equipment level constraints. Traditional mathematical algorithms such as Lambda iterative method, gradient method, linear programming and Newton's method can solve the convex optimization problem better, but these algorithms are more sensitive to the initial value and are prone to form local optimum. Intelligent algorithms such as genetic algorithms and particle swarms can make up for the shortcomings of traditional algorithms and can be used to solve non-convex optimization problems. However, these algorithms are implemented in centralized controllers and require the physical quantities of all adjustable devices, and the reliability of the system is relatively low. In traditional power systems, the economic operation of a distributed energy grid cluster can be represented as a Lagrangian optimization model containing constraints, which can be solved by the equal consumed energy ratio approach, but it requires the calculation of the system's power flow, and the traditional power flow calculations are only suitable for power systems containing balancing nodes. However, in droop-controlled distributed energy grid clusters, all DG containing regulation capabilities are involved in load regulation, and balancing nodes do not necessarily exist, and thus cannot be directly applied in distributed energy grid clusters. To address these problems, Chen et al. (2016) and

Zhang and Yu (2015) proposed a distributed optimization method based on a multi-agent first-order consistency protocol, which coordinates the active outputs of the DG to make the incremental costs of the DG the same, so as to achieve the overall optimization of the system cost. Based on this, literature (Binetti et al. 2014) proposed a distributed optimization algorithm considering network losses. However, in the droop controlled distributed energy grid cluster, the regulation of the active power of DG needs to be achieved by adjusting the droop parameter, while the above literature, although it achieves the optimization of the active power, does not make a detailed analysis of the specific implementation in the droop controlled distributed energy grid cluster; and it does not optimize the control of the system's voltage and frequency either.

To address the above problems, this chapter proposes a layered optimization approach, which is implemented at each layer by a distributed consistency protocol and the system control is fully distributed. To address the problem of active power regulation for DG in droop control, this chapter proposes a new second-order consistency protocol for coordinating the droop parameters of DG, so as to realize the precise control of active power by DG. Meanwhile, the hierarchical optimization method proposed in this chapter is asymptotically stable at each layer, and thus avoids the problem of the influence of the droop coefficients on the stability of the system in the traditional distributed energy grid clusters.

(2) Optimization Model

The cost of DG consists of fuel cost and maintenance cost, where maintenance cost is generally a linear function of active power and fuel cost is a quadratic function of active power. Therefore, the generation cost of DG can be expressed as:

$$C_i(P_i) = a_i P_i^2 + b_i P_i + c_i \tag{5.60}$$

where a_i, b_i and c_i are the cost function coefficients, and $a_i, b_i, c_i > 0$. For a distributed energy grid clusters containing the number of DG n_G, the objective of economic dispatch is to minimize the total cost of the distributed energy grid clusters, which is: Set the multi-DG decision portfolio as $\mathbf{x} = [x_1, x_2, \ldots, x_n]$. The payment function of the seller and buyer of DG at a certain point in time is:

$$\min c = \sum_{i=1}^{n_G} C_i(P_i) \tag{5.61}$$

Simultaneously satisfying the constraints at the system level and the DG itself, include:

$$\sum_{i=1}^{n_G} P_i = P_t \tag{5.62}$$

$$\sum_{i=1}^{n_G} Q_i = Q_t \tag{5.63}$$

$$U_{\min} \leq U \leq U_{\max} \tag{5.64}$$

$$f_{\min} \leq f \leq f_{\max} \tag{5.65}$$

$$P_{i\_\min} \leq P_i \leq P_{i\_\max} \tag{5.66}$$

$$\sqrt{P_i^2 + Q_i^2} \leq S_{i\_\max} \tag{5.67}$$

where P_t and Q_t are the total power demand of the system including network losses; $U_{\min}$ denotes the minimum value of the node voltage and $U_{\max}$ denotes the maximum value of the node voltage. $f_{\min}$ denotes the minimum value of system frequency and $f_{\max}$ denotes the maximum value of system frequency. $P_{i\_\min}$ denotes the minimum value of node injected active power, $P_{i\_\max}$ denotes the maximum value of node injected active power and $S_{i\_\max}$ denotes the maximum value of node injected apparent power. Limit values of active and apparent power. Equations (5.62)–(5.65) denote the system level constraints, where Eqs. (5.62) and (5.63) denote the supply–demand balance of the distributed energy grid cluster, and Eqs. (5.64) and (5.65) denote that the node voltages and the system frequency need to be within the required range. Equations (5.66) and (5.67) describe the generation capacity constraints of the generation units. Combined with the associated droop regulation characteristics, it can be seen that the optimization function in Eq. (5.61) can be equated to the optimization of the droop parameter, which is:

$$\min c = \sum_{i=1}^{n} C_i(f_{0i}, U_{0i}, m_i, n_i) \tag{5.68}$$

From Eq. (5.68), it can be seen that the optimal operation of distributed energy grid cluster can be translated into the optimization of droop parameters. Analysing the regulation characteristics of droop control, it can be seen that optimizing the operation cost of distributed energy grid clusters will change the operating characteristics of DG, affecting the system frequency and node voltage at the same time.

(3) Hierarchical Real-time Optimization Based on Multi-agent Consistency

The process of DG participation in system optimization is divided into three layers: active power optimization, active power control and system frequency and voltage optimization, and the realization of each layer requires the coordination of the corresponding control parameters of the DG. In the following, the coordination of the system is achieved at each of the three levels of the hierarchical optimization strategy. In a MAS, an agent can interact with neighbouring agents through a distributed communication system, and then judge its own operating state and adjust its output based on this information so that it has the same amount of target features as the neighbouring agents, thus achieving overall consistency.

1. Cost Optimization

In the hierarchical optimization strategy proposed in this chapter, the main task of the top layer is to minimize the total cost of the system by adjusting the output active power of the distributed power sources while ignoring the dynamic response of the system. In conventional power systems, the incremental cost is commonly used to solve the optimization function. By deriving the cost function for the active power P while considering the power limit of the DG, the incremental cost of the DG is:

$$\lambda = \begin{cases} 2a_i P_{i\_\min} + b_i, P_i = P_{i\_\min} \\ 2a_i P_i + b_i \\ 2a_i P_{i\_\max} + b_i, P_i = P_{i\_\max} \end{cases} \tag{5.69}$$

From the conventional equal consumed energy ratio, the total cost of the system is minimum when the incremental cost of the DG is the same, as $\lambda_i = \lambda_j$. At this point, the optimized active power of the system P_{refi} is:

$$\lambda^* = \frac{P_t + \sum_{i=1}^{n_G} \frac{b_i}{2a_i}}{\sum_{i=1}^{n_G} \frac{1}{2a_i}} \tag{5.70}$$

$$P_{refi} = \begin{cases} P_{i\_\min} \\ \frac{\lambda_i^* - b_i}{2a_i} \\ P_{i\_\max} \end{cases} \tag{5.71}$$

Equations (5.70) and (5.71) are the optimization solution obtained by the centralized approach. From the previous analysis results, it is clear that this optimization solution is difficult to obtain directly because the demand of the system is related to the operating point of the system. This section applies the first-order consistency protocol of Eq. (5.72) to solve the problem.

$$\lambda_i = \beta_i \sum_{j=1, j \neq i} (\lambda_i - \lambda_j) - \alpha_i \left(\frac{\lambda_i - b_i}{2a_i} - P_i \right) \tag{5.72}$$

where α_i and β_i are non-negative control parameters. When the system operates stably, there is $\lambda_i = \lambda_j$, when the total cost of the system is minimized. Meanwhile, the stability analysis from the Zheng et al. (2009) shows that the distributed energy grid cluster under the control of the applied Eq. (5.72) is stable.

2. Distributed Power Tracking Control

This section focuses on the analysis of the control of the active power of the DG so that it can accurately track the reference value of the medium optimization proposed in the previous sections.

By applying the linear feedback control method, the active power is derived twice consecutively to obtain the active power dynamic response model of the distributed power supply.

$$\begin{cases} \dot{P}_i = w_i \\ \dot{w}_i = u_{Pi} \end{cases} \tag{5.73}$$

where w_i is defined as the first-order differential value of the active power. Based on the above dynamic model, Yu et al. (2017) proposes the consistency protocol of Eqs. (5.74)–(5.80) to realize the regulation of the active power of the DG.

$$u_{Pi} = u_{P\alpha i} + u_{P\beta i} + u_{P\gamma i} \tag{5.74}$$

$$u_{P\alpha i} = -k_{P1} \sum_{j=1, j\neq i}^{n} a_{ij} \frac{P_{ij}^2 - P_{(ref)ij}^2}{P_{(ref)ij}^2 P_{ij} + P_{ij}^3} \tag{5.75}$$

$$u_{P\beta i} = \sum_{j=1, j\neq i}^{n} a_{ij} sign(w_{ji}) \left| w_{ji} \right|^{k_{P2}} \tag{5.76}$$

$$u_{P\gamma i} = k_{P3} \left[-k_{P1} \frac{P_{iL}^2 - P_{(ref)iL}^2}{P_{(ref)iL}^2 P_{iL} + P_{iL}^3} + sign(w_{Li}) \left| w_{Li} \right|^{k_{P2}} \right] \tag{5.77}$$

$$P_{ij} = \left| P_i - P_j \right| \tag{5.78}$$

$$P_{(ref)ij} = \left| P_{refi} - P_{refj} \right| \tag{5.79}$$

$$w_{ij} = w_j - w_i \tag{5.80}$$

where L denotes the virtual leader in active power control;k_{p1}, k_{p2} and k_{p3} are control parameters, k_{p1} and k_{p3} are non-negative. $u_{P\alpha i}$, $u_{P\beta i}$ and $u_{P\gamma i}$ denote the three independent control components in the consistency protocol, respectively, and the role of each part is: $u_{P\alpha i}$ is used to control the active power deviation between DG to be equal to the deviation from the target, $u_{P\beta i}$ is used to have its differentials equal, and $u_{P\gamma i}$ is used to control the active deviation between DG_i and the virtual leader to be equal to the deviation from the target. Combined Eq. (5.73) gives the equilibrium point of the system.

The active power tracking target value of the DG can be controlled by the consistency agreement of Eqs. (5.74)–(5.80). While the control objective is achieved, the stability of the system is further analyzed later on and Lyapunov direct method is used in this chapter. For the system consisting of Eqs. (5.73)–(5.80), the energy function of Eq. (5.81) is considered in this chapter.

$$V(x)=\frac{1}{2}\sum_{i=1}^{n}(w_{iL})^2+\frac{1}{2}\sum_{i=1}^{n}\sum_{j=1,j\neq i}^{n}k_{P1}a_{ij}\ln\frac{P_{(ref)ij}^2+P_{ij}^2}{2P_{(ref)ij}P_{ij}}$$
$$+\sum_{i=1}^{n}k_{P1}k_{P3}\frac{P_{(ref)iL}^2+P_{iL}^2}{2P_{(ref)iL}P_{iL}} \tag{5.81}$$

Analysis of the components in Eq. (5.81) shows that the first three requirements are satisfied by Lyapunov's direct method. The analysis shows that for all $x\neq 0$, $\dot{V}(x)<0$ is satisfied. Combined with the stability law of Lyapunov, it can be seen that the control in Eqs. (5.73)–(5.80) consistency protocol can achieve accurate tracking of the optimized active power while the system is asymptotically stable.

3. Optimal Control of System Frequency and Voltage

After the tracking of active power is achieved, the system frequency and voltage need to be further controlled so that the system operates in the desired state. From the droop control characteristics, it can be seen that the system frequency and nodal voltage are affected by the droop parameters, and too large active droop coefficient m_i may make the system unstable (Yu and Ai 2017). In this book, the optimization of system frequency and voltage is mainly achieved by adjusting f_{0i} and V_{0i}. In this chapter, the optimization objective of the system frequency and voltage is the nominal value. This chapter uses the consistency protocol proposed in Chap. 2 of the literature to achieve the optimal control of the system voltage and frequency:

$$u_{yi}=u_{y\alpha i}+u_{y\beta i}+u_{y\gamma i} \tag{5.82}$$

$$u_{y\alpha i}=-k_{y1}\sum_{j=1,j\neq i}^{n}a_{ij}\frac{2y_{ij}}{1+y_{ij}^2} \tag{5.83}$$

$$u_{y\beta i}=\sum_{j=1,j\neq i}^{n}a_{ij}sign(x_{yji})|x_{yji}|^{k_{y2}} \tag{5.84}$$

$$u_{y\gamma i}=k_{y3}\left[-k_{y1}\frac{2y_{iL}}{1+y_{iL}^2}+sign(x_{yLi})|x_{yLi}|^{k_{y2}}\right] \tag{5.85}$$

$$y_{ij}=y_i-y_j \tag{5.86}$$

$$y_{iL}=y_i-y_L \tag{5.87}$$

$$x_{yji} = x_{yj} - x_{yi} \tag{5.88}$$

$$x_{yLi} = x_{yL} - x_{yi} \tag{5.89}$$

where y denotes the optimized system quantity and in this chapter denotes the frequency f and voltage v. From the analytical results in the previous chapters, it can be seen that under the control of the consistency protocol shown in Eqs. (5.82)–(5.89), the system frequency and voltage can be operated at the reference values; at the same time, the system is asymptotically stabilised, and the control equations of the system sag parameter are as follows:

$$f_{oi} = \int \left(m_i \int u_{cPi} dt + \int u_{cwi} dt \right) dt + f_{oi0} \tag{5.90}$$

$$V_{oi} = \int \left(n_i \dot{Q}_i + \int u_{cvi} dt \right) dt + V_{oi0} \tag{5.91}$$

where f_{oi0} and V_{oi0} are the initial values of the droop control parameters. Combining the active power optimization of Eqs. (5.74)–(5.80), the distributed hierarchical optimal control of distributed energy grid clusters can be achieved by active power tracking control, and its control structure is shown in Fig. 5.16.

5.2.2 Real-Time Optimal Adjustment Model for Bi-level Games

The introduction of large-scale DG into the grid poses a challenge to the energy grid and requires the study of new DG control technologies. DG is able to rationally adjust internal optimization strategies in conjunction with the prevailing demand of DG or the market demand to achieve the optimal environmental and economic benefits while achieving internal cooperative operation and meeting the market demand. In this chapter, in conjunction with the joint operation of renewable energy sources, the multi-distributed power supply layered control structure based on MAS is established by taking the DG containing WT, PV, micro GTs and DR resources as an example, and the multi-distributed power supply optimization problem is transformed into a bi-level coordinated optimization model. The upper level is the multi-DG game bidding model, which formulates the multi-DG direct trading strategy according to the specific values of demand and supply. The lower level is a cooperation model between distributed energy grid cluster within DG to obtain the optimal response power with the objective of the lowest total cost. In order to improve the ability of DG to consume clean energy and reduce the impact of prediction error, a multi-timescale optimization strategy is proposed. Through the coordination and interaction of multiple DG, revenue maximization and intraregional power balance are achieved.

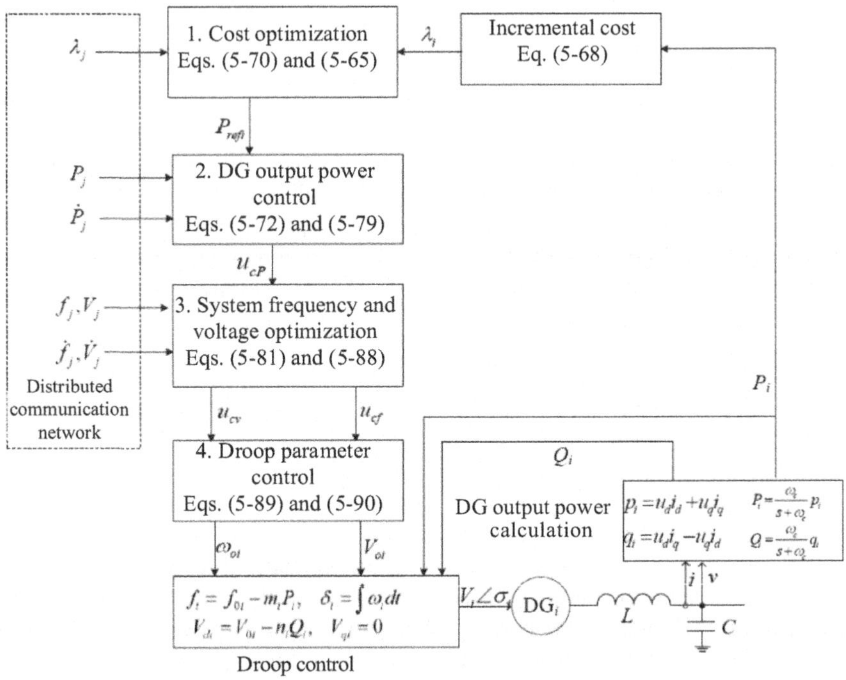

Fig. 5.16 Hierarchical optimization control block diagram

5.2.2.1 Multi-agent Based DG Mechanism Design

(1) Hierarchical Control Framework

The distributed energy grid clusters integrate multiple DG and participates in grid operation as an energy carrier in a form different from the traditional grid. The hierarchical control framework in a DG market environment based on a MAS constructed in this chapter is shown in Fig. 5.17, which includes three levels, namely, the power market agent, the DG/power company agent, and the device agent, with optimized coordination between the levels achieved through information and energy exchange.

(2) Hierarchical Control Strategy

(a) Device agent. The power generation agent is responsible for providing the DG agent with information on operation status and power forecast, etc. The load agent prioritises the loads according to their importance, and sends information on different types of loads to the upper-level agent. It dynamically adjusts the amount of power generation/load according to the control instructions of the upper-level agent and its own constraints, and provides timely feedback of its own power information to the upper-level agent.

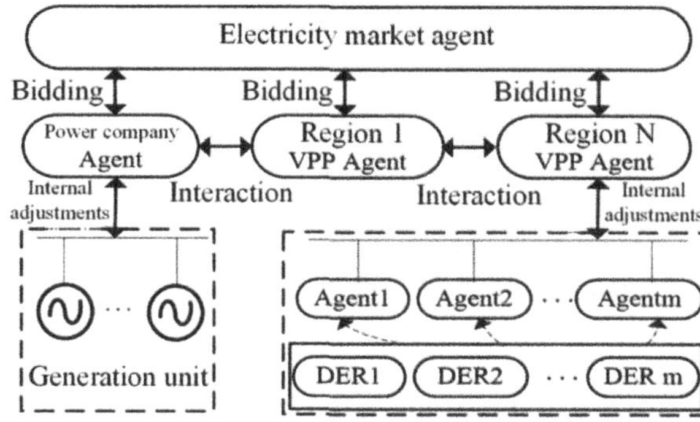

Fig. 5.17 MAS-based control structure diagram

(b) DG/power company agent. Combining its own power generation forecast and load forecast in the region and other information, it adjusts the operation plan of the lower-level agent; it receives tariffs issued by the power market agent and bidding information of other DG intelligent bodies, adjusts the bidding strategy with the goal of maximizing its own benefits, determines the amount of power to be sold or purchased from neighbouring agents, and feeds the strategy back to the upper-level agent.

(c) Power market agent. Aggregate the information of willingness to purchase and sell electricity and power output range reported by the lower-level agent, consider the system security, carry out the calculation of tariff decision, and send the tariff and other DG agent bidding information to each DG agent, and at the same time, monitor the situation of the power of each DG agent, evaluates whether Nash equilibrium conditions are met, and if so, stop the process of gaming.

Negotiation of aggregation members within DG. When participating in the market, optimization within the DG is carried out with the goal of minimizing the cost or maximizing the benefit of the DG. In the node level, the DG control centre is used as a dynamic agent, which is responsible for the state monitoring, scheduling and control of the equipment in the region, and establishes a dynamic cooperation mechanism for the component agent in the region.

Negotiation between DG. DG can achieve cross-regional coordination by negotiating with neighbouring DG. If the neighbouring DG proposes to be more favourable than the market, it chooses to trade part of its power with the neighbouring DG.

Participation of DG in the energy market. Enables interaction between DG and the energy market for optimal global dispatch.

5.2.2.2 Bi-level Game Model for DG

(1) Lower Level Optimization Model

The information about the output of each DG is used as a decision variable, and the combination of its strategies is $\mathbf{a} = [a_1, a_2, \ldots, a_n]$. Using the opposite of the cost of each agent as a payment function, the cost minimization problem is transformed into a game problem of maximizing the payment. The payment function at a certain moment is:

$$v_i(a_i, \mathbf{a}_{-i}) = \begin{cases} -\left(a_{\mathrm{MT}} a_i^2 + b_{\mathrm{MT}} a_i + c_{\mathrm{MT}}\right), i \in n_{\mathrm{MT}} \\ -\left(a_{\mathrm{IL}} a_i^2 + b_{\mathrm{IL}} a_i\right) - \lambda_{\mathrm{load}}^t a_i, i \in n_{\mathrm{IL}} \\ -\lambda_{\mathrm{wt}}(P_{\mathrm{WT}}^t - a_i), i \in n_{\mathrm{WT}} \\ -\lambda_{\mathrm{pv}}(P_{\mathrm{PV}}^t - a_i), i \in n_{\mathrm{PV}} \\ \lambda_{\mathrm{load}}^t a_i, i \in n_{\mathrm{D}} \end{cases} \tag{5.92}$$

where a_i is the decision of the i-th DG source, $\mathbf{a}_{-i}$ is the decision aggregation of the remaining DG excluding the i-th DG. $n_{\mathrm{MT}}, n_{\mathrm{IL}}, n_{\mathrm{WT}}$ and n_{PV} represent the aggregation of MT, IL, WT and PV, respectively, and n_{D} represents the aggregation of load agents. The controllable DG takes MT as an example, a_{MT}, a_{MT} and c_{MT} are the operating cost coefficients, and the interruptible load is considered to achieve DR, where the two parts of the equation are the cost of interruption and the revenue from energy sales due to the reduction of interruptible loads, a_{IL} and b_{IL} are the cost coefficients of interruption, and $\lambda_{\mathrm{load}}^t$ is the power purchase tariff of loads. λ_{wt} and λ_{pv} are the cost coefficients of wind and solar rejection, and P_{WT}^t, P_{PV}^t are the values of the planned output of WT, and PV, respectively, in the moment of t. The generation cost of WT and PV is not considered.

(2) Upper Level Optimization Model

Set the multi-DG decision portfolio as $\mathbf{x} = [x_1, x_2, \ldots, x_n]$. The payment function of the seller and buyer of DG at a certain point in time is:

$$u_i(x_i, \mathbf{x}_{-i}) = \begin{cases} \lambda_{\mathrm{vpp}}^t x_i + \lambda_{m,s}^t (P_{m,i}^t - x_i) - \lambda_{ser} \sum_{j=1}^{n_p} x_{ij}, \ i \in n_s \\ \lambda_{\mathrm{vpp}}^t x_i + \lambda_{m,p}^t (P_{m,i}^t - x_i) - \lambda_{\mathrm{load}}^t P_{m,i}^t, \ i \in n_p \end{cases} \tag{5.93}$$

where x_i is the direct trading power decision of DG source i, n_s and n_p are the energy sale and purchase aggregation, respectively. λ_{vpp}^t, $\lambda_{m,s}^t$ and $\lambda_{m,p}^t$ are the direct trading price, market selling and purchasing price, respectively, λ_{ser} is the unit price of grid service charge, and x_{ij} is the amount of traded between the i-th DG and the j-th DG. The premise of the successful transaction between the sale and purchase of energy is that the revenue gained from participating in direct trading is greater than that from trading with the energy market. When the direct trading of multiple DG is unsuccessful or there is a surplus or shortage of electricity, the energy market will cooperate with the trading. According to the characteristics of the energy market, when the supply exceeds the demand, the direct trading tariff is low, and as the demand increases, its competitive use of energy will lead to higher tariffs. Construct the tariff function as follows:

$$\lambda_{\text{vpp}}^{t} = k\left(\sum_{i=1}^{n_p} x_i\right)^2 \Big/ \sqrt{\sum_{j=1}^{n_s} x_j} + \lambda_{m,s}^{t} \tag{5.94}$$

where k is the unit cost factor. In order to prevent the scalping behaviour of DG, it is stipulated that in the same period, the DG only has the authority to either sell or purchase energy, and will not simultaneously purchase energy from the party with a low energy price and resell it to the party with a high energy price. Therefore, the constraint should be satisfied:

$$P_{m,i}^{t} \in [P_{m,i,\min}^{t}, P_{m,i,\max}^{t}] \tag{5.95}$$

$$x_i \in [0, P_{m,i}^{t}],\ i \in n_s \tag{5.96}$$

$$x_i \in [P_{m,i}^{t}, 0],\ i \in n_p \tag{5.97}$$

$$\sum_{j}^{n_s} x_j \geq \left|\sum_{i}^{n_p} x_i\right| \tag{5.98}$$

In addition, the model needs to satisfy the DG supply–demand balance constraints, MT agent power and ramp rate constraints, IL agent interruption constraints, and grid contact line security constraints.

(3) Modelling of the Game

In view of the stochastic nature of wind and solar power generation, which requires the cooperation of controllable power generation/energy-using agents in order to improve the level of consumption, the lower-level participants can form a DG, adopting the operation principle of "internal coordination and external coordination and competition", and the internal agents form a cooperative alliance to achieve energy complementarity and improve the total return. For the Nash equilibrium in the non-cooperative game, it is defined as follows:

Definition 5.1 In an n-player game $\mathbf{G} = \{\mathbf{S}_1, \mathbf{S}_2, \ldots, \mathbf{S}_n;\ u_1, u_2, \ldots, u_n\}$ formulated in the standard form, for each participant $i \in \mathbf{N}$, an optimal payment s_i^* can be obtained when the other game decision makers each adopt a strategy $s_1^*, s_2^*, \ldots, s_{i-1}^*, s_{i+1}^* \ldots, s_n^*$ and the combination of strategies is $\mathbf{s}^* = (s_1^*, s_2^*, \ldots, s_i^*)$ a Nash equilibrium for that game $\mathbf{G}$. This can be expressed as

$$\begin{aligned}&u_i(s_1^*,\ldots,s_{i-1}^*,s_i^*,s_{i+1}^*\ldots,s_n^*)\geq u_i(s_1^*,\ldots,s_{i-1}^*,s_i',s_{i+1}^*\ldots,s_n^*),\\&\forall s_i\in\mathbf{S}_i,s_i^*\neq s_i',\forall i\in\mathbf{N}\end{aligned}\tag{5.99}$$

When a Nash equilibrium is satisfied, no rational participant will choose to leave the Nash equilibrium situation. Therefore, if there is a Nash equilibrium in the game, the participants must spontaneously stabilize in the Nash equilibrium. It is important to note that a Nash equilibrium is not necessarily an optimal solution for each participant in the game, but rather a stable solution obtained by equilibrating the various participants in the game. There is the following theorem about the existence of Nash equilibrium.

Theorem 5.1 *Examine a strategy-based game whose strategy space is a nonempty tightly convex set of Euclidean space. A pure strategy Nash equilibrium exists for this game if pairs of its payoff functions are continuously fitted concave.*

According to the theorem, each payoff function of the lower level game is convex and there exists a Nash equilibrium a_i^* such that $v_i(a_i^*,\mathbf{a}_{-i}^*)\geq v_i(a_i,\mathbf{a}_{-i}^*),\ i\in N$. The optimal strategy of the lower level co-operative game is:

$$\mathbf{a}^*=\arg\ \max\sum_{i\in n}v_i(a_i,\mathbf{a}_{-i})\tag{5.100}$$

Similarly, there exists a Nash equilibrium x_i^* in the upper level game such that $u_i\left(x_i^*,\mathbf{x}_{-i}^*\right)\geq u_i\left(x_i,\mathbf{x}_{-i}^*\right)$.

5.2.2.3 Real-Time Optimization of Regulation Strategies

(1) Uncertainty Treatment

For WP prediction, the wind speed probability distribution is often described by the Weibull distribution model, and the WT output power is calculated by combining the relationship equation between the WT output power and the wind speed; for PV power prediction, the irradiance intensity is often described by the Beta distribution, and the PV output power is calculated by combining the relationship equation between the PV output power and the irradiance intensity. Therefore, the wind and solar output data can be generated by Monte Carlo simulation. For load forecasting, an artificial neural network method is used to introduce historical data of similar days and load and temperature data of the first 2 h and the first 1 h of the same day into the database as training objects, and obtain the load forecast data of the next moment. It is set that the forecast error and load fluctuation of wind and solar power generation are portrayed by normal distributions with zero mean and independent of each other, and since the forecast accuracy increases with the decrease of the forecast time scale, the coordinated operation strategy of the day-ahead, intra-day, and real-time multi-timescales is investigated to alleviate the pressure of the grid operation and the potential abandonment of wind and solar power, and to increase the economic efficiency of the system and the orderly operation.

$$\begin{aligned}\delta_{\mathrm{WT}}^{t} &\sim N(0, \sigma_{\mathrm{wt},t}^{2}) \\ \delta_{\mathrm{PV}}^{t} &\sim N(0, \sigma_{\mathrm{pv},t}^{2}) \\ \delta_{\mathrm{load}}^{t} &\sim N(0, \sigma_{L,t}^{2})\end{aligned} \tag{5.101}$$

where δ_{WT}^{t}, δ_{PV}^{t}, $\delta_{\mathrm{load}}^{t}$ are wind and solar output fluctuations and load fluctuations at time t. Since the prediction accuracy increases with the decrease of the prediction time scale, the coordinated operation strategy of the day-ahead, intra-day, and real-time multi-timescales is investigated to reduce the pressure of grid operation and the potential wind and solar power abandonment, and to improve the economic efficiency and orderly operation of the system.

(2) Real-Time Optimization

Real-time optimization is used to correct the deviation of the operating plan from the forecast result by adjusting the size of the controllable unit output and the DR (denoted as DDR) used for DLC in the DG. The cost function of the DDR is the same as the constraints as the IL. The scenery and the load forecast error are generated by the simulation of the normal distribution. The objective function is:

$$\begin{aligned}\min F_3 = {} & (a_{\mathrm{MT}}(P_{\mathrm{MT}}^{t})^2 + b_{\mathrm{MT}} P_{\mathrm{MT}}^{t} + c_{\mathrm{MT}}) + \lambda_{\mathrm{wt}} P_{w}^{t} + (a_{\mathrm{DDR}}(P_{\mathrm{DDR}}^{t})^2 + \\ & b_{\mathrm{DDR}} P_{\mathrm{DDR}}^{t}) + \lambda_{\mathrm{pv}} P_{p}^{t} + (a_{\mathrm{IL}}(P_{\mathrm{IL}}^{t})^2 + b_{\mathrm{IL}} P_{\mathrm{IL}}^{t}) - V_{\mathrm{VPP}}\end{aligned} \tag{5.102}$$

Need to satisfy opportunity constrained planning as follows:

$$\begin{aligned}P\{(P_{\mathrm{WT}}^{t} + \delta_{\mathrm{WT}}^{t} - P_{w}^{t}) + (P_{\mathrm{PV}}^{t} + \delta_{\mathrm{PV}}^{t} - P_{p}^{t}) + P_{\mathrm{MT}}^{t} + P_{\mathrm{IL}}^{t} + P_{\mathrm{DDR}}^{t} \\ \geq P_{m}^{t} + (P_{\mathrm{load}}^{t} + \delta_{\mathrm{load}}^{t})\} \geq \alpha\end{aligned} \tag{5.103}$$

The tasks in this phase include: determination of MT output, determination of DDR and wind and solar abandonment.

5.2.2.4 Solution Steps

Since the above model involves predicting random variables and opportunity constraints in the intraday and real-time optimization phases, although it can be transformed into an equivalent deterministic form and then solved using linear/nonlinear programming problems, the transformation is more difficult in practical applications, especially when there are more random variables. Therefore, genetic algorithms are uniformly used in this section to solve them.

For the bi-level optimization model, since the upper and lower levels can reach equilibrium after a finite number of games, a nested genetic algorithm is used to

Fig. 5.18 Algorithm flow chart

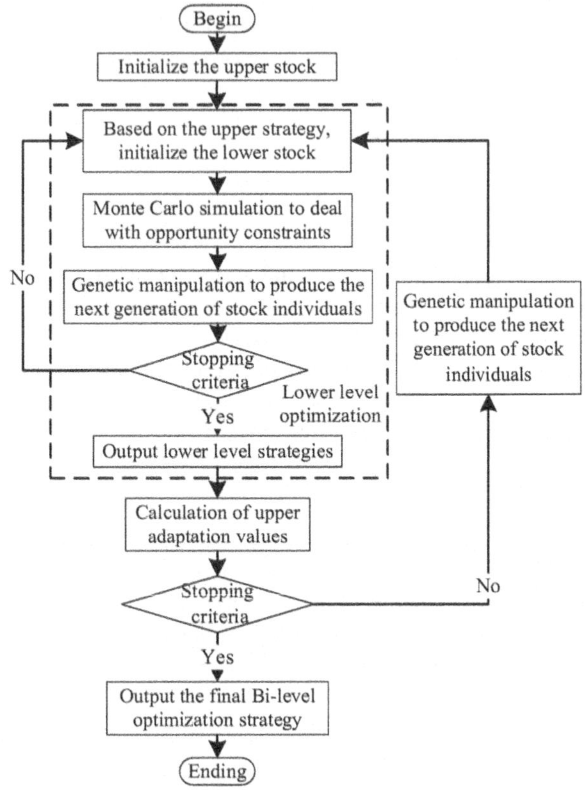

solve the model, and the process is shown in Fig. 5.18. For the intra-day and real-time coordinated optimization model, a combination of Monte Carlo simulation and genetic algorithm is used for direct solution, which is simpler and more effective and more widely applicable, and the process is shown in the lower level optimization part of Fig. 5.18.

5.2.2.5 Case Analysis

(1) Description of the case

Taking the test system of Fig. 5.19 containing three DEGs as an example. the capacity of WT1/WT3/PV2 are 500 kW, 300 kW, 200 kW respectively, using the same model MT with 200 kW capacity and 2 kW/min creep rate. Taking the maximum call rate of IL as 10%, $a = 0.008$ yuan/(kW^2·h), $b = 0.4$ yuan/(kW^2·h), and the DDR as 5% of the total load, $a = 0.009$ yuan/(kW^2·h), $b = 0.85$ yuan/(kW^2·h). The energy market sale price and purchase price of energy are 1 yuan/(kW h) and 0.4 yuan/(kW h) respectively, the load price is 0.6 yuan/(kW h), and the distributed energy bidding price range [0.4,1] yuan/(kW h).

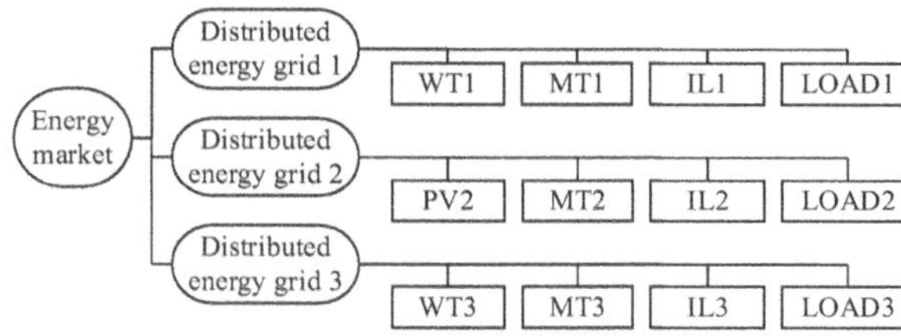

Fig. 5.19 Test system diagram

Set the variance of day-ahead, intraday and real-time forecast fluctuations of WT and PV output to be 0.2, 0.05 and 0.02, respectively, and the variance of load forecast fluctuations to be 0.02, 0.005 and 0.002, respectively. The curves of output forecasts and load forecasts of the three distributed energy sources at multiple time scales are shown in Fig. 5.20. The population size of 50 and the number of iterations of 200 are taken in the genetic algorithm of this chapter.

(1) Analysis of multi-scale optimization results

For the purposes of this discussion, using the multiple distributed energy co-operation model mentioned in the previous sections as an example, it is assumed that there are only multiple distributed energy direct transactions. Two models are considered here:

(a) The results of each distributed energy dispatch are determined directly from the day-ahead optimisation, i.e. both DDR and standby capacity are determined from the day-ahead optimization.
(b) The multi-timescale model proposed in this chapter is used.

The results of the two modes of scheduling and the required spinning reserve capacity are shown in Fig. 5.21, both for distributed energy source 1.

As seen in Fig. 5.21, when the day-ahead forecast data is used, the operation results are smoother than Strategy 2 due to few sampling points (24 data points for 24 h). Where the DR is mainly provided by the IL and does not call the DR which is capable of DLC. In strategy 2, in the 1st and 5th hour, the real-time forecast update is not able to consume all the WP after the internal output adjustment of the distributed energy sources, resulting in the wind abandonment phenomenon. The real-time optimisation in Strategy 2 is updated once in 5 min, and the DR in real-time optimisation is mainly implemented by the direct load-controlled DR due to the slow IL response. In addition, from the distributed energy 1 interaction power change in the figure, it can be seen that the interaction power of strategy 1 is slightly larger relative to that of strategy 2 at 1–8, which is due to the larger error of WP forecast and load forecast in the previous day, and in the rolling optimisation phase of strategy 2 intra-day, when the dynamic adjustment of each controllable distributed energy within distributed power is based on the forecast data with smaller error, the game model is updated for that time period due to the inability to satisfy the trading power determined in the previous day. The game model for the time period is updated. The

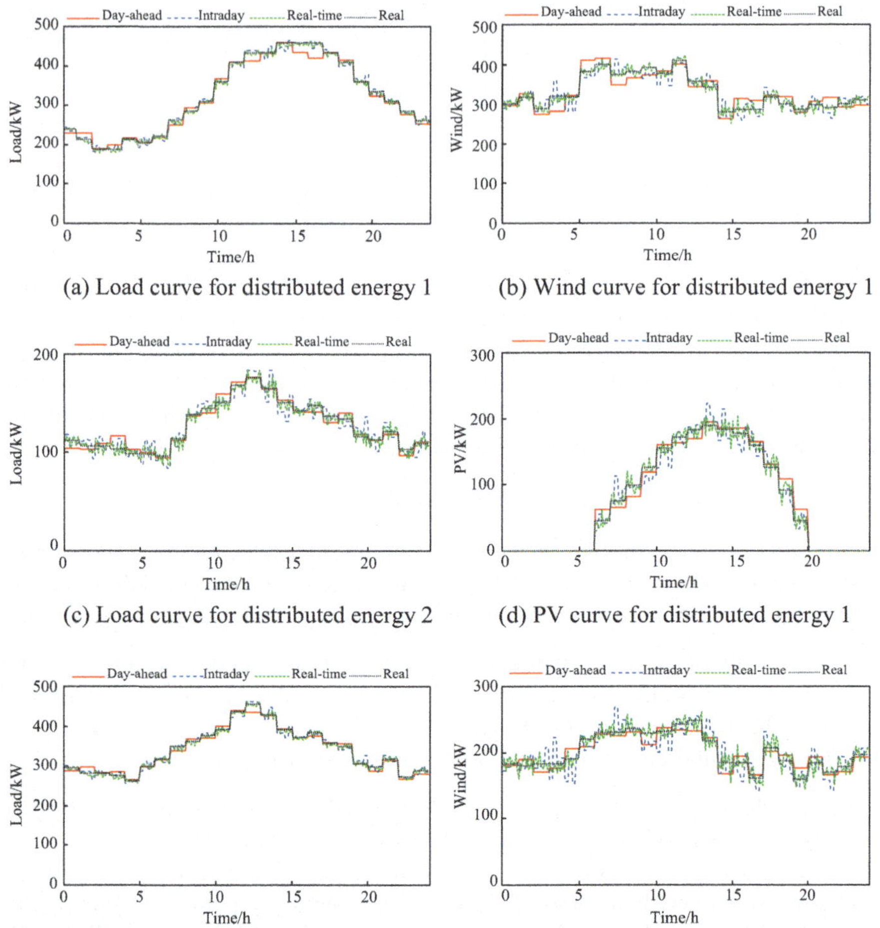

(a) Load curve for distributed energy 1 (b) Wind curve for distributed energy 1

(c) Load curve for distributed energy 2 (d) PV curve for distributed energy 1

(e) Load curve for distributed energy 3 (f) Wind curve for distributed energy 3

Fig. 5.20 Distributed energy output forecasting and load forecasting

figure also gives the quantified value of lost load due to unsatisfied chance constrained planning for both strategies. Since the real-time prediction accuracy is higher than the day-ahead prediction value and its lost load is smaller, only a small amount of spare capacity is needed to smooth out the output fluctuation and improve the stability of the system. Including lost load penalties in the cost:

$$C_{risk} = \sum_{t=1}^{T} \lambda_e^t E_{ens}(t) \tag{5.104}$$

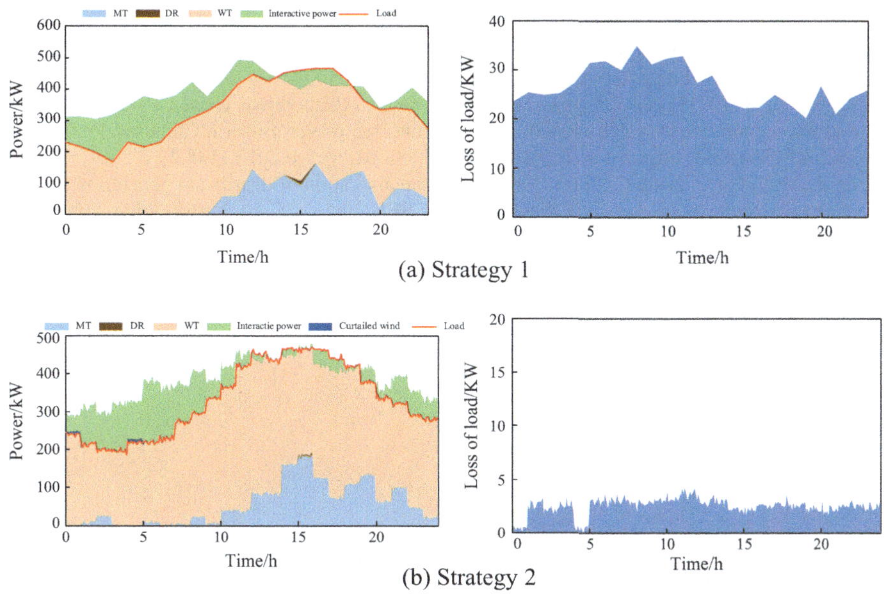

Fig. 5.21 Scheduling results for distributed energy source 1 under strategy 1–2

Table 5.6 Loss of load and gain situation

Operational strategy	Loss of load/(kW)	Total gain/(yuan)
Strategy 1	1435.74	5420.53
Strategy 2	113.92	6755.73

where $E_{ens}(t)$ is the value of lost load at time period t and λ_e^t is the lost load penalty cost (usually higher than the energy grid sale price). From the optimization results in Table 5.6, it can be seen that the multi-timescale optimization operation reduces the risk of loss of load, and is able to improve the system reliability while increasing the economic efficiency.

References

Binetti G, Davoudi A, Lewis FL (2014) Distributed consensus-based economic dispatch with transmission losses. IEEE Trans Power Syst 29(4):1711–1720. https://doi.org/10.1109/TPWRS.2014.2299436

Chen G, Li Z, Zhao Z (2016) Distributed optimal droop control in micro-grid systems (in Chinese). Control Theory Appl 33(8):999–1006. https://doi.org/10.7641/CTA.2016.50893

Yu Z, Ai Q (2017) Adaptive droop control strategy for microgrid based on consensus of multi-agent system (in Chinese). Electr Power Autom Equip 37(12):150–156. https://doi.org/10.16081/j.issn.1006-6047.2017.12.021

Yu Z, Ai Q, Xiong W (2017) Hierarchical distributed real-time optimization strategy for microgrid based on consistency protocol of multi-agent system (in Chinese). Autom Electr Power Syst 41(18):31–37+94. https://doi.org/10.7500/AEPS20170105009

Zhang X, Yu T (2015) Virtual generation tribe based collaborative consensus algorithm for dynamic generation dispatch of AGC in interconnected power grids (in Chinese). Proc CSEE 35(15):3750–3759. https://doi.org/10.13334/j.0258-8013.pcsee.2015.15.002

Zheng Z, Ai Q, Gu C, Jiang C (2009) Multi-objective allocation of distributed generation considering environmental factor (in Chinese). Proc CSEE 29(13):25–30. https://doi.org/10.3321/j.issn:0258-8013.2009.13.004

Chapter 6
Distributed Coordinated Scheduling of Distributed Energy Grid Clusters

The distributed energy grid mainly has two development directions: one is to connect the distributed energy grid with the superior energy grid; the other is to interconnect multiple neighboring DEGs, forming distributed energy grid clusters. With the gradual popularization of DEGs (Lasseter 2011; Saad et al. 2012; Bie et al. 2012), distributed energy grid clusters are also moving from theory to practical engineering applications. In an energy grid area, multiple adjacent DEGs may coexist, forming the distributed energy grid clusters system (Hatziargyriou 2009; Kumar and Doolla 2013; Nunna and Doolla 2012; Martini 2015a). As an important supplement to the large energy grid, distributed energy grid clusters play an increasingly important role in improving power supply reliability, security, and addressing energy supply issues in remote areas and islands. These clusters also provide vast development potential for renewable energy, which can be better integrated through interaction with electricity, heating, and hydrogen networks. By coupling these energy systems, the overall flexibility and efficiency of energy distribution and consumption can be significantly enhanced.

This chapter, based on the study of multiple time scales within DEGs, adds a distributed energy cluster optimization layer, and further proposes a coordinated scheduling framework for distributed energy grid clusters. This framework not only coordinates the scheduling of power grids but also integrates heat networks and hydrogen grids into the optimization process. The chapter also provides methods for the distributed coordinated scheduling of multiple heterogeneous distributed energy grid clusters, emphasizing the interactions between electricity, heat, and hydrogen as fundamental components in enhancing system resilience and flexibility.

However, the interconnection of DEGs leads to an increase in distributed devices, complicating information exchange and global control and presenting unprecedented challenges to traditional centralized control strategies. The shortcomings of traditional scheduling strategies can be summarized as follows:

T. Lu et al., *Operation Modeling and Collaborative Regulation of Distributed Energy Grid Clusters*, Smart Energy Systems, https://doi.org/10.1007/978-981-96-9942-1_6

(1) The wide-area distributed energy access requires the deployment of global optimization, point-to-multipoint communication, and vertical control for all distributed components, including electricity, heat, and hydrogen systems. This increases the complexity of management and coordination, especially when coordinating energy flows between these different grids.
(2) Due to the strong randomness and time-varying characteristics of distributed energy grid clusters scheduling, rolling correction and dynamic optimization are key to its scheduling. However, under centralized scheduling strategies, due to communication delays and server capacity limitations, it is difficult to achieve satisfactory tracking performance and control effects across the integrated electricity, heat, and hydrogen systems.
(3) Centralized decision-making systems may fail to protect the privacy of controlled entities, especially when information from heterogeneous grids (electric, thermal, hydrogen) is shared among multiple stakeholders. This raises concerns about the protection of sensitive data.
(4) Centralized optimization frameworks often struggle to allocate benefits fairly among all participants, particularly when multiple energy sources, such as electricity, heat, and hydrogen, interact in a complex multi-energy environment, leading to potential inefficiencies and conflicts of interest.

Therefore, to overcome these issues, the algorithm proposed in this chapter is fully decentralized, suitable for collaboration among multiple entities across electricity, heating, and hydrogen grids. It also supports plug-and-play functionality and online expansion for distributed energy grid clusters, providing new research ideas for optimization at the distributed energy grid clusters level while ensuring system flexibility, privacy protection, and fair resource allocation.

6.1 Operation Modeling of Distributed Energy Grid Clusters

6.1.1 Dynamic Cooperative Game Model

Under the premise of DEGs cooperative game, the MAS of distributed energy grid clusters needs to transmit their respective power deficits or surpluses as well as location information to the superior energy grid. At the same time, they also receive corresponding trading instructions from the superior energy grid to complete the trading process. This trading model will generate transaction costs and communication costs. Therefore, the payment function of the distributed energy grid system clusters should include the electricity purchase cost, grid loss cost, communication cost, and compensation cost for interruptible loads of each distributed energy grid.

The goal of the distributed energy grid clusters cooperative game is to find the optimal clusters substructure, thereby minimizing the payment function of each distributed energy grid. This problem can be divided into three parts.

(1) Minimizing the Clusters Payment Function

An algorithm for determining the purchase and sale combinations between DEGs is used for solving. For a given clusters S, which consists of multiple DEGs, for each buyer distributed energy grid, the seller DEGs are sorted based on their distance to the buyer energy grid, and energy trading is carried out according to the proximity principle.

(2) Benefit Distribution Among Members of the Clusters

The concept of the Shapley value is used to represent the contribution of each member within the clusters to the overall payment function. For the distributed energy grid within the Clusters S, its Shapley value can be expressed as:

$$\phi_i = \sum_{S \subseteq N \backslash \{i\}} \frac{|S|!(|N| - |S| - 1)!}{|N|!}(v(S \cup \{i\}) - v(S)) \tag{6.1}$$

where N represents the total number of participants, $v(S)$ represents the total payment function of clusters S, and S represents the number of participants in clusters S.

(3) Finding the Optimal Substructure of the Clusters

A three-stage centralized optimization framework based on merging and splitting is adopted, introducing a Pareto optimal approach based on the payment function to achieve the maximum benefit of the clusters. The three-stage optimization framework consists of request exchange, merging-splitting, and joint trading: In the first stage, the agents of each distributed energy grid will send their location, power deficit, and power surplus information to the superior energy grid control center (SC). After receiving information from all the DEGs, the SC will perform the merging-splitting operation to form the optimal clusters structure. The Pareto condition will be used as the criterion for judging whether the merging or splitting process is successful. When the merging-splitting process finally converges, the resulting clusters structure will ensure that all the members of the clusters have minimized payment functions. Based on this clusters structure, the SC will send corresponding instructions (including clusters pairing, joining order, etc.) to the agent system of each energy grid. After receiving the instructions, the DEGs will form clusters with other DEGs based on the joining order and proceed with the energy trading.

6.1.2 Bi-level Game Model

Assume that the superior energy grid and each distributed energy grid can participate in market operations as independent entities, and consider the existence of both dispatchable and non-dispatchable distributed energy sources. The goal is to achieve a balance of interests among all market participants while considering the output uncertainty of distributed energy sources. A bi-level game model for the distributed energy grid system is established, where the upper-level problem involves the energy trading among multiple energy grids, with the control objective being the maximization of transaction profits for each participant (energy grid) while ensuring system energy balance. The lower-level problem involves the operational scheduling within the energy grids, determining the optimal cost scheduling plan within each energy grid while meeting the upper-level market demand and individual declared quantities.

(1) Upper-Level Direct Trading Model Between DEGs

The upper-level game model adopts a coalition cooperative game model, where the decision variables are the declared energy sale or purchase quantities of each distributed energy grid. The objective function is the total trading profit of distributed energy grid clusters, and the constraint is the system's energy balance.

First, each distributed energy grid forecasts its energy deficit or surplus for the current period based on the predicted output of distributed energy sources and load demand data. The feasible range for the declared quantities is set as the forecasted values with a 20% fluctuation margin (i.e., the range of the decision variables). On this basis, each distributed energy grid uses the coalition cooperative game model to solve for the optimal feasible solution, aiming to maximize the total coalition trading profit.

Considering the coalition cooperative game among DEGs, the elements of the participant set are divided into microsource-type and load-type grids. The distinction between the two is whether the total generation capacity exceeds the total predicted load. In the game, the decision variable for the microsource-type is the tradable energy, while the decision variable for load-type is the purchased energy. The decision quantities of each participant must satisfy the declared decision variable limits and the system energy balance constraint.

For a given strategy selection x_i, the total payment function of all DEGs can be defined as:

$$u(x_i, X_{-i}) = \begin{cases} p_{dp}(X)x_i + (\mathrm{Req}_i - x_i)p_{pp} - C_C\mathrm{Req}_i, x_i \le \mathrm{Req}_i, i \in N_s \\ p_{dp}(X)x_i + (\mathrm{Req}_i - x_i)p_{wp} - C_C\mathrm{Req}_i, x_i > \mathrm{Req}_i, i \in N_s \\ s(x_i) + p_{dp}(X)x_i + (\mathrm{Req}_i - x_i)p_{wp}, x_i \ge \mathrm{Req}_i, i \in N_b \\ s(x_i) + p_{dp}(X)x_i + (\mathrm{Req}_i - x_i)p_{pp}, x_i < \mathrm{Req}_i, i \in N_b \end{cases} \tag{6.2}$$

where X is an $N \times 1$ vector representing the strategy choices of all DEGs. Req_i^w represents the energy deficit or surplus of the distributed energy grid. X_{-i}^w represents the strategy choices of DEGs excluding participant i, and p_{dp}^w denotes the trading energy price determined by the declared strategies of the DEGs. $s(x_i^w)$ is the income function of the energy selling, C_C represents the energy cost of the distributed energy grid, and p_{wp} and p_{pp} are the energy purchase price and sale price for the distributed energy grid.

Considering the cooperative game among DEGs, the objective function is the total payment of the clusters. For the entire coalition structure G(N) = {DEG_1, DEG_2, ..., DEG_N}, the total payment function for the entire coalition is as follows:

$$\max_{x \in X} u_N = \max_{x \in X} \sum_{w} \sum_{i \in N} u(x_i^w, X_{-i}^w) \tag{6.3}$$

(2) Lower-Level Internal Optimization Scheduling Model for Energy Grids

The lower-level game model takes the output of distributed energy sources within the cluster as the decision variables, including the output of distributed energy sources over a 24-h period (daily operating curve). The objective function is the total operating cost of the cluster, which includes fuel costs, operation and maintenance costs, environmental management costs, and wind/solar curtailment costs. The constraints include system power balance, output constraints for each distributed energy source, and the declared electricity quantities for the superior energy grid.

The participants within the distributed energy grid clusters can include WT, PVs, ES devices, or micro GTs. For each participant's decision variables, power balance constraints and output constraints for the distributed energy sources must be satisfied.

For a given strategy choice a_i ($i \in N$), the payment function for different distributed energy sources can be defined as follows:

$$v_i(a_i, A_{-i}) = \begin{cases} -c_{wt}(g_i - a_i), i \in N_{wt} \\ -c_{pv}(g_i - a_i), i \in N_{pv} \\ -c_{bat} a_i, i \in N_{bat} \\ -(k_{fuel} + k_{env} + k_{om}), i \in N_{mt} \end{cases} \tag{6.4}$$

where, the payment function is defined as the negative of the costs for each distributed energy source, thus transforming the cost minimization problem into a payment maximization game problem. c_{wt} and c_{pv} represent the cost coefficients for wind and solar curtailment, respectively; c_{bat} represents the cost coefficient for ES device charging and discharging; and k_{fuel}, k_{env}, k_{om} represent the fuel cost coefficient, environmental management cost coefficient, and operation and maintenance cost coefficient for GTs, respectively.

For the cooperative game among distributed energy sources considered in this section, the objective function should similarly be defined as the total payment of the clusters. Considering the full coalition structure G(N) = {n_1, n_2, ..., n_N} of all distributed energy units, the total payment for the entire clusters is defined as follows:

$$\max_{a \in A} v_N = \max_{a \in A} \sum_{i \in N} v(a_i, A_{-i}) \tag{6.5}$$

6.1.3 Bi-level Evolutionary Game Decision Model for Superior Energy Grid and Distributed Energy Grid Clusters

The superior energy grid uses net losses and voltage stability levels as their benefit objectives and decides the output of the grids. The distributed energy grid clusters use operating costs as their benefit objective and decide the output of each distributed energy source. The power at the common connection point is jointly decided by both the distributed energy grid clusters and the superior energy grid. The benefit objectives of the distributed energy grid clusters and the superior energy grid are treated as virtual participants in the game process. A cooperative evolutionary game algorithm (CEGA) is proposed to solve the benefit model, yielding the optimal operating point for both the distributed energy grid clusters and the superior energy grid, which can be used to guide their joint operation.

During the interactive operation between the distributed energy grid and the superior energy grid, energy exchange occurs between them, creating an inseparable relationship. On the one hand, the superior energy grid schedules energy for the distributed energy grid, affecting the economic performance of the energy grid; on the other hand, the output of the energy grid also influences the safety and stability of the superior energy grid's operation, and there is a certain conflict of interests between the two. Using game theory, the distributed energy grid and the superior energy grid, with their conflicting interests, can be treated as game participants and optimized according to certain mathematical rules. The optimization result depends on their respective benefit objectives. Based on this, using an appropriate algorithm, a stable strategy can be found in which all game participants reach a global optimum. Therefore, the superior energy grid and the distributed energy grid can jointly formulate a scheduling strategy through the game. The specific virtual game participants are shown in Fig. 6.1.

(1) Optimization Benefit Function of the Superior Energy Grid

The goal of the superior energy grid is to minimize the total production cost, including the cost of equipment operation, the cost of interacting with the distributed energy grid clusters, and the cost of pipe grid operation and maintenance.

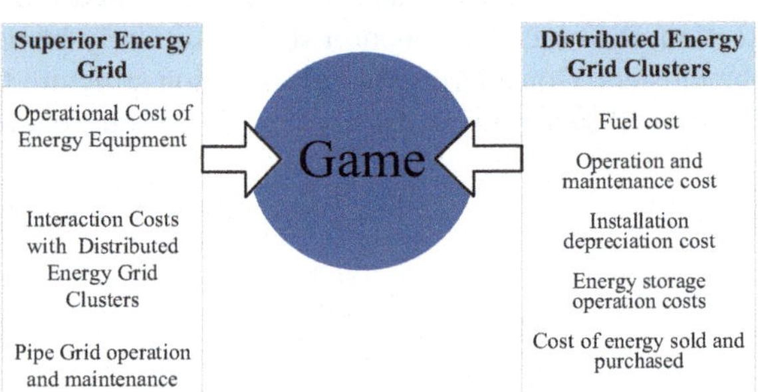

Fig. 6.1 Game structure of superior energy grid and distributed energy grid clusters

$$
\begin{cases}
\min F^{\text{DIES}} = \sum \Delta t \left(C_t^{\text{om}} + C_t^{\text{mg}} + C^{\text{line.op}} \right) \\
C_t^{\text{om}} = \delta^{\text{CHP}} P_{\text{elec},t}^{\text{CHP}} + \delta^{\text{GB}} \sum\limits_{s \in \Omega_{\text{CB}}} H_{s,t}^{\text{GB}} \\
C_t^{\text{mg}} = \sum\limits_{m} \left(c_{\text{e}}^{\text{mg}} P_m^{\text{BP}} - c_{\text{h}}^{\text{mg}} H_m^{\text{DB}} \right) \\
C^{\text{line.op}} = \lambda^{\text{line}} \cdot C^{\text{line}} \cdot \left[\dfrac{r(1+r)^n}{(1+r)^n - 1} \right]
\end{cases}
\tag{6.6}
$$

The F^{DIES} represents the total cost of the superior energy grid operator. C_t^{om} is the equipment operating cost of power generation and heat production equipment during period t; C_t^{mg} is the energy interaction cost between the superior energy grid and the distributed energy grid clusters during period t; $C^{\text{line.op}}$ is the pipe grid operation and maintenance cost; r represents the annual discount rate; n represents the pipeline's service life in years; δ^{CHP} and δ^{GB} represent the operation and maintenance cost per unit output of the combined CHP unit and GBs; $P_{\text{elec},t}^{\text{CHP}}$ and $H_{s,t}^{\text{GB}}$ represent the electricity and thermal power outputs of the CHP unit and the s-th GBs located at the superior energy grid during period t; c_{e}^{mg} and c_{h}^{mg} represent the interaction cost of electricity and thermal energy between the superior energy grid operator and the distributed energy grid clusters; m is the set of upper-level GBs; Δt is the time span of each scheduling period. P_m^{BP} represents the active power transferred from the superior energy grid to the distributed energy grid clusters, which is defined as positive in this context. H_m^{DB} represents the heat power transferred from the superior energy grid, with its positive direction defined as flowing from the thermal grid to the distributed energy grid clusters.

(2) Optimization Constraints of the Superior Energy Grid

The constraints of the superior energy grid optimization model mainly include the constraint of heating network power flow equations, interconnection line capacity limits, and the power transmission capability constraints of the energy grid.

(a) Constraint of thermal grid

According to the law of energy conservation, the temperature mixing constraint of the working fluid from different pipelines at a node is:

$$
T_{n,t}^{\text{sp}} = \frac{\sum_{p \in r_b^{p+}} \left(m_{p,t}^{\text{sp}} T_{p,t}^{\text{sp,out}} \right) + T_{n,t}^{\text{ks}} m_{n,t}^{+}}{\sum_{p \in r_n^{p+}} m_{p,t}^{\text{sp}} + m_{n,t}^{+}} \tag{6.7}
$$

$$
T_{n,t}^{re} = \frac{\sum_{p \in r_u^{n+}} \left(m_{p,t}^{\text{re}} T_{p,t}^{\text{re,out}} \right) + T_{n,t}^{\text{tl}} m_{n,t}^{-}}{\sum_{p \in r_x^{(r+}} m_{p,t}^{\text{re}} + m_{n,t}^{-}} \tag{6.8}
$$

$$
\begin{cases}
T_{n,f}^{sp} = T_{p,t}^{sp\ \text{in}} & \forall p \in \Gamma_n^{sp-} \\
T_{n,f}^{\text{re}} = T_{p,f}^{\text{re,in}} & \forall p \in \Gamma_n^{\text{re}-}
\end{cases}
\tag{6.9}
$$

where $T_{n,t}^{\text{sp}}$ and $T_{n,t}^{re}$ represent the temperatures at node n in the supply and return pipelines of the district thermal grid, $T_{p,t}^{\text{sp,out}}$ and $T_{p,t}^{\text{re,out}}$ represent the temperatures at the outlet of supply and return pipeline p; $T_{p,t}^{sp\text{ in}}$ and $T_{p,f}^{\text{re,in}}$ represent the temperatures at the inlet of supply and return pipeline p, $m_{p,t}^{\text{sp}}$ and $m_{p,t}^{\text{re}}$ represent the mass flow rates of supply and return pipeline p in the grid, $T_{n,t}^{\text{ks}}$ and $T_{n,t}^{\text{tl}}$ represent the supply temperature of the heat source and the return temperature after passing through the heat exchange station at node n, respectively; $m_{n,t}^{+}$ and $m_{n,t}^{-}$ represent the mass flow rates of the heat source at node n for supply and return, Γ_n^{sp-} and $\Gamma_n^{\text{re}-}$ are a set of pipes starting from node n in the water supply network and the return water network.

(b) Power flow constraints

$$\begin{cases} P_i + P_{DEGi} - P_{Li} - V_i \sum\limits_{j \in i} V_j (G_{ij} \cos\theta_{ij} + B_{ij} \sin\theta_{ij}) = 0 \\ Q_i + Q_{DEGi} - Q_{Li} - V_i \sum\limits_{j \in i} V_j (G_{ij} \sin\theta_{ij} + B_{ij} \cos\theta_{ij}) = 0 \end{cases} \tag{6.10}$$

where P_i and Q_i are the active and reactive power generated by the generator at node i; P_{DEGi} and Q_{DEGi} are the active and reactive power transmitted from the distributed energy grid clusters to node i, P_{Li} and Q_{Li} are the active and reactive power of the load at node i; V_i and V_j are the voltage magnitudes at nodes i and j; G_{ij} and B_{ij} are the conductance and susceptance of the line between nodes i and j; θ_{ij} is the voltage phase angle between nodes i and j, $\theta_{ij}=\theta_i - \theta_j$.

(c) Interconnection line capacity constraints

$$P_{\min} \le P_{PCCt} \le P_{\max} \tag{6.11}$$

where P_{PCCt} is the power of the interconnection line; $P_{\min}$ and $P_{\max}$ are the lower and upper power limits of the interconnection line.

(d) Constraint of superior energy grid transmission capability

$$\sum P_{it\min} - P_{Lt} \le P_{PCCt} \le \sum P_{it\max} - P_{Lt} \tag{6.12}$$

where $\sum P_{it\max}$ and $\sum P_{it\min}$ are the sums of the maximum and minimum output powers of all DG sources at time

(3) Optimization Benefit Function of the Distribution Energy Grid Clusters

The distribution energy grid clusters aims to minimize the total economic cost, which includes gas costs, operation costs of distributed units, ES operation costs, and energy interaction costs with the superior energy grid.

$$\min F_m^{\text{MEMG}} = \sum_t \Delta t \left(C_{m,t}^{\text{e}} + C_{m,t}^{\text{DG}} + C_{m,t}^{\text{BU}} + C_{m,t}^{\text{DESS}} \right) \tag{6.13}$$

$$
\begin{cases}
C_{m,t}^{\mathrm{e}} = c^{\mathrm{gas}} P_{\mathrm{gas},m,t}^{\mathrm{EXT}} \\
C_{m,t}^{\mathrm{DG}} = a_m^{\mathrm{g}} \left(P_{m,t}^{\mathrm{CGU}}\right)^2 + b_m^{\mathrm{g}} P_{m,t}^{\mathrm{CGU}} + c_m^{\mathrm{g}} + \delta_m^{\mathrm{PV}} P_{m,t}^{\mathrm{PV}} + \delta_m^{\mathrm{DCHP}} P_{m,t}^{\mathrm{DCHP}} + \delta_m^{\mathrm{HP}} P_{m,t}^{\mathrm{HP}} \\
C_{m,t}^{\mathrm{BU}} = \delta_m^{\mathrm{BU}} \left(P_{\mathrm{ch},m,t}^{\mathrm{BU}} + P_{\mathrm{dc},m,t}^{\mathrm{BU}}\right) \\
C_{m,t}^{\mathrm{DIES}} = c_{\mathrm{h},m}^{\mathrm{mg}} H_{m,t}^{\mathrm{BM}} - c_{\mathrm{e},m}^{\mathrm{mg}} P_{m,t}^{\mathrm{MB}} + c_{i,t}^{\mathrm{G}}
\end{cases}
\tag{6.14}
$$

where F_m^{MEMG} represents the total production cost of the m-th distribution energy grid; $C_{m,t}^{\mathrm{e}}$ is the energy purchase cost; $C_{m,t}^{\mathrm{DG}}$ is the operation cost of distributed power sources; $C_{m,t}^{\mathrm{BU}}$ is the operation cost of ES; $C_{m,t}^{\mathrm{DIES}}$ is the energy interaction cost between the distributed energy grid and the superior energy grid; a_m^{g}, b_m^{g}, and c_m^{g} are the cost coefficients of small diesel generators in the m-th distribution energy grid; δ_m^{PV}, δ_m^{DCHP}, δ_m^{HP}, and δ_m^{BU} represent the operation and maintenance costs per unit power of PV, CHP units, HPs, and ES, $P_{m,t}^{\mathrm{CGU}}$, $P_{m,t}^{\mathrm{PV}}$, $P_{m,t}^{\mathrm{DCHP}}$, and $P_{m,t}^{\mathrm{HP}}$ represent the electric power outputs of small diesel generators, PV, CHP units, and HPs in the m-th distributed energy grid at time t, respectively; $P_{\mathrm{ch},m,t}^{\mathrm{BU}}$ and $P_{\mathrm{dc},m,t}^{\mathrm{BU}}$ represent the charging and discharging power of ES in the m-th distributed energy grid at time t; $c_{\mathrm{h},m}^{\mathrm{mg}}$ and $c_{\mathrm{e},m}^{\mathrm{mg}}$ are the interaction costs of thermal and electrical energy per unit between the m-th distributed energy grid and the superior energy grid, $H_{m,t}^{\mathrm{BM}}$ represents the thermal power obtained by the m-th distributed energy grid from the centralized heating network; $P_{m,t}^{\mathrm{MB}}$ represents the active power flowing into the superior energy grid from the m-th distributed energy grid. $c_{i,t}^{\mathrm{G}}$ represents the cost of buying natural gas from the superior energy grid.

(4) Optimization Constraints of the Distributed Energy Grid Clusters

The constraints for the operation of the distributed energy grid clusters include system power and thermal balance constraints, DG output constraints, and ramp-up constraints for CHP units, micro GTs, and HP systems, among other limitations.

(5) Model Fuzzy Processing

Due to the different nature of the benefit objectives between the distributed energy grid clusters and the superior energy grid, direct game-theoretic interaction between the two is not feasible. Therefore, this chapter uses fuzzy mathematics to handle the benefit functions, transforming them into a unified value for easier comparison of the optimization results after the game. First, the optimal solutions for the benefit objectives of the distributed energy grid clusters and the superior energy grid are determined under their respective constraints, and then these optimal solutions are used to fuzzify their respective benefit functions.

Fuzzy mathematics is a precise mathematical method used to represent and process the fuzzy phenomena that exist in reality. To achieve this, the membership function u must first be determined. The value of u reflects the degree of optimization, where

$u = 1$ represents the best optimization and $u = 0$ represents the worst optimization. After comparative analysis, this chapter selects the decreasing half-Γ distribution. For the above benefit functions, the membership function of the decreasing half-Γ distribution is given by:

$$u_k = \begin{cases} 1, & F_k \leq F_{k\min} \\ \exp\left(\frac{F_{k\min} - F_k}{F_{k\min}}\right), & F_k > F_{k\min} \end{cases} \tag{6.15}$$

where $F_{k\min}$ represents the minimum value of the benefit function F_k under the constraints; $k = 1, 2$.

(6) Model Solution Based on the Cooperative Game Algorithm (CGA)

Evolutionary game theory is a branch of game theory that assumes evolutionary changes are caused by natural selection within a population. It evolves through selection behavior with frequency-dependent effects to search for evolutionary stable strategies and studies the evolutionary process. The co-evolutionary algorithm introduces the concept of an ecosystem into traditional evolutionary algorithms, mapping the problem to an ecosystem composed of species that interact and influence each other. The problem is solved through the evolution of the ecosystem. This chapter, combining evolutionary game theory and co-evolutionary algorithms, proposes a CEGA to solve the model.

In the CEGA, the distributed energy grid clusters and the superior energy grid act as game participants, generating two populations, P1 and P2, respectively. P_1 and P_2 represent the population distribution probabilities in the initial population. Each population contains a certain number of individuals, each recording a decision vector and the population to which it belongs. P1 and P2 aim to maximize the benefits with objectives u_1, u_2, respectively. The population game structure is shown in Fig. 6.2.

In the evolutionary game theory, each individual is randomly selected from the population to participate in repeated games, with the assumption that each participant possesses only limited rationality. The algorithm draws on the concepts and selection mechanisms of evolutionary games. When two individuals in the population encounter each other, they engage in a game for the same resource.

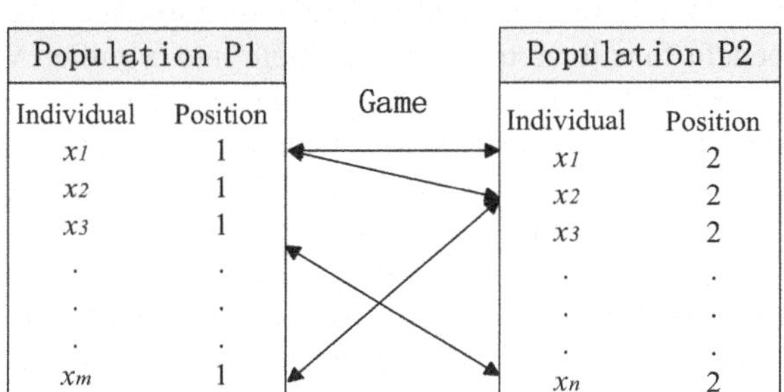

Fig. 6.2 Population game structure

(7) Dynamic Optimization Submodel Based on Maximum Dissatisfaction

Due to the significant influence of environmental and climatic conditions on WTs and PVs in the energy grid, it is necessary to account for the stability of decision satisfaction caused by randomness to prevent their randomness and volatility from having a substantial impact on the operation of the superior energy grid. This requires dynamically adjusting the algorithm parameters during the evolutionary game to account for the randomness.

In each generation of evolution, based on the consideration of the randomness and impact of distributed energy sources, the dissatisfaction is calculated through the dynamic optimization submodel. This allows for the adjustment of the distribution probabilities of the two populations, changing the game positions of the distributed energy grid clusters and superior energy grid. The next evolution is then carried out, ultimately yielding the optimal stable scheduling decision.

The solution steps are as follows (Fig. 6.3):

Step 1: Input the original data and parameters, set the population distribution probabilities, randomly generate the initial population, and create the initial feasible strategy combinations.

Step 2: Randomly select two individuals from the population to engage in a game, and calculate the payoffs for each individual.

Step 3: Repeat Step 2 until the number of repetitions reaches the maximum allowed number of repeated games.

Step 4: Calculate the fitness of each individual.

Step 5: Generate random scenarios and calculate the dissatisfaction of the evolutionary scheduling decision for the current generation. If the dissatisfaction exceeds the maximum tolerance value, dynamically adjust the population distribution probabilities.

Step 6: Perform selection, crossover, and mutation for the two populations to generate the next generation.

Step 7: Repeat Steps 2–6 until the entire population reaches evolutionary stability or the maximum evolutionary generations are reached.

6.2 Distributed Scheduling Strategy

With the advancement of renewable energy utilization technologies, an increasing number of distributed energy units are being integrated into energy systems, bringing significant economic and social benefits. Distributed energy grid clusters encompass a local cluster of loads, distributed resources, BESSs, and increasingly, multi-energy interactions such as electricity, heat, and gas. It provides an effective means to integrate various distributed resources, particularly renewable energy systems. Currently,

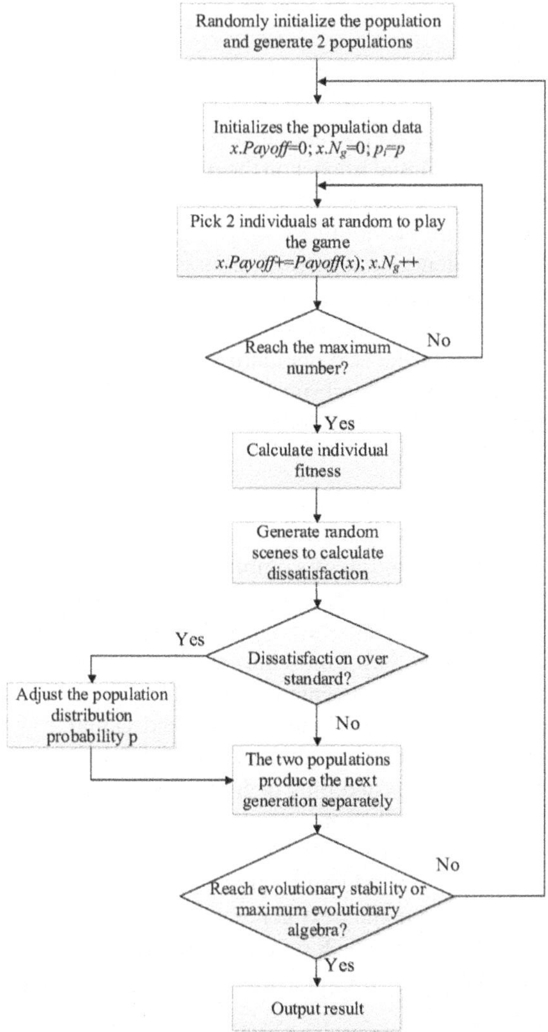

Fig. 6.3 Algorithm flow chart

DEGs have been widely adopted in many countries, and distributed energy grid clusters have inevitably become a key approach for scaling up the utilization of distributed resources across wide areas.

This section proposes a distributed autonomous scheduling strategy based on a virtual leader agent. This strategy combines the contact line power optimization, distributed consensus iteration, and reference point tracking control technologies under the MAS framework, coordinating the distributed energy grid with distributed resources. By utilizing the synchronous Alternative Direction Multiplier method (ADMM), optimization problems with boundary coupling constraints are transformed into decoupled sub-problems. With the help of the virtual leader, the scheduling of distributed resources and interconnected DEGs is integrated. The distributed optimization results proposed in this section can approach the global optimal solution obtained by traditional centralized algorithms. On the other hand, this method avoids sharing participants' private information and requires a control center with dedicated wide-area communication.

6.2.1 Multi-agent-Based Hierarchical Control Framework for Distributed Energy Grid Clusters

In recent years, distributed collaborative scheduling strategies characterized by plug-and-play features have received increasing attention. The European Integrated Research Program introduced the concept of the "web-of-cell" at the International Conference on Electrical Engineering (Martini 2015b). In this framework, a "cell" is defined as a flexible combination of distributed resources within a certain voltage level or geographic boundary (Morch et al. 2015; Syed et al. 2015). Through information sharing between cells, point-to-point decentralized decision-making is performed, thus achieving global optimization. Furthermore, it allows for timely and effective adjustments to respond to the internal fluctuations of renewable energy sources and loads, thereby achieving dynamic collaboration within the cell.

On the other hand, scalable architectures not only provide a plug-and-play interface for distributed resources but also significantly enhance system reliability. The sudden interruption of a communication line does not affect the system's stability, avoiding the issue where centralized controllers may cause the control system to collapse due to grid attacks (Xu and Li 2014). Therefore, multi-agent distributed scheduling with autonomous capabilities is also the preferred method for multi-energy grid scheduling.

Wang et al. (2014) only proposed a method for adjusting the contact line power for ESSs, while further coordination of other adjustable distributed resources can be achieved through rolling correction. Lv and Ai (2016) proposed centralized optimization for active superior energy grids, where reference point tracking control can improve the response speed and control accuracy in this centralized scheduling mode. Binetti et al. (2014) considered transmission losses in distributed economic dispatch. This chapter deeply analyzes the decentralized scheduling problem for interconnected energy grids. The impact of transmission losses is considered in the contact line optimization, and a distributed consensus lambda algorithm is designed. Combined with the virtual leader agent, it achieves coordination of contact line power tracking and rolling optimization objectives. Additionally, a second-order consensus algorithm is designed to accelerate reference tracking.

This study adopts a MAS architecture to design a hierarchical control framework. The distributed energy grid clusters system, consisting of interconnected DEGs, provides an effective means for seamless integration of wide-area loads, DGs, BESS, and other energy forms, such as heating and gas networks. Through energy interactions, significant benefits can be achieved. The target system is a medium-scale energy system. With the proliferation of distributed energy grid clusters, many small distribution systems on the user side are equipped with adjustable distributed resources. Uneven load distribution leads to a wide distribution of DGs, necessitating autonomous optimization of energy interactions and distributed scheduling across distributed energy grid clusters.

A MAS is a computational system composed of multiple interacting intelligent agents whose intelligence is manifested in systematic processes or algorithm searches. Each agent is adaptive and can be used to solve problems that are difficult or impossible for individual agents or the overall system to solve (Azevedo et al. 2017; Shafiee et al. 2016). This section proposes a decentralized, self-regulated scheduling strategy based on multi-agent consistency, which can effectively coordinate all adjustable resources automatically. The specific functions of the four types of agents are:

(A) Management Agent (MA): A coordination agent for superior energy grid and distributed energy grid clusters. As a connection agent for large-capacity energy systems, it has three functions: First, it can directly receive control signals from the administrator and switch modes via interconnected switches. Second, it serves as a virtual distributed energy grid clusters' control center agent, collecting price signals from the grid and broadcasting them as fixed marginal costs to other DEGs. Third, MA has a fault knowledge base and the ability to actively manage emergency events.

(B) Distributed energy grid Control Center Agent (DEGCCA): The coordinator for local and neighboring DEGs. λ iteration takes place between the local and neighboring DEGs. Additionally, during the solution process of the subproblems, boundary state information is exchanged between the local distributed energy grid control center and the neighboring distributed energy grid control centers. Using this communication method, a distributed optimization algorithm optimizes contact line power.

(C) Dominate Distributed Resource Agent (DDRA): The virtual leader agent within a distributed energy grid. Highly flexible distributed resource can be set as DDRA. Generally, Battery Energy Storage Systems (BESS) have a large adjustable capacity and fast regulation capability. BESS is typically configured at the PCC, making it suitable for obtaining contact line power deviations. As such, BESS is ideal as DDRA. Based on the actual contact line power and the optimized contact line power deviation from DEGCCA, DDRA introduces global iterative correction terms in the consistency iteration to track the reference contact line power of the distributed energy grid clusters. In addition, DDRA also has the same functions as a distributed resource agent (DRA).

(D) DRA: DRA participates in consistency iteration within the distributed energy grid. The reference energy of distributed resource can be calculated based on the iterative micro-increase rate. At the same time, λ is updated based on real-time generation. Once the consistency is reached, the second-order consistency algorithm automatically tracks the power output of distributed resource.

6.2.2 Multi-agent-Based Distributed Scheduling Model

The distributed optimization of contact line power between distributed energy grid clusters is based on the ADMM (Boyd et al. 2011). This algorithm is suitable for distributed optimization of large-scale distributed network physical systems. ADMM does not require the configuration of a centralized computing center; its computation process can be fully distributed across local agents (Ma et al. 2018).

This section first establishes a cost model for adjustable distributed resources and clusters. Then, the sub-gradient descent algorithm is used to update the contact line power. Additionally, this study synchronizes the traditional ADMM algorithm to achieve asynchronous updates of boundary states.

The cost of a typical dispatchable unit is generally defined as a quadratic function. The marginal cost of the distributed resource unit i can be expressed as:

$$\lambda_i = \begin{cases} 2a_i P_{i\_\min} + b_i, P_i = P_{i\_\min} \\ 2a_i P_i + b_i \\ 2a_i P_{i\_\max} + b_i, P_i = P_{i\_\max} \end{cases} \tag{6.16}$$

where $P_{i\_\min}$, $P_{i\_\max}$ represents the operating range of the distributed resource unit i, and a_i, b_i are the coefficients of the quadratic and linear terms of the function.

For distributed energy grid clusters with m distributed resource units in the system n, the operation cost can be expressed as:

$$C_n(x) = \sum_{m=1} C_{DG,n}^m(x_{DG,n}^m) + C_{BESS,n}(x_{BESS,n}) + \sum_{j=1}^{n} C_{\text{heat},j} + \sum_{k=1}^{p} C_{\text{fuel},k} \tag{6.17}$$

$$\begin{cases} C_{DG,n}^m(x_{DG,n}^m) = a_n^m (x_{DG,n}^m)^2 + b_n^m (x_{DG,n}^m) + c_n^m \\ C_{BESS,n}(x_{BESS,n}) = \lambda_*[1 + \sigma(Soc_{BESS,n}, x_{BESS,n})]x_{BESS,n} + c_{BESS,n} \end{cases} \tag{6.18}$$

where σ represents the depreciation rate of the ESS, λ_* denotes the real-time average marginal cost of all distributed resources in the clusters, and c_{BESS} represents the average cost of the BESS. $C_{\text{heat},j}$ is the heating cost of the heating unit j, $C_{\text{fuel},k}$ is the fuel cost of the k-th energy source.

6.2.3 Case Study Verification Based on Multi-agent Design

A case study was proposed to verify the effectiveness of the proposed approach. The optimization results of the distributed computation for the contact lines were compared with the optimal results of centralized computation, where the results from the distributed computation were found to be very close to the optimal outcomes. Following this, joint simulations of the discrete consensus algorithm for λ iteration and the second-order consensus algorithm for reference tracking were performed. This significantly improved the convergence. Finally, the impact of delay on the consensus algorithm was studied.

To test the feasibility of the proposed decentralized scheduling strategy, a study strategy based on decentralized scheduling was constructed as an early proof of concept. The numerical example is based on a distributed energy grid clusters system composed of 5 DEGs which includes various types of equipment such as DG, GB, HP, FC, CHP units, Energy Aggregation and Conversion Equipment (EACE). In this system, the default distributed energy grid 1 serves as the leader DDRA for all the corresponding DEGs. The architecture of the MAS is as described earlier (Fig. 6.4).

The specific operating cost parameters for all distributed resource units are shown in Table 6.1.

The decentralized scheduling strategy is compared with the traditional scheduling methods, and the decentralized optimization mentioned above is made equivalent to a centralized optimization problem.

Eight scenarios were designed to compare the decentralized method proposed in this chapter with the traditional centralized method. These 8 scenarios (see Table 6.2) were optimized using both decentralized and centralized methods.

The economic costs of the two methods are compared in Table 6.3.

From the optimization results, under high-precision convergence criteria, the results of the distributed computation method can approximate the optimal solution, with the maximum error rate being only 0.036%.

The simulation results demonstrate the application prospects of this algorithm in optimizing the internal operations of DEGs. In the simulation, the first distributed energy grid was used as the research object. The power adjustment settings were made for different time periods: $C_{adjust,1} = 320¥$ during $t \in [0, 30)$, $C_{adjust,1} = 420¥$ during $t \in [30, 60)$, and $C_{adjust,1} = 370¥$ during $t \in [60, 90]$.

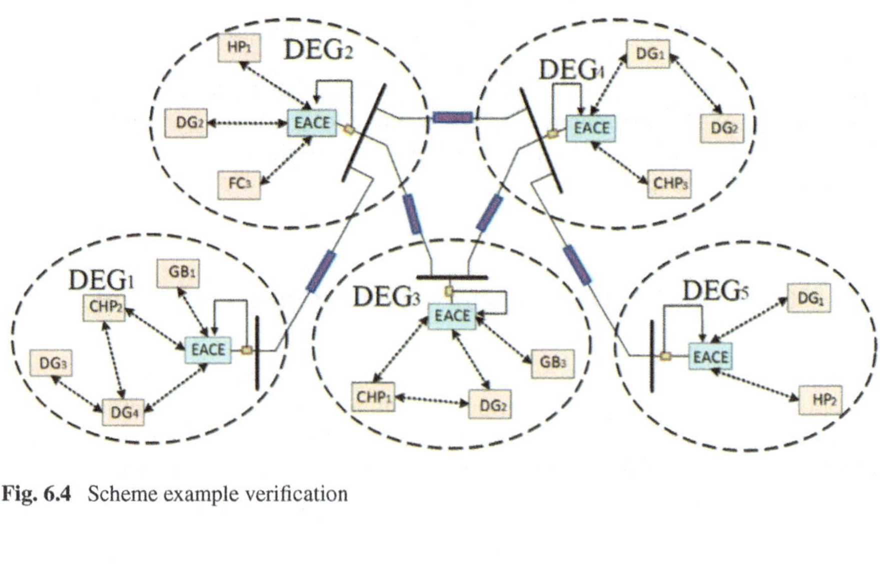

Fig. 6.4 Scheme example verification

Table 6.1 Operating cost parameters for distributed energy grid clusters

DEG	DG	a_i	b_i	c_i
DEG_1	EACE	0.000862	0.342	25
	GB_1	0.00111	0.462	5
	CHP_2	0.00156	0.323	60
	DG_3	0.00119	0.275	40
	DG_4	0.00134	0.329	20
DEG_2	EACE	0.00147	0.342	25
	HP_1	0.00111	0.462	5
	DG_2	0.00156	0.173	60
	FC_3	0.00134	0.329	20
DEG_3	EACE	0.00147	0.342	25
	CHP_1	0.00111	0.462	5
	DG_2	0.00156	0.173	60
	GB_3	0.00134	0.329	20
DEG_4	EACE	0.00147	0.342	25
	DG_1	0.00158	0.319	10
	DG_2	0.00120	0.275	60
	CHP_3	0.00134	0.329	20
DEG_5	EACE	0.00158	0.319	10
	DG_1	0.00120	0.329	20
	HP_2	0.00111	0.462	5

Table 6.2 8 Typical running scenario parameters

$C_{adjust,i}$	DEG_1	DEG_2	DEG_3	DEG_4	DEG_5
1	460	360	260	280	180
2	420	380	350	190	250
3	370	340	320	240	210
4	320	220	170	220	200
5	430	350	200	270	170
6	380	420	300	400	320
7	490	410	240	310	290
8	380	290	190	440	220

Distributed consensus control is an effective tool to alleviate the burden on distributed energy grid clusters. However, transmission delays are ubiquitous in almost all communication means. In practical control systems, if the matrix diagonal is not zero, discrete consensus will converge. The convergence speed depends on

Table 6.3 Comparison of centralized and distributed optimization results

Scene	Centralized method	Decentralized method	Percentage error (%)
1	1002.2147	1002.0847	0.01297
2	1029.1589	1029.0385	0.01166
3	970.3164	970.1765	0.01443
4	793.7721	793.5827	0.02386
5	938.9354	938.8454	0.009585
6	1157.3389	1156.9589	0.03283
7	1111.9276	1111.8376	0.008094
8	991.6069	991.2469	0.03630

the spectral radius of the matrix, i.e., the second-largest eigenvalue. The smaller the radius, the faster the convergence, and the fewer iterations required.

Based on the topology of the example system, a dual-random matrix is constructed. The delay situation is discussed (for $\tau = xT_s x = 0, ..., 60$). The sampling interval T_s is set to 200 ms.

From the eigenvalues of the dual-random matrix, as the arrows along the direction of the arrow gradually increase, the spectral radius of the matrix increases. With the increase in zero delay, the corresponding eigenvalue changes from real conjugate complex numbers to gradually approach the origin of the unit circle, ultimately converging to the point (1,0).

As the average spectral radius increases, periodic forced oscillations are more likely to occur. When there is no communication delay, the system behaves normally. However, as communication delay increases, the average spectral radius increases, and oscillations are enhanced. Although the control performance deteriorates, discrete iterations can still converge under time delays. According to Fig. 6.5, small delays also significantly reduce the system's stability margin. Therefore, it is crucial to minimize delays as much as possible.

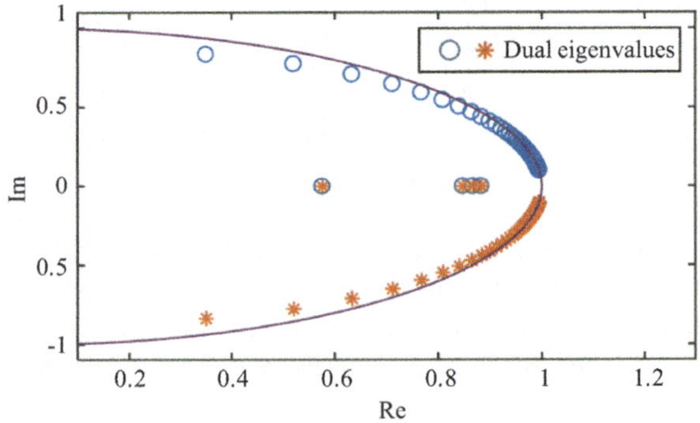

Fig. 6.5 Zero locus of a double random matrix

To reduce the impact of delays, the following measures can be taken in practical systems:

(a) Use Modbus communication or reliable Point-to-Point (P2P) communication to reduce delays and packet loss.
(b) Appropriately increase the sampling/iteration time interval.
(c) Through graphical analysis, reasonably plan the network structure and the locations where distributed resource units are connected.
(d) Construct dual-random matrices that are insensitive to delays and optimize the eigenvalue distribution of matrices with large delays.

This section presents a decentralized, self-regulating scheduling strategy for interconnected energy grids based on a virtual leader agent. The fully decentralized strategy avoids the sharing of participants' private information and truly achieves the goal of "decentralization." Under virtual leadership, rolling corrections ensure the economic operation of the system under renewable energy and load fluctuations. This solution can effectively reduce total generation costs and maintain the supply–demand balance within the system in real time.

Compared to existing literature, this chapter has several advantages and contributions:

(a) Transmission line losses are considered, and the contact line power of distributed energy grid clusters is constrained by the boundary coupling relationship.
(b) A discrete consensus algorithm is used to design distributed λ iteration, forming a team response to scheduling.
(c) Under the leadership of DDRA, a high-sensitivity second-order consensus iteration algorithm is adopted for reference point tracking.
(d) Numerical simulations show that this method supports plug-and-play functionality and analyzes the impact of delays on convergence performance.
(e) However, the complexity of this method also brings some limitations. The decentralized scheduling strategy in this chapter requires proper coordination and scheduling of all agents, and the saturation of power electronic systems and communication system failures can pose potential risks.

6.3 Multi-energy Demand Response

With the increasing popularity of large-scale intermittent renewable energy, regional energy systems are moving towards faster development, denser interactions, and more flexible transactions. However, existing energy systems face problems such as poor scalability, difficult interaction, and single transaction modes. The DR in distributed energy grid clusters provides a new idea to solve these problems by utilizing flexible multi-energy flow and information flow interactions.

DR aims to motivate and encourage users to dynamically adjust their energy consumption based on the balance between energy supply and demand, and is seen as an effective way to tap into the flexibility of the demand side. DR aggregators (DRA) can achieve scheduled or real-time demand reduction through different methods, including the use of real-time electricity prices, peak prices, demand-side bidding, and interruptible loads, which can fully leverage the complementary advantages of multiple regions. Current DR applications can be divided into two categories: DLC and Dynamic Price Incentives.

The increase in distributed devices has made centralized DR interaction and global control difficult due to issues like adjustment delays, privacy protection, and incentive price formulation. Therefore, in order to effectively utilize flexible resources on the user side, it is necessary to aggregate their response capabilities to increase adjustable capacity and coordinate distributed DR through local DRAs to coordinate the multi-energy demand of users across a wide area. Since DR programs will ultimately involve a large number of residential, industrial users, and distributed energy resources on the user side, the DR center needs to collect all necessary information from the distribution terminals and optimize the response plan. The limited computational capability and high system expansion costs make centralized response strategies unsuitable for large-scale, decentralized regional energy systems.

To address the distributed DR problem, Zhu and Martínez (2012) develops a distributed adaptive diffusion algorithm. Latifi et al. (2017) uses a randomized ADMM to achieve a fully distributed DR solution through low-cost wireless networks.

This section mainly covers the following four aspects:

(a) An online inverse optimization algorithm is developed to dynamically perceive user elasticity parameters using historical data. This method is more flexible.
(b) The proposed DR strategy does not require initialization and is fully distributed, avoiding node blocking.
(c) General multi-energy system applications based on MASs, such as those in Morstyn et al. (2018) and Hao et al. (2018), are designed based on rules but lack convergence and optimality proofs. This chapter provides proofs for convergence and optimality in such cases.
(d) Conventional dual decomposition or ADMM distributed algorithms (Xu et al. 2018; Ma et al. 2018) fail to solve global inequality constraints. The D-PPDS algorithm can better solve problems with global inequality constraints in a distributed manner.

6.3.1 Design of Distributed Demand Response Framework

To enable multiple DRAs to fully leverage their collaborative and integration advantages, this chapter adopts a distributed DR framework. For comparative analysis, the traditional centralized and the distributed DR frameworks are discussed in this chapter.

Unlike the centralized framework where each DRA must delegate scheduling authority to a central control center, the distributed framework does not require the deployment of a central control center. By designing an appropriate distributed

DR algorithm, the superiority of the distributed interactive framework can be fully leveraged to achieve a globally optimal response solution.

In the distributed DR framework presented in this chapter, the MA in the distributed framework is treated as the leader, while the DRAs are considered as followers. Distributed DR is achieved through hierarchical optimization between the two. The distributed MA manages only real-time prices and DR constraints directly communicated with any DRA. To reduce communication costs and improve system scalability, this agent does not directly communicate with each DRA.

Additionally, multiple DRAs interact with each other. The DRA directly connected to the MA first changes the price variable upon receiving the signal, and then broadcasts this change to all DRAs via distributed singular consensus communication. Meanwhile, each DRA performs local optimization asynchronously based on the price, which gradually alters the interconnecting line power. The leader's optimization takes rescheduling costs into account and gradually adjusts the reference price. This process addresses the issue of global line deviation within the distributed algorithm.

The optimization problem of the DRA directly connected to the MA is reformed, cleverly introducing the upper-level problem constraints into the lower-level solution process through perturbation points and primal–dual gradient methods. Ultimately, this results in a distributed solution for the DR optimization problem with global variables and global inequality constraints.

6.3.2 Demand Response Modeling and Model Online Identification

The uncertainty of the user price-demand elasticity model has always been a challenge in the practical application of DR. To accurately describe the load characteristics and identify the corresponding model parameters, this section designs a data-driven process as a bi-level programming problem. The lower-level objective is to minimize the model identification error, while the upper-level objective is the DRA optimization problem. The complementary slackness conditions in the lower-level problem are considered as constraints for the upper-level problem.

(1) DRA Optimization

Define c_t as the real-time price. The consumers within the DRA maximize individual benefits, with the objective function being the total economic benefit of the region. It is assumed that the aggregated consumers adjust their energy usage plan by balancing the DR settlement benefits $\sum_{t\in\tau} c_t \Delta P_{i,t}^l$ and the changes in energy utility $\sum_{t\in\tau} \Delta F_{i,t}^l$.

$$\begin{cases} \max\limits_{\Delta P_{i,t}} U_i = \sum_{t\in\tau} c_t \Delta P^l_{i,t} - \Delta F^l_{i,t} \\ \qquad = \sum_{t\in\tau} -\alpha_{i,t} \Delta P^{l2}_{i,t} + (c^h_{i,t} - 2\alpha_{i,t} - \beta_{i,t}) \Delta P^l_{i,t} \\ s.t.\ g_j\left(\Delta P^l_{i,t}\right) \le 0,\ j \in \Omega \end{cases} \tag{6.19}$$

The decision variables of DRA i are influenced by the load growth rate and the consumption constraint set $g_j \le 0,\ j \in \Omega$. The decision variables are subject to load ramp-up rates and energy consumption limits, which define a feasible region for the user response model.

$$\begin{aligned} s.t.\ \Delta P^l_{i,t} \le \Delta \bar{P}^l_{i,t},\quad \Delta P^l_{i,t} \ge \Delta \underline{P}^l_{i,t}\ & t \in \tau = \{t : t = 1, \ldots, T\} \\ \Delta P^l_{i,t} - \Delta P^l_{i,t-1} \le r^u_{i,t} \qquad & t \in \tau^{-1} = \{t : t = 2, \ldots, T\} \\ \Delta P^l_{i,t-1} - \Delta P^l_{i,t} \le r^d_{i,t} \qquad & t \in \tau^{-1} = \{t : t = 2, \ldots, T\} \end{aligned} \tag{6.20}$$

where $\Delta \overline{P}^l_{i,t}$ and $\Delta \underline{P}^l_{i,t}$ are the maximum multi-energy increment and decrement, of all users in DRA i at time t. $r^u_{i,t}$ and $r^d_{i,t}$ are the corresponding ramp-up and ramp-down limits.

(2) DRA Online Inverse Optimization

The DR parameters of users can be obtained online through an inverse optimization process, meaning that the DRA can simulate the decision-making process of consumers to identify the users' response parameters. In inverse optimization, the Kuhn-Tucker conditions of the primal problem serve as the nonlinear complementary slackness conditions in the model, where μ^*_j is the Lagrange multiplier of the constraint g_j. Using this method, an inverse optimization model is constructed to determine the parameters of the users' DR behavior.

The online inverse optimization estimates the parameter vector $\phi_{i,t} = \{\alpha^l_{i,t}, \beta^l_{i,t}, \Delta \overline{P}^l_{i,t}, \Delta \underline{P}^l_{i,t}, r^u_{i,t}, r^d_{i,t}\}$ for the users in the DRA, which describes the decision-making behavior of the users. The time series of price and consumption data $(c^h_{i,t}, \Delta P^h_{i,t})$ maintain a fixed length and are updated automatically. Based on the least-squares principle, the optimal solution $\Delta P^*_{i,t}$ of this problem minimizes the Euclidean distance from the historical data $\Delta P^h_{i,t}$. The inverse optimization problem is described as follows:

$$\min_{\phi_{i,t}, \Delta P^*_{i,t}} \sum_{t\in\tau} \left|\Delta P^*_{i,t} - \Delta P^h_{i,t}\right|^2$$

$$\begin{cases} s.t.\ \nabla U_i(\Delta P^*_{i,t}) + \sum\limits_{j\in\Omega} \mu^*_j \nabla g_j\left(\Delta P^*_{i,t}\right) = 0 \\ \mu^*_j \ge 0, \qquad\qquad j \in \Omega \\ \mu^*_j = 0 \perp g_j(\Delta P^*_{i,t}) = 0, j \in \Omega \end{cases} \tag{6.21}$$

where $g_j(\Delta P^*_{i,t}) \leq 0$ represents all the constraints in the user response model. The constraints in the above equation are used to provide the optimality conditions for the DR aggregation optimization problem $\Delta P^*_{i,t}$.

The online inverse optimization proposed in this section combines the advantages of both model-driven and data-driven approaches. It can output data models with practical significance and also perform online data rolling. In contrast, data-driven "black-box" models cannot provide a determined data model, making further DR optimization impossible. Additionally, traditional offline model identification introduces significant errors in practical applications due to its inability to adapt to model changes.

6.3.3 Multi-agent-Based Hierarchical Optimization Design for Distributed Demand Response

The distributed DR hierarchical optimization proposed in this section proceeds as follows: The leader agent first adjusts the price signals based on the reallocated costs and broadcasts them to the followers via distributed communication. Then, each follower determines the optimal strategy as the best reflection of the leader's strategy and executes its own strategy. Afterward, the leader perceives the interaction capabilities between the DRA agents and the power grid. Based on this feedback, the leader updates its strategy and broadcasts the price signal again. This interactive process occurs in real time. Under the perturbation of external price signals, all strategies are readjusted. This game can be formulated as a bi-level optimization problem, which will be described in the following subsection.

(1) Leader Optimization Model

Based on real-time energy price provided by the grid, it encourages users to participate in DR and formulates the scheduling plan for the previous day based on the all types of energy price signals. Considering the rescheduling costs, each DRA gives a correction price λ_t is the decision variable in this stage. $\Psi_L = \{\lambda_t, t \in \tau\}$. Adjustable equipment includes: GT, WB, EB, hydrogen electrolytic tank (HE), FC, hydrogen-storage tank (HT), thermal storage tank (TT); and the energy balance constraints of electricity, heat and hydrogen can be described as

$$\begin{cases} P^0_{\mathrm{GT}}(t) + P^0_{\mathrm{FC}}(t) + P_{\mathrm{EG}}(t) + P_{\mathrm{W}}(t) + P_{\mathrm{PV}}(t) + P^0_{\mathrm{LB}}(t) \\ \quad = P_{\mathrm{eload}}(t) + P^{\mathrm{I}}_{\mathrm{HE}}(t) + P^{\mathrm{I}}_{\mathrm{LB}}(t) + P^{\mathrm{I}}_{\mathrm{EB}}(t) \\ P^0_{\mathrm{WB}}(t) + P^0_{\mathrm{EB}}(t) + P^0_{\mathrm{TT}}(t) = P_{\mathrm{tL\,oad}}(t) + P^{\mathrm{I}}_{\mathrm{TT}}(t) \\ P^0_{\mathrm{HE}}(t) + P^0_{\mathrm{HT}}(t) = P_{\mathrm{hload}}(t) + P^{\mathrm{I}}_{\mathrm{FC}}(t) + P^{\mathrm{I}}_{\mathrm{HT}}(t) \end{cases} \tag{6.22}$$

The upper corner marks I and 0 indicate the input and output energy flows of the corresponding device respectively Energy flow; $P_{\mathrm{eload}}(t)$, $P_{\mathrm{tL\,oad}}$ and P_{hload} are respectively the electrical, thermal and hydrogen loads at time t. The goal of the DRA leader is to maximize total profit:

$$\max_{\lambda_t} U_{Leader} = \sum_{t\in\tau} \lambda_t \Delta P_t - \sum_{t\in\tau} c_t \Delta P_t - \sum_{t\in\tau} \kappa \Delta P_t^2 \tag{6.23}$$

where ΔP_t is the variation in the interconnection multi-energy between the DRA and the grid. $\Delta P_{i,t}^g$ is the generation electricity and heat change of the dispatchable units in DRA i. The first term of the objective function represents the DR revenue. The second term represents the cost of purchasing energy from the real-time market. The third term is the square of the interconnection power deviation, representing the penalty for rescheduling. κ is a coefficient indicating the unit cost of power rescheduling.

When the objective function is maximized, the linear optimization result can be obtained:

$$\lambda_t = c_t - 2\kappa \Delta P_t \tag{6.24}$$

After the leader agent optimization, it sends the information λ_t to the DRA agents that are connected to the leader agent via communication links. Additionally, the total response amount ΔP_t must satisfy the global DR constraint, as specified in the following equation, which defines the demand for the total response.

$$\begin{cases} \Delta P_t \geq \Delta P_{t,-} \ \ if\ \Delta P_t < 0 \\ \Delta P_t \leq \Delta P_{t,+} \ \ if\ \Delta P_t \geq 0 \end{cases} \tag{6.25}$$

The maximum energy reduction of the grid is denoted as $\Delta P_{t,-}$, and the maximum demand increase of the grid is denoted as $\Delta P_{t,+}$. When the energy price is high, the grid aims to inject more energy through the interconnection line $\Delta P_t < 0$. When the electricity price is low, the grid wants the users to consume more energy $\Delta P_t \geq 0$.

(2) Follower Optimization Model

For each follower of the leader, let the strategy set for the followers be $\Psi_i = \{\Delta P_{i,t}^l, \Delta P_{i,t}^g, t \in \tau\}$. The follower's objective function consists of four parts: the cost of dispatchable distributed resource, the load utility function, the DR payment to the DRA, and user satisfaction evaluation. The objective is to maximize the total utility:

$$\max_{\Psi_i} \boldsymbol{u}_i = \sum_{t\in\tau} \lambda_t \Delta P_{i,t} + \Delta F_{i,t}^l(\Delta P_{i,t}^l) - \Delta C_i^g(\Delta P_{i,t}^g) + Q_i(d_{i,t}^l, \Delta P_{i,t}^l) \tag{6.26}$$

With reference to load utility, the operating cost of the dispatchable distributed resource can be modeled as a quadratic function with polynomial coefficients $(\alpha_{i,t}^g, \beta_{i,t}^g)$.

$$\Delta C_i^g = \alpha_{i,t}^g \Delta P_{i,t}^{g2} + (2\alpha_{i,t}^g + \beta_{i,t}^g)\Delta P_{i,t}^g \tag{6.27}$$

where $Q_i(d_{i,t}^l, \Delta P_{i,t}^l)$ is the user satisfaction evaluation. It quantitatively describes the user's satisfaction with the difference between multi-energy demand and actual supply. As the actual load reduction increases, the function value increases. When the actual load is lower than the predicted demand, the user is more satisfied and $Q_i(d_{i,t}^l, \Delta P_{i,t}^l)$ is positive. When $Q_i(d_{i,t}^l, \Delta P_{i,t}^l)$ is negative, it indicates the user is dissatisfied with the multi-energy supply limit. As the supply limit increases, satisfaction tends to saturate, and as the actual multi-energy demand continues to increase, the rate of decrease of satisfaction slows down. The satisfaction function used in this book is as follows:

$$Q_i\left(d_{i,t}^l, \Delta P_{i,t}^l\right) = d_{i,t}^l m_i \left[\left(\frac{\Delta P_{i,t}^l + d_{i,t}^l}{d_{i,t}^l}\right)^{r_i} - 1\right] \tag{6.28}$$

where $r_i > 0$ and $r_i m_i < 0$ are user-specific parameters. Therefore, this function satisfies the three conditions listed below:

(1) if $\Delta P_{i,t}^l = 0$, $Q_i(d_{i,t}^l, \Delta P_{i,t}^l) = 0$.
(2) $\partial Q_i / \partial \Delta P_{i,t}^l < 0$.
(3) $\partial^2 Q_i / \partial \Delta P_{i,t}^{l2} > 0$.

6.3.4 Distributed Communication and Algorithm Considering Global Predefined Constraint

This section designs a distributed communication framework and a fully distributed algorithm to solve the above optimization problem. The Singular Consensus Algorithm is used to broadcast key global information. The Singular Consensus Algorithm broadcasts price changes to each DRA, and the DRAs use the D-PPDS algorithm to consider global inequality constraints, causing all DRAs to converge to the optimal solution that satisfies all global constraints. The local DR based on the optimization results will change the interconnection line power and other energy change, the leader will then adjust the price signal in real time. These steps can all proceed asynchronously without requiring timing coordination.

(1) Distributed Communication of Price Signals

A directed weighted graph $\zeta = (\upsilon, E, \boldsymbol{W})$ is constructed for DRA agent communication. E is the set of all valid communication links in the graph. For each $i = 1, 2, \ldots, N$, every DRA interacts with its neighbors to exchange price signals λ_t to update the real-time price signals.

A doubly stochastic matrix is constructed to broadcast price signals to all follower agents (DRAs) in a distributed manner. When the eigenvalues of the matrix are less than or equal to 1, the system's state converges to the average value. This chapter uses the Metropolis construction method of the doubly stochastic matrix (Meng et al. 2016). In the discrete Singular Consensus Algorithm, the marginal cost is the iterative state variable.

(2) Distributed Perturbation Primal–Dual Sub-gradient (D-PPDS) Method

Due to the absence of a centralized control center, conventional algorithms face challenges in achieving parallel distributed processing while simultaneously satisfying global inequality constraints through limited information exchange. It is impossible to directly satisfy the global constraints, which are composed of nearly all decision variables. To overcome this, the chapter introduces the D-PPDS algorithm, which estimates global perturbation points using local consistency. The algorithm ensures global convergence and accurate perturbation point estimation. The following steps outline the D-PPDS process:

(a) Average Consensus Estimation: The communication graph of each DRA is represented by a directed weighted graph. D-PPDS updates auxiliary variables and price signals through distributed communication. Each DRA interacts with its neighbors to estimate average values of the auxiliary variables, despite not having complete global inequality constraint information.
(b) Perturbation Point Calculation: Since global constraints are linear and smooth, each DRA updates its local perturbation point using a specific formula, projecting out-of-feasible points to the boundary of the feasible region.
(c) Primal–Dual Perturbation Sub-gradient Update: Based on the updated perturbation points, each DRA updates its primal and dual variables, using a fixed step size and initialization within a given range.
(d) Auxiliary Variable Update: Each DRA updates its auxiliary variables based on changes in local objective functions and global inequality constraints, with initial values specified.
(e) These steps enable the D-PPDS algorithm to efficiently compute the optimal solution while addressing global constraints in a distributed manner.

6.3.5 Case Study Analysis

An example is constructed based on a real industrial park in Shandong, China. Historical load data from five major user groups (enterprises) in this region at different multi-energy price levels were surveyed. Using online inverse optimization, a real-time multi-energy elasticity model was established. Based on the distributed resource costs obtained from the survey and historical load data from a specific day, the performance of the D-PPDS, ADMM, and C-PDS algorithms was compared within the framework of distributed DR, considering global inequality constraints. The optimization results and response benefits were analyzed, verifying the effectiveness of the data-driven inverse optimization algorithm for online parameter identification in DR models (Fig. 6.6).

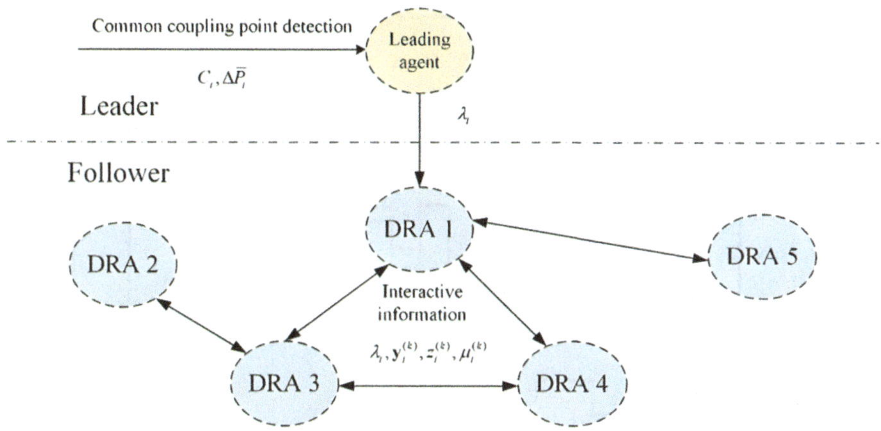

Fig. 6.6 Schematic diagram of multi-agent distributed communication links

(1) Distributed DR Optimization Process

To verify the superiority of the incomplete information global optimization algorithm proposed in the book, the distributed scheduling results of D-PPDS were compared with the general distributed algorithm (ADMM), as shown in Fig. 6.7.

D-PPDS estimates the pre-set DR constraints using local perturbation point estimates. In contrast, ADMM can only handle incomplete local constraints and cannot

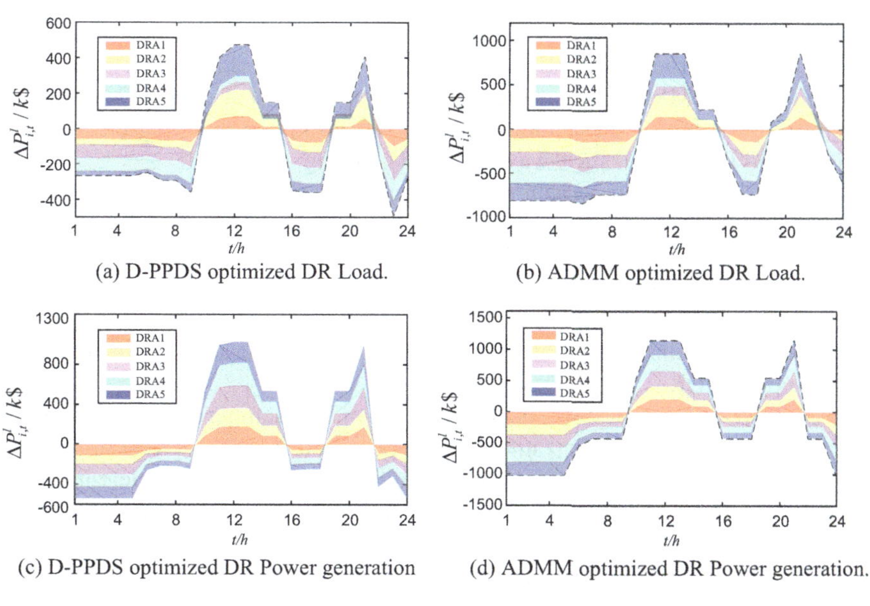

Fig. 6.7 Distributed scheduling results of D-PPDS and ADMM algorithms

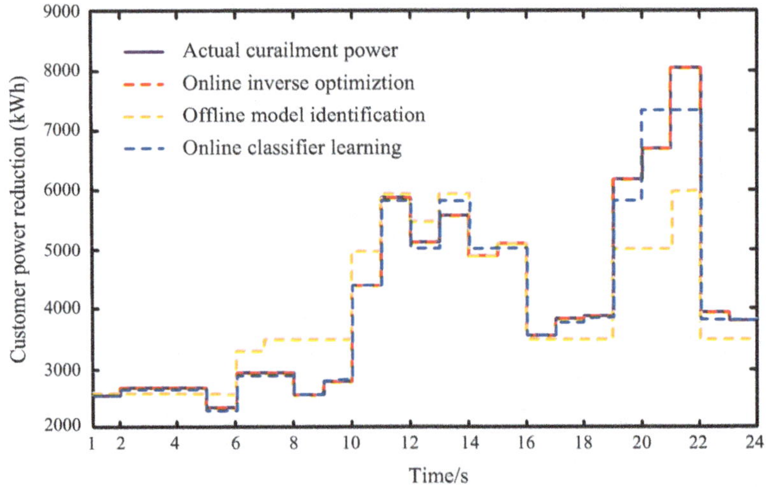

Fig. 6.8 Comparison of identification methods of customer power reduction and price model

address global constraints. Therefore, during certain time periods, the load and generation responses optimized by D-PPDS differ from those optimized by ADMM. The following section provides a detailed analysis of the causes and an economic comparison of the results.

(2) Comparative Analysis of Distributed DR Optimization

To verify the effectiveness of D-PPDS in distributed DR optimization, a comparison was made between the optimization results and DR revenues of D-PPDS, C-PDS, and the traditional ADMM algorithm (Xu and Li 2014; Wang et al. 2014). Furthermore, the impact of online inverse optimization identification on improving the accuracy of distributed DR optimization was also validated (Fig. 6.8).

Table 6.4 shows a comparison of the distributed DR optimization results of C-PDS and D-PPDS in five scenarios. Since centralized methods have access to all information and obtain the complete optimization problem, they can search for the optimal solution. In fact, across the 24 simulation scenarios, the optimization results of distributed DR were very close to the optimal solution, with the maximum error in all scenarios being only 1.11%. This means that D-PPDS can serve as an alternative for near-optimal DR.

To compare the accuracy of model identification, real sample data were identified using the online inverse optimization method proposed in this chapter, offline least-squares fitting model identification, and data-driven online classifier learning. The identified models were then applied to a real operational day. The results showed that offline identification, due to its inability to update over time, resulted in larger model errors. Online identification had higher accuracy, with the online inverse optimization model being slightly more accurate than the online classifier learning method. However, machine learning models, being "black-box" models, cannot be used for

Table 6.4 Comparison of DR results between C-PDS and D-PPDS

Items	DRA1	DRA2	DRA3	DRA4	DRA5
C-PDS	−148.26	−113.29	−191.38	−179.16	−161.04
D-PPDS	−147.16	−112.90	−190.82	−177.22	−159.94
Errors	0.65%	0.43%	0.29%	0.11%	0.68%
C-PDS	−150.36	−114.91	−193.82	−180.25	−159.62
D-PPDS	−148.96	−114.51	−191.75	−178.33	−158.07
Errors	0.93%	0.36%	1.07%	1.07%	0.97%
C-PDS	161.499	181.118	200.938	216.127	302.404
D-PPDS	161.714	179.739	199.715	215.782	300.382
Errors	0.13%	0.76%	0.61%	0.16%	0.67%
C-PDS	−88.183	−115.278	−129.367	−128.162	−139.631
D-PPDS	−88.259	−114.928	−129.904	−128.287	−139.009
Errors	0.086%	0.30%	0.42%	0.12%	0.46%
C-PDS	−85.794	−124.491	−119.351	−115.739	−154.596
D-PPDS	−85.258	−123.538	−119.045	−115.553	−153.189
Errors	0.62%	0.77%	0.26%	0.16%	0.91%

subsequent distributed DR optimization since they do not provide a deterministic data model. Additionally, offline identification did not meet the required accuracy, making both offline identification and online machine learning methods unsuitable for the application scenario in this chapter.

Although local constraints can be handled by the ADMM algorithm, ADMM cannot address global inequality constraints under incomplete information. D-PPDS, however, estimates the execution of global inequality constraints using local perturbation point estimates, achieving excellent results.

To analyze the DR scheduling results of D-PPDS with online inverse optimization, D-PPDS without online inverse optimization, and ADMM, a comparison of the optimized DR scheduling power and benefits of these algorithms was made. The multi-energy that does not exceed the response demand is settled according to real-time prices, while the multi-energy exceeding the demand is settled according to the previous day's prices. Detailed comparisons are shown in Figs. 6.9 and 6.10.

From Fig. 6.10, it is evident that ADMM can only decompose the global equality constraints into multiple boundary-coupling constraints based on the auxiliary problem principle. As a result, it not only has low iteration efficiency but also struggles to handle global inequality constraints. This causes the computed response power to sometimes exceed the target demand reduction/increment. On the other hand, D-PPDS, which optimizes the response power, never violates the inequality constraints provided by the scheduling. The difference between D-PPDS with online inverse optimization and D-PPDS with offline identification is relatively small.

On the other hand, in Fig. 6.10, compared to traditional offline identification, D-PPDS with online inverse optimization achieves the maximum DR profit. The

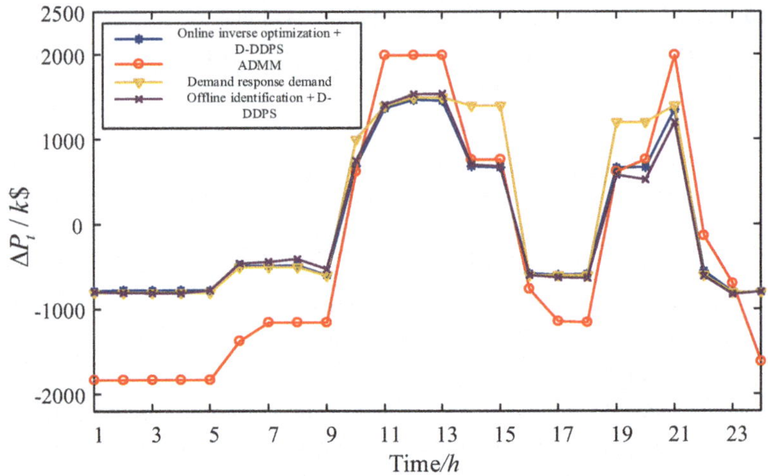

Fig. 6.9 Comparison of scheduling results between D-PPDS and ADMM algorithm

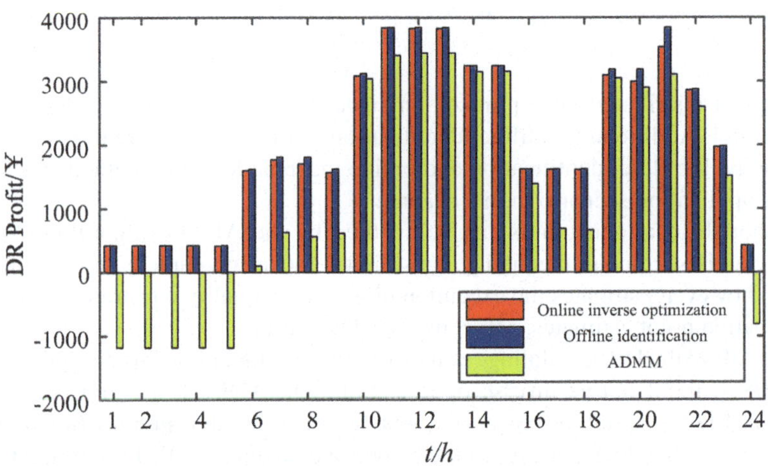

Fig. 6.10 Comparison of D-PPDS and ADMM algorithm scheduling benefits

increase in elastic parameter identification accuracy leads to a 1.83% higher DR profit when online optimization is considered, compared to scenarios without online optimization. This improvement arises because traditional model-driven algorithms generally cannot be updated in real-time, and incomplete information analysis results in a 39.03% decrease in DR profit. This demonstrates the superiority of online inverse optimization in parameter identification.

Furthermore, due to the failure to meet the pre-set constraints, any energy exceeding the response demand is settled at the previous day's price, which results in significantly poorer performance of the ADMM algorithm. The profit of the ADMM algorithm even drops to negative values when it deviates significantly from the constraints at certain times.

The incentive subsidies for adjustable loads and generation on the user side can bring significant benefits such as peak shaving, valley filling, improved stability, delayed investments, and enhanced utilization of renewable energy. This chapter proposes a multi-agent-based DR optimization method with global inequality constraints. By performing inverse optimization on the user distributed DR model, the method allows for online acquisition of more accurate user elasticity parameters. In the distributed optimization process, a discrete singular consensus algorithm is used to update the price auxiliary variables, and the D-PPDS algorithm is designed to ensure that the DRAs in the distributed optimization satisfy the global inequality constraints.

The distributed DR strategy proposed in this book has been verified through practical system simulations. According to the simulation results, online inverse optimization improves the parameter identification accuracy and brings a 1.83% increase in response. The distributed DR optimization problem, which incorporates global states and global inequality constraints, can be solved in a distributed manner without complete information exchange. The computational error compared to centralized optimal solutions is no more than 1.11%.

References

Bie Z, Zhang P, Li G et al (2012) Reliability evaluation of active distribution systems including distributed energy grids. IEEE Trans Power Syst 27(4):2342–2350

Binetti G, Davoudi A, Lewis FL, Naso D, Turchiano B (2014) Distributed consensus-based economic dispatch with transmission losses. IEEE Trans Power Syst 29(4):1711–1720

Boyd S, Parikh N, Chu E, Eckstein J et al (2011) Distributed optimization and statistical learning via the alternating direction method of multipliers. Found Trends Mach Learn 3(1):1–122

de Azevedo R, Cintuglu MH, Ma T, Mohammed OA (2017) Multiagent-based optimal distributed energy grid control using fully distributed diffusion strategy. IEEE Trans Smart Grid 8(4):1996–2008

Hao R, Jiang Z, Ai Q et al (2018) Hierarchical optimisation strategy in distributed energy grid based on the consensus of multi-agent system. IET Gener Transm Distrib 12(10):2444–2451

Hatziargyriou N (2009) Advanced architectures and control concepts for more distributed energy grids. In: ICCS

Kumar N, Doolla S (2013) Multiagent-based distributed energy-resource management for intelligent distributed energy grids. IEEE Trans Ind Electron 60(4):1678–1687

Lasseter R (2011) Smart distribution: coupled distributed energy grids. Proc IEEE 99(6):1074–1082

Latifi M, Khalili A, Rastegarnia A et al (2017) Fully distributed demand response using the adaptive diffusion–stackelberg algorithm. IEEE Trans Ind Inform 13(5):2291–2301

Lv T, Ai Q (2016) Interactive energy management of networked distributed energy grids-based active distribution system considering large-scale integration of renewable energy resources. Appl Energy 163:408–422

Ma WJ, Wang J, Gupta V, Chen C (2018) Distributed energy management for networked distributed energy grids using online ADMM with regret. IEEE Trans Smart Grid 9(2):846–856

Martini L (2015a) Trends of smart grids development as fostered by European research coordination: the contribution by the EERAJP on smart grids and the ELECTRAIRP. In: Proceedings of international conference on power engineering, energy and electrical drives, Milan, Italy, 8p
Martini L (2015b) Trends of smart grids development as fostered by European research coordination: the contribution by the EERA JP on smart grids and the ELECTRA IRP. In: 2015 IEEE 5th international conference on power engineering, energy and electrical drives (POWERENG), Riga, pp 23–30
Meng FL, Xi JX, Shi ZY et al (2016) Admissible output consensus control for singular swarm systems. Int J Syst Sci 47(7):1734–1744
Morch AZ et al (2015) Future control architecture and emerging observability needs. In: 2015 IEEE 5th international conference on power engineering, energy and electrical drives (POWERENG)
Morstyn T, Savkin AV, Hredzak B et al (2018) Multi-agent sliding mode control for state of charge balancing between battery energy storage systems distributed in a DC distributed energy grid. IEEE Trans Smart Grid 9(5):4735–4743
Nunna H, Doolla S (2012) Demand response in smart distribution system with multiple distributed energy grids. IEEE Trans Smart Grid 3(4):1641–1649
Saad W, Zhu H, Poor H et al (2012) Game-theoretic methods for the smart grid: an overview of distributed energy grid systems, demand-side management, and smart grid communications. IEEE Signal Process Mag 29(5):86–105
Shafiee Q, Nasirian V, Vasquez JC et al (2016) A multi-functional fully distributed control framework for AC distributed energy grids. IEEE Trans Smart Grid (99):1–3
Syed MH et al (2015) Demand side participation for frequency containment in the web of cells architecture. In: IEEE 2015 international symposium on smart electric distribution systems and technologies (EDST)
Wang D, Ge S, Jia H et al (2014) A demand response and battery storage coordination algorithm for providing distributed energy grid tie-line smoothing services. IEEE Trans Sustain Energy 5(2):476–486
Xu J, Sun H, Dent CJ (2018) ADMM-based distributed OPF problem meets stochastic communication delay. IEEE Trans Smart Grid
Xu Y, Li Z (2014) Distributed optimal resource management based on the consensus algorithm in a distributed energy grid. IEEE Trans Ind Electron 62(4):2584–2592
Zhu M, Martínez S (2012) On distributed convex optimization under inequality and equality constraints. IEEE Trans Autom Control 57(1):151–164

Chapter 7
Market Trading Strategies for Distributed Energy Grid Clusters

With the restructuring of the country's economic and industrial structure in the past few years, the demand for energy by enterprises and residents is increasing. Under the background of the rapid development of the country, energy in the form of electricity, heat, and gas has become a huge engine of the country's economic growth and an irreplaceable part of the daily life of the residents. In this environment, the development and expansion of energy companies have ushered in a favorable space for growth. In the past few decades, large energy grids have developed rapidly by their unique advantages, and have become the main supply channels for electricity, heat, and gas. With China's energy market reform, the separation of distribution and sales, the liberalization of the incremental distribution energy grid business, and the involvement of a variety of capital in the construction and operation of power generation projects, the participation of distributed energy grid clusters in the market will be an inevitable trend in the future development of the market. After the distributed energy grid cluster joins the electricity, heat, and natural gas market, the role of the distributed energy grid cluster is also converted into a market entity, and how to maximize its profit is the primary issue that each self-financing market entity must consider, so it is necessary to analyze and research on the market transaction, competitive offer and cooperative offer of the distributed energy grid cluster participating in the electricity, heat and natural gas market.

7.1 Analysis of the Current Status of Market Transactions

At present, many countries in the world are involved in the research and development of distributed energy grid clusters. Europe, the United States, Japan, and other developed countries and regions have completed the basic theoretical study of distributed energy grid clusters, completed the distributed energy grid cluster control and protection strategies, operation and economic analysis and verification

T. Lu et al., *Operation Modeling and Collaborative Regulation of Distributed Energy Grid Clusters*, Smart Energy Systems, https://doi.org/10.1007/978-981-96-9942-1_7

through experimental testing and on-site demonstration projects. With the maturity of distributed energy grid cluster technology, distributed energy grid cluster will not only be able to flexibly and reasonably carry out economic scheduling of internal distributed power supply but also participate in the competition of electricity, heat, and natural gas markets to solve the problem of a surplus of its power, heat, and natural gas energy, to achieve the reasonable and effective use of resources.

7.1.1 Electricity Market

7.1.1.1 Electricity Market Basic Concepts

Before conducting a study on the participation of distributed energy grid clusters in the electricity market, it is necessary to understand some basic concepts of microeconomics to understand the participation of distributed energy grid clusters in the electricity market competition. The following study and analysis are mainly based on demand and supply, and some basic economics terms are introduced to provide a theoretical basis for the participation of distributed energy grid clusters in the market competition.

(1) Balance Between Supply and Demand

Demand is the quantity of a good that a consumer is willing and able to purchase at each price level in a given period. The demand referred to in economic analysis is effective demand, which is the unity of desire and ability to buy. Demand is affected by several factors, including the price of the good itself, the price of related goods, the income level of the consumer, the consumer's hobbies or preferences, the consumer's expectations of the future, and socio-demographic conditions.

Similarly, supply is the product that producers are willing and likely to sell at a particular time and under certain price conditions, including newly supplied goods and own stock. Supply is the unity of desire and ability to supply. The factors affecting supply are more complex than those affecting demand and are both economic and non-economic. It is not only affected by the price of the commodity itself but also by the production objectives of the manufacturer, the price of related products, the cost of factors of production, the level of production technology, the government's macro-control policies, and so on.

Usually, the quantity demanded falls with the price increase, and the quantity supplied increases with the price increase. When the price rises when the demand is less and less, the price decreases when the supply also decreases. This phenomenon is called the law of supply and demand, i.e. demand and supply are linked to each other and influence each other. When demand is greater than supply, the price will rise and vice versa, the price will fall. However, the market will eventually reach an equilibrium point that the market equilibrium and the market supply will fluctuate around this equilibrium point up and down through regulation so that the market irregularity of the automatic regulation of a certain degree of regularity. When other conditions change, the market will establish a new equilibrium point (Li 2016).

(2) Market Clearing Price

For power companies, the tariff structure can be divided into supply tariff structure and regulation tariff structure. Supply price refers to the price between power companies, such as the price between power plants and power grid companies, power grid companies, and power companies. Regulated tariffs generally include seasonal tariffs, peak, and valley tariffs, and power factor adjustment tariffs. Market clearing refers to a market in which commodity prices are sufficiently flexible to allow demand and supply to equilibrate quickly. With this flexibility, there is no oversupply in either the electricity sales market or the electricity supply market. Each market is in or tends towards a general equilibrium where supply and demand are equal. There are two main types of clearing methods, namely, uniform price clearing MCP and offer-based clearing PAB. Under offer-based clearing, different generators or purchasers sell or buy electricity at different prices, which are related to the price they offer. Under uniform price clearing, all generators sell electricity at the same price and all power purchasers buy at the same price, with a kind of marginal clearing price settlement. Under both price mechanisms, market transactions are based on the offers of units or generators, allocating generation loads from low to high until the system supply–demand balance is met. The difference is that the offer settlement is based on the actual offer price, whereas the marginal clearing price is based on the offer price of the last unit that meets the system load balance, which is taken as the marginal price for settlement. In the electricity market settled by marginal clearing price, no matter how high or low the power generation company's offer is, once it is selected, it will be settled according to the marginal clearing price (Cui et al. 2016).

At present, most of the world's electricity markets have adopted a uniform price mechanism for settlement at the marginal clearing price (Ma et al. 2014), because the marginal clearing price is more efficient than the offer mechanism, more favorable to small and medium-sized power producers, and better able to prevent the exercise of monopoly power by large power producers to improve the stability of electricity prices in the electricity market. Under the marginal clearing price mechanism, generating units in a mature market will offer prices based on their marginal costs, so that the market clearing result can meet the constraints of power system security, thus realizing economic and optimal dispatch of the power system. In addition, the use of marginal clearing price offer also helps to improve the efficiency of clean energy consumption, because at present, the operating cost of conventional power sources such as thermal power is mainly the cost of fuel, which is becoming increasingly expensive and volatile, and with the improvement of the level of technology, the marginal cost of clean energy power generation has been close to zero. Therefore, in the spot market with marginal clearing price offers, clean energy will occupy an absolute advantage and become the first choice for peaking and dispatching in power spot trading (Peng 2014).

(3) Distributed Energy Grid Cluster Market Transactions Background

In a distributed energy grid cluster, the existence of electricity trading between DEGs will have an impact on the operation and management mode of the region. If there is (or tends to be) electricity trading between DEGs, on the one hand, the amount of electricity directly traded between DEGs and distribution grids will be reduced, and the content of the work of distribution grid scheduling and management may be changed; on the other hand, there is a need to have a specific organization to provide a platform for the trading of DEGs as well as the supporting hardware and software facilities, and the DEGs involved in the trading need to bear the corresponding costs at the same time. On the other hand, there is a need for a specific organization to provide a platform and supporting hardware and software facilities for transactions between DEGs, for which the participating DEGs will have to bear the corresponding costs. If there is no (or no tendency to) electricity trading among DEGs, individual DEGs can be regarded as distributed power sources with certain regulatory functions, and the distribution grid is responsible for the unified operation and management of DEGs. The existence of electricity trading among DEGs or all DEGs only have transactions with the distribution grid are the two most dominant and basic modes of distributed energy grid cluster operation (Bai et al. 2016).

DEGs use highly intermittent renewable energy sources as the main form of energy, which predisposes DEGs to have different power demands or supply at different moments. At the same time, the distributed energy grid may have power surplus at one moment and power demand at another moment, with the increase of distributed energy grid how to solve the power demand and power supply of distributed energy grid seems to be very urgent. At the same time, due to renewable energy power generation and weather and climate correlation being relatively large, different regions of the distributed energy grid generation characteristics are not the same, the same region of the distributed energy grid in different seasons and even different weather power generation characteristics of the difference between the characteristics are also large, to take full advantage of the distributed energy grid of the surplus power, efficient use of resources to protect the environment, the need for interconnection of the distributed energy grid, that is, the formation of a distributed energy grid cluster.

Interconnected DEGs can sell or buy power from other DEGs to meet their load demand. Through the interconnection of large power grids, DEGs can be complementary to each other, thus constituting the prototype of the electricity market with DEGs. In summary, it can be seen that the effective demand and supply of electricity between DEGs, with the conditions to participate in the market.

(4) Types of Distributed Energy Grid Cluster Participation in Market Transactions

In the electricity market environment, electrical energy is an electricity trading commodity. However, due to the special characteristics of electric energy commodities, under the current level of technology, it is difficult to economically realize large-scale storage of electric energy, and the production and consumption of electric energy can be said to occur almost simultaneously. At present, the electricity market mainly has contract trading, spot trading, futures trading, and other transaction types (Chen 2011).

(a) Contract Trading: Contract trading is often referred to as medium- and long-term contract trading, which is a forward physical delivery contract. Usually, the supply and demand sides through private negotiations, to reach a simultaneous satisfaction of the needs and objectives of both sides of the terms of the transaction involve a large amount of electricity, and a long period is generally used in this way of trading. The organization of the medium- and long-term contract market is to prevent market risks, reduce the price and grid security risks arising from complete reliance on the spot market, and ensure the orderliness of the electricity market, security, price stability, and the continuity of power production. It gives the grid company ample time to consider how to ensure grid security power generation companies have enough time to organize fuel procurement plans to maintain long-term stability of fuel supply, which is conducive to the development of generating unit overhaul plans trading centers have enough time to increase the depth of competition in power trading to facilitate the optimal allocation of power generation resources on a wider scale.
(b) Spot Trading: most of the electricity market in the electricity market can be traded through the decentralized open market, but the market cannot guarantee the reliability of the power system, to make up for the shortcomings of the open electricity market, the need to establish a centrally managed spot market. The centralized spot market can regulate the output of the more flexible power generation companies and reduce the demand of voluntary load-shedding users, to achieve the matching of power generation and load, and the centralized spot market must also be able to deal with some sudden or unpredictable accidents. Spot transactions in the centralized spot market are divided into day-ahead transactions and real-time transactions. Day-ahead trading is a day ahead of the future hours of power trading, the market completed most of the spot market power transactions while providing economic incentives to the power generation side of the prompt by real-time scheduling requirements for scheduling. However, at the same time, the safe and stable operation of the power system and power quality needs to be adjusted and balanced in real-time, thus requiring real-time trading of the balancing market, i.e. real-time market.
(c) Futures Trading: The power futures market refers to the trading of power commodities in the form of power futures contracts that are traded at a specific price, with delivery starting at a specific time in the future and completed within a specific period. In the power futures trading model, power futures prices and power spot prices can achieve unity, market participants can hedge through the power futures, avoiding the existing power trading in the price of electricity fluctuations on both sides of the impact of the transaction and the power producers to determine the stability of the production plan, and power consumers to ensure a stable supply of electricity. Stable supply and demand can weaken the market participants to conduct power of market speculation. It is conducive to stabilizing the price of electricity in the spot market to ensure stable supply, and demand at

the same time, both sides of the transaction can production, consumption plans for flexible adjustment, in the external environment changes, to better adapt to the market. Therefore, futures trading can well avoid the price risk of the electricity market and is conducive to the healthy and stable development of the electricity market.

7.1.1.2 Economic Behavioral Decisions for Distributed Energy Grid Clusters

In the economic and behavioral decision-making of distributed energy grid clusters, the establishment of feature extraction and behavioral analysis decision-making of distributed energy grid clusters is an important link. Intelligent analysis has already covered a variety of fields, data mining and analysis technology are one of the basic links, computer technology, communication technology, measurement technology, information control, and so on are cross-applied, and in practice to achieve the visualization operation. In terms of the technical application of the power consumption information collection system, the interaction between the electricity supply system and the electricity users has been realized through the intelligent technology information network. However, the research on data mining and behavioral analysis for distributed energy grid clusters is still in the preliminary stage due to its many challenges such as complex data structure and diverse sampling time (Zakariazadeh and Jadid 2014).

The basis of data mining and analysis is the information acquisition system. The establishment of the information acquisition system can improve the traditional manual meter reading operation mode, forming a fully closed innovation mode, that is, it can realize the real-time collection of electricity consumption of energy cells and online detection through the metering device. Systematic management is implemented from meter reading to accounting of electricity bills.

Based on the collection of system (big) data, combined with external system data to study the economic behavior of distributed energy grid clusters is a relatively large research topic. According to the characteristics of the collection system data, foreign scholars divided the research direction into two categories: predictive type and optimization type (Gkatzikis et al. 2013). Predictive type is mainly through the collection system historical data and the periphery of a variety of possible factors for multi-dimensional data analysis and modeling, from a wide range of factors to find the main factors affecting economic behavior, and at the same time based on the data analysis of the law of the future economic behavior of a limited number of reasonable predictions. For example: through the study of a certain range of relatively stable distributed energy grid electricity consumption data will be time and climate temperature change factors for multi-dimensional analysis, can be more accurate to get the distributed energy grid electricity consumption and temperature correlation between the law, then through the historical temperature data trends, can be predicted in the future within a year of the monthly electricity consumption of the user group. Optimization type is mainly through the study of big data, the traditional analysis methods cannot quantitatively analyze certain characteristics of the power grid or there is no data to support the management of the work, to find the exact data, to help business managers to carry out management optimization. For example, by analyzing the total meter power value of the collected public transformer stations, it is easy to get the curve of power and time, and by reasonably setting the threshold value, it can quickly get the number and details of normal transformer load rate, key monitoring and overloading of the selected samples, which will help the managers to formulate a reasonable inspection plan, and at the same time, it also provides a detailed data basis for the division of the station area to expand the capacity, which greatly improves the management efficiency. The following are some examples of how this system can be used to improve management efficiency.

In the process of commercial operation of the cluster, it faces the problem of coordinating the economic interests of different investors in distributed energy, which needs to be guided by market mechanisms such as competitive bidding. The formation of a regional bidding system by multiple autonomous DEGs will be conducive to the market-based management and economic operation of the power grid. Based on autonomous operation, the bidding function is realized, reflecting the interests of each distributed energy grid through market means and guiding the optimal operation of clusters.

Taking the multi-cluster data-based integrated operation system as the object of analysis, the self and interactive behaviors of each unit in the system are data-based modeled. In Fig. 7.1, each cluster is equipped with a local data center with its server and data management system. The data center regularly collects cluster data according to system requirements, and the data are divided into three types: distributed new energy generation data collected through the power prediction system, user electricity consumption data collected through the electricity consumption information collection system, and market transaction data collected through the electricity sales platform. These three types of data constitute the data analysis as well as the training samples for the behavior of each cluster. The whole operation system adopts a decentralized, distributed, and discrete data management mode, which allows the upper-layer operators to analyze and generalize the group economic behaviors of the clusters from the perspective of biomimetic analysis based on the group characteristics reflected in the lower-layer data network and provides theoretical support for the boundary modeling of different market gaming strategies.

For a single cluster, the new energy generation data, user data, and transaction data are used as the basis for data mining and analysis using noise reduction inverse optimization techniques. Firstly, the new energy generation model, the user's electricity consumption model, the cluster's external transaction model, and the upper layer supervisory model are established, and then, to reduce the prediction error and the overfitting of historical data, the electricity consumption model and the external transaction model are state-optimized and trained to eliminate the boundaries of different types of electricity consumption and transactions, and finally, the upper layer supervisory model summarizes the optimized and trained state models into behavioral information using collected cluster data and applies it to the subsequent research.

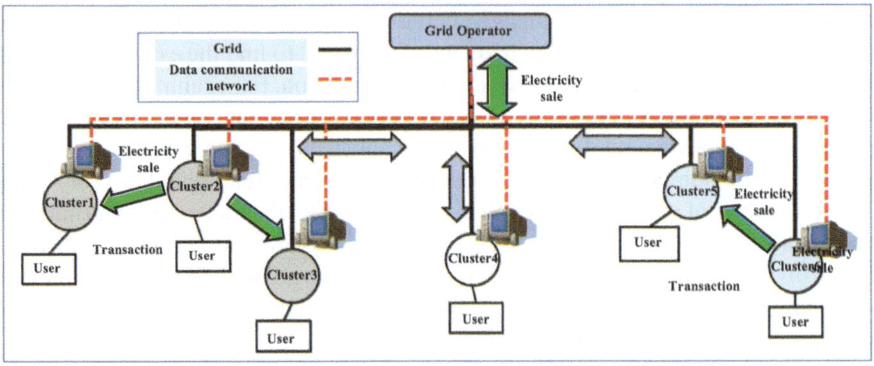

Fig. 7.1 Cluster data-enabled integrated operating system

For the whole system network, the discrete behavioral information stored in the local server of each cluster is analyzed holistically according to the group-related behavioral characteristics in anthropomorphic biology (e.g., migratory birds' guided flights, beehive migrations, etc.), to ensure the completeness and the group nature of the information analysis, and to summarize the interactive operational behavioral patterns of the distributed energy grid clusters, as well as the gaming framework. At the same time, according to the cooperative and non-cooperative behavioral characteristics of distributed energy grid clusters under competitive market boundary conditions, the market game strategy based on the Stackelberg model and the cooperative economic operation strategy based on dynamic decision-making are established respectively. The logical relationship is shown in the following Fig. 7.2.

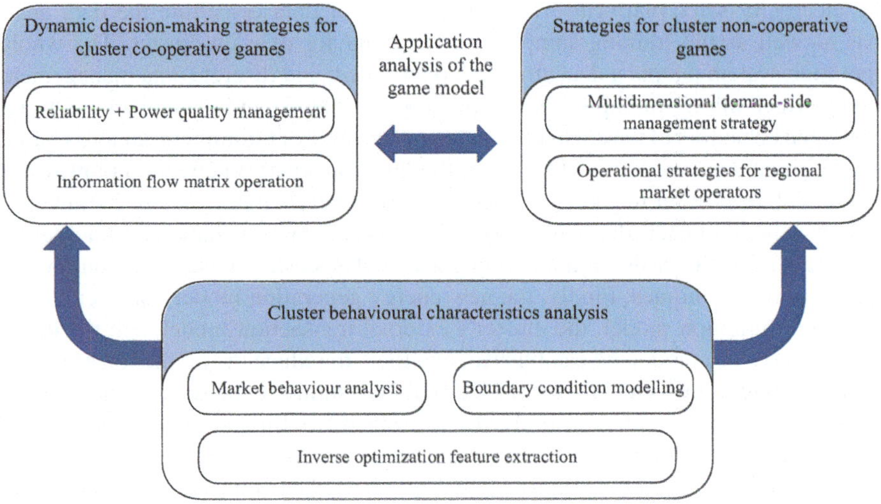

Fig. 7.2 A framework for analyzing multi-cluster market game strategies

In the modelling of Stackelberg's game theory, an energy cell can either choose to conduct its market and power activities independently, maintaining only a basic soft connection to the grid, or it can choose to form coalitions with other cells to support each other within the coalition while externally obtaining greater economic benefits and demand for service production. Based on the Shapley value, the multi-cells can find the optimal mode of interaction as well as the network structure based on the optimal individual or coalition utility function:

$$\phi_i(v) = \sum_{S \subseteq N} \frac{|S|!(|N| - |S| - 1)!}{|N|}(v(S \cup \{i\}) - v(s)) \tag{7.1}$$

where $v(S \cup \{i\}) - v(s)$ is the utility corresponding to a coalition or individual S, and represents the average payment that the set receives under all possible network structures (individual or coalition).

Based on the above theory, boundary models based on different behavioral interaction patterns and game strategies can be established. For the behavioral analysis and boundary modeling framework of the multi-energy cell data-based integrated operation system, distributed data extraction and analysis algorithms are formulated to improve the computational efficiency and increase the prediction and modeling accuracy.

In the non-cooperative game behaviour, the market management framework is divided into management and tracking layers; the management (grid operator) of the model, aiming at obtaining economic benefits and matching the dispatch demand, influences the power consumption activities and market behaviour of the aggregates in the tracking layer through price signals based on the information related to the power activities provided by the DEGs; and the DEGs in the tracking layer adjust the power consumption behaviors of the managed users and adjust their power consumption behaviors with the price signals issued by the management; and the DEGs in the tracking layer adjust their power consumption behaviors with the price signals issued by the management. The distributed energy grid in the tracking layer adjusts the power consumption behaviour of the managed users according to the price signals issued by the management and conducts a bidding game with the distribution market operator and other DEGs to minimize costs and maximize user satisfaction; the behaviour of the tracking layer adjusted according to the price of the power sale and the bidding game inversely affects the formulation of the price of the power sale at the decision-making level, thus forming a cyclic dynamic game between the upper, lower and the same layer (the tracking layer) and finally reaching dynamic equilibrium.

Under the compensatory market boundary conditions, the interactive behaviour between DEGs is based on the cooperative game of information interaction, so it is suitable to establish the market game strategy based on the information flow matrix. Firstly, the game variables of DEGs are established according to the market cooperation agreement and redundant electricity resources, and the variables are linked to the spatial and temporal scales and operation parameters to form information units;

then, the information units with different directional attributes are integrated into the information flow matrix specific to each distributed energy grid according to the directionality of the energy, information, and economic flows among the DEGs, which contains all the information needed for participating in the cooperative game; finally, the information flow matrix is integrated into the information flow matrix of each distributed energy grid, and the information flow matrix of each distributed energy grid is integrated into the information flow matrix of each distributed energy grid. Finally, all DEGs and their information flow matrices are substituted into the game model for dynamic decision-making.

7.1.1.3 MASs in Distributed Energy Grid Cluster Electricity Markets

With the deepening of the study of DEGs, the degree of practicality of DEGs is getting higher and higher, the number of DEGs will also be more and more of its installed capacity is also getting bigger and bigger, and all the output of power generation components within the distributed energy grid will be greater than the total demand of the local load, at this time, to make full use of the surplus power in the distributed energy grid, DEGs will participate in the power transmission grid. Therefore, with the advancement of technology, cost reduction, and capacity increase, the distributed energy grid will participate in the bidding system of the electricity market while ensuring its consumption. However, due to the high cost of the distributed energy grid with renewable energy as the main source of energy, the smaller volume of power trading, and the more obvious intermittency of renewable energy, the distributed energy grid is not yet able to fully compete with the traditional fossil energy generation (Wu et al. 2015).

The agent role can be divided into four levels of structure: electricity market agent, generator agent, user agent, and distributed energy grid agent (Ai and Zhang 2010; Ai et al. 2016).

(1) Electricity Market Agent: classifies power generators, users, and DEGs and bids or competes for them respectively. According to the bidding or tender data declared by each agent and the system security level, the trading plan is prepared, the market price is determined, and real-time market trading is realized. At the same time, the final market pricing and the output power of each power generator and distributed energy grid are fed back to the agents at all levels, so that their members can make the next trading decisions based on the information provided. At the same time, the scheduling of power generators and DEGs is carried out to ensure the safe and stable operation of the system.
(2) Generator Agent: applies for bidding to the higher-level agent according to its situation, and learns and adjusts the bidding strategy based on the data provided by the higher-level agent. When the power generator obtains the bidding power, it generates power according to the scheduling information of the electricity market agent, and at the same time carries out real-time monitoring of the generating units to ensure the safe operation of the system.

(3) User Agent: apply for bidding to the electricity market agent according to its load demand, and learn and adjust the bidding strategy from the data provided by the superior agent.
(4) Distributed Energy Grid Agent: the data provided by the component agent for distributed energy grid generation and load forecast, according to the forecast data to the electricity market agent to report its operating role in the sale of electricity or purchase of electricity distributed energy grid and then, the higher-level agent for bidding or bidding applications, and the higher-level data provided by the agent to learn and adjust the bidding strategy. After successful bidding, the distributed energy grid agent schedules component agents with a cost-minimizing bidding strategy. At the same time, real-time monitoring of the distributed energy grid, when the distributed energy grid fails should be quickly excised from the grid into the island operation state, to ensure the safe operation of the large system.

MAS-based distributed energy grid bidding model, which can be divided into the electricity market agent, distributed energy grid agent, power generation components agent, and distributed energy grid user agent four-level structure, they both compete with each other and are interconnected to ensure the safe and stable operation of the distributed energy grid, the structure is shown in Fig. 7.3.

The distributed energy grid generation element agent predicts its output power according to the characteristics of its generation element and environmental factors, and reports the prediction data, operating costs, generating unit offer, and various operating constraints to the distributed energy grid agent; at the same time, it always monitors the operating status of the generation element and reports it to the distributed

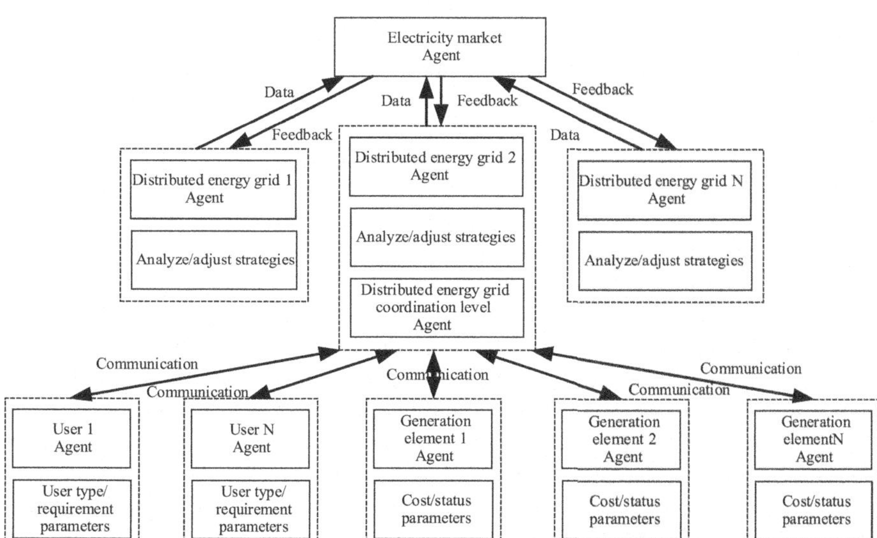

Fig. 7.3 MAS-based bidding model for distributed energy grid clusters

energy grid agent in time to carry out real-time control of the generation element to ensure the safe and stable operation of the distributed energy grid. Real-time control of power generation components to ensure the safe and stable operation of the distributed energy grid.

The distributed energy grid user agent is based on the importance of the user into sensitive loads, interruptible loads, and adjustable loads and reported to the distributed energy grid agent, not only to the user's power load demand forecast, and forecast data reported to the distributed energy grid agent, at the same time, the user's power consumption stability and security of real-time monitoring, coordination between the user and distributed energy grid operation.

7.1.2 Heat/Gas Market

7.1.2.1 Heat Market

The heat market is driven by factors such as climate, policy, cost, and customer demand, and is an energy supply industry operated by a coordinated group of stakeholders, including producers, suppliers, and service providers. In many countries, heat systems are an important part of modern infrastructure and markets are highly competitive. The organization of heat markets may vary from country to country, or from urban area to urban area, but they also follow certain common features (Ai et al. 2016): first, thermal energy is considered as a partially substitutable commodity in heat market transactions. Secondly, heat markets need to consider the physical structure and economics of the heating system in their operation. Thirdly, heat markets and heat prices remain highly regulated due to limited competition from suppliers and retailers.

The heat market suffers from a lack of competitiveness compared to the electricity and natural gas markets. Electricity and natural gas systems are usually large-scale systems with cross-regional or even cross-border interconnections, providing a favorable environment for market competition. However, unavoidable heat losses during the transport of thermal energy greatly limit the size of thermal systems. As a result, usually, the heating system only provides heat supply services in one city, and the number of competing heat supply companies and consumers within the system is limited. In the future, it is expected that large-scale heat systems will be constructed to introduce market competition on the supply side by integrating more energy supply equipment. However, depending on the size of the system, the advantages and disadvantages of liberalized and regulated markets have to be carefully weighed. Regulated markets lack incentives to invest in cost-effective new technologies; the competitive pressures of liberalized markets have the potential to improve the efficiency of the system but do not necessarily reduce energy prices for consumers. For example, in parts of Sweden, a liberalized heating market has caused heating prices to rise.

7.1.2.2 Natural Gas Market

The natural gas market is an energy supply industry with strong upstream and downstream linkages due to the nature of natural gas, which must be discharged without interruption after extraction. The development of the natural gas market is constrained by the high price of natural gas and the high feed-in tariffs for natural gas-fired power generation, which are not competitive with coal-fired power generation. Natural gas-fired power generation is cleaner than coal-fired power generation and removes fewer pollutants. However, its environmental and peaking benefits have not been translated into tangible economic benefits and its development has been limited. The establishment of a better pricing mechanism for environmental benefits and peak-to-valley tariffs will help improve the competitiveness of natural gas-fired power generation. The implementation of categorized gas prices and incentives will help promote the development of the natural gas market.

In the United States natural gas market, the purchase, sale, and transmission of natural gas are usually handled by separate markets. The signing and trading of physical contracts for natural gas volumes are completed in the commodity market, while the transmission contracts for natural gas are completed in the transmission capacity market (Jenkins et al. 2015). Compared to the electricity market, the current natural gas market in the United States is less competitive and less flexible. Firstly, the natural gas spot market is based daily and its natural gas price is fixed throughout the day; secondly, there are two types of natural gas transmission services, fixed and interruptible. Natural gas units are interruptible natural gas transmission services. In the current natural gas market environment, the coupled operation of the electricity-gas network makes the shortage of natural gas for gas units a very serious problem. In Japan, Russia, the United Kingdom, and several other European countries, the natural gas market has also developed more maturely, basically going through a process from state regulation of prices to gradual marketization.

The interactions between multiple energy markets, such as electricity, heat, and natural gas, are becoming more and more significant, and risk mitigation needs to be achieved through effective coordination mechanisms between multiple markets.

As a multi-energy interconnected distributed energy grid cluster with large power grids, natural gas networks, and cooling and heating pipeline networks, the interaction and synergy between the power, heat, and natural gas markets will be a key focus of future research.

7.2 Competitive Pricing Strategy

Distributed energy grid clusters have received widespread attention and development in recent years due to their reliable power supply, flexible operation mode, and environmental friendliness. With the increasing maturity of electric hot gas synergy technology, multiple types of resources such as electricity, heat, and gas will be integrated into the comprehensive energy system in the future. The future distributed energy grid clusters will become an emerging market entity, participating in the competition of integrated energy system electricity market and heat/gas market. The development of distributed energy grid clusters enables it to participate in market transactions as a decision-making subject, gradually transforming into an open trading mode of multi-party participation and supply relationship regulation market. Therefore, the competitive bidding strategy of distributed energy grid clusters has been widely studied.

7.2.1 Non Cooperative Game Model

Game theory refers to the discipline of studying the strategies of multiple individuals or teams in a game under specific constraints, using the strategies of relevant parties to implement corresponding strategies. Behavior with a competitive or confrontational nature is called game behavior. In such behavior, the parties involved in the struggle or competition each have different goals or interests. The bidding in the electricity/heat/gas market with DEGs is a process of mutual game (Duan et al. 2024).

The key to solving a game lies in finding a combination of strategies that each player is unwilling or unwilling to change individually. As long as this combination of strategies exists and is unique, the game has an absolutely certain solution. The strategy combination in which each player is unwilling or unwilling to individually change their strategy is the most important concept in game theory, the Nash equilibrium (Syothert and MacLeod 2000).

Based on the previous description of the models of electricity, heat, and gas energy systems, the goal of each system is to maximize its own interests, and the coexistence of competition and cooperation conforms to the Nash equilibrium theory. This section considers the electricity, heat, and gas systems as three independent entities and establishes a non cooperative game model as follows:

$$G = \{N, S, u\} = \{E, G, H; s_E, s_G, s_H; C_E, C_G, C_H\} \tag{7.2}$$

The above game model includes three elements: *N* is the set of game participants, representing individuals who participate in the game process and can adjust their strategies. *S* is a set of game strategies, representing a set of strategies that game participants can make. *U* is a set of utility functions used to evaluate the profits of participants and adjust their next game strategies.

(1) Game Participants

The participants in the game are the power system, thermal system, and natural gas system, which can be represented as:

$$N = \{E, G, H\} \tag{7.3}$$

(2) Game Strategy

Game strategy can be seen as a set of output plans for each unit that satisfy the constraint conditions, that is, the output constraints of each system. Among them, s_E, s_G and s_H are the constraint conditions of the preceding electrical, thermal, and gas systems.

$$S = \{s_E, s_G, s_H\} \in R^3, S \neq \varnothing \tag{7.4}$$

(3) Utility Function

The utility function of game participants refers to the total operating cost of each participant. Therefore, the utility functions of the power system, thermal system, and natural gas system are:

$$u = \{C_E, C_G, C_H\} \tag{7.5}$$

(4) Nash Equilibria

The strategy set $\left(P_E^*, P_G^*, P_H^*\right)$ is the equilibrium strategy in this section. When three systems satisfy this strategy set, the income function of the participants reaches the maximum value in the Nash equilibrium sense, and the following conditions need to be met:

$$u_i\left(s_1^*, \ldots, s_{i-1}^*, s_i^*, s_{i+1}^*, \ldots, s_n^*\right) \geq u_i\left(s_1^*, \ldots, s_{i-1}^*, s_i', s_{i+1}^*, \ldots, s_n^*\right) \tag{7.6}$$

7.2.2 Market Bidding Strategy Based on Demand Response

With the opening up of the market sales side, the traditional centralized purchasing and sales model has been broken. And by introducing competitive sales entities, users gradually gained the right to choose energy usage and have the power to independently choose sales entities.

However, when distributed energy and small residential load users are used as demand-side resources, they cannot be directly traded in the energy market due to their dispersed distribution, small individual capacity, and small response volume. They are usually consumed within a certain power supply range. When a large number of distributed energy and residential users participate in DR, traditional network architecture is difficult to effectively manage and reasonably consume these demand-side resources.

Distributed energy grid clusters can utilize advanced information and communication technologies and control architectures to aggregate and coordinate demand-side resources, enabling them to freely participate in energy market transactions. Distributed energy and the participation of small residential users in DR can generate additional benefits, and by aggregating and managing these demand-side resources, distributed energy grid clusters can to some extent smooth out output fluctuations caused by uncertain power generation units such as wind and solar power, improve their disturbance resistance, and increase the energy sales of the energy network group, thereby improving the stability and economy of the system.

7.2.2.1 The Impact of Sales Side Liberalization on Demand Side Resource Participation in the Energy Market

With the gradual opening up of the energy market to the sales side, opportunities for distributed energy sources such as distributed power sources and demand side resources to participate in competition in the electricity/heat/gas market are gradually increasing and becoming more fair. The opening up of the sales side market has two basic connotations: firstly, breaking the monopoly of the sales side market and creating a situation where multiple sales entities participate in the competition of the sales side market; Secondly, load users gradually have the freedom to choose their energy usage. Within the scope permitted by policies, load users can compare the advantages and disadvantages of energy sales contracts provided by various sales entities, and thus freely choose the sales entity that is more suitable for themselves. They can also change the seller after the contract ends.

The relaxation of the sales side will have a certain impact on the participation of demand-side resources in the energy market, providing them with a more fair competitive environment and platform. On the one hand, with the gradual opening up of the sales side market, more demand-side resources have gained channels for fair participation in energy market competition, promoting fair competition in the energy market and adding vitality to it. The traditional energy market mainly focuses on demand-side management for large industrial and commercial users, and cannot effectively manage a large number of dispersed distributed energy and small-scale energy users. The gradual opening up of the sales side market has broken the traditional market model. DEGs, load aggregators, or power sales companies have become new sales side market entities to participate in the energy market. Through reasonable electricity pricing strategies or incentive policies, they guide distributed energy and small-scale energy users to participate in energy market competition and provide DR services for the energy system. On the other hand, the gradual opening up of the sales side market can provide effective guarantees for the benefits of demand side resources participating in energy market competition, thereby stimulating the initiative and enthusiasm of demand side resources to participate in the energy market. In the traditional energy market, the utilization of demand side resources mainly relies on policy based mandatory energy load users to undertake scheduling tasks, which does not conform to market rules. With the gradual opening up of the sales side market, energy prices can better reflect changes in costs and market demand. The price policies and incentive mechanisms related to DR are gradually improving, and the participation of demand side resources in the energy market provides services such as peak shaving and valley filling for the system. This not only reduces the overall operating costs of the system, but also brings corresponding benefits to demand side resources, thereby increasing the enthusiasm of demand side resources to actively participate in the energy market.

7.2.2.2 Distributed Energy Grid Participation in Market Process Considering Demand Response in the Context of Sales Side Liberalization Environment

As the sales side market gradually opens up, distributed energy grid clusters that take into account DR participate in energy market competition. Firstly, it is necessary to organize and optimize the combination of various distributed energy sources aggregated internally, predict and collect energy consumption information habits of load users, determine the mode and type of electricity sales services that the distributed energy grid can provide as a whole, and formulate targeted sales contracts. On this basis, users can independently choose energy sellers based on their own energy consumption habits by comparing the advantages and disadvantages of various energy sales contracts. The distributed energy grid clusters receives the autonomous selection results of load users, optimizes the combination of internal resources, and conducts daily optimization of each distributed unit with the goal of maximizing overall revenue. According to the optimization results, the distributed energy grid will participate in the energy market bidding plan, and the market will be cleared through the dispatch center. Based on the dispatch instructions, a daily plan will be issued, and each distributed unit will execute it within the day and settle it in the future. After the end of the electricity sales contract, the load user can choose another electricity retailer based on their actual situation. The process of participating in energy market competition for DEGs that take into account DR is shown in Fig. 7.4.

7.2.2.3 Distributed Energy Grid Clusters Bidding Model Taking into Account Demand Response

When participating in electricity market bidding for DEGs, it is necessary to consider the cost of internal distributed energy and the uncertain factors of uncontrollable unit output such as WP and PVs. The distributed energy grid clusters bidding model proposed in this section takes WP as an example, considering its internal DR and the uncertainty of renewable energy output, with the goal of maximizing the consumption of renewable energy and achieving the lowest overall cost of the distributed energy grid group.

(1) DR Cost

According to different response mechanisms, DR can be divided into two categories: incentive based demand response (IBDR) and price sensitive demand response (PSDR).

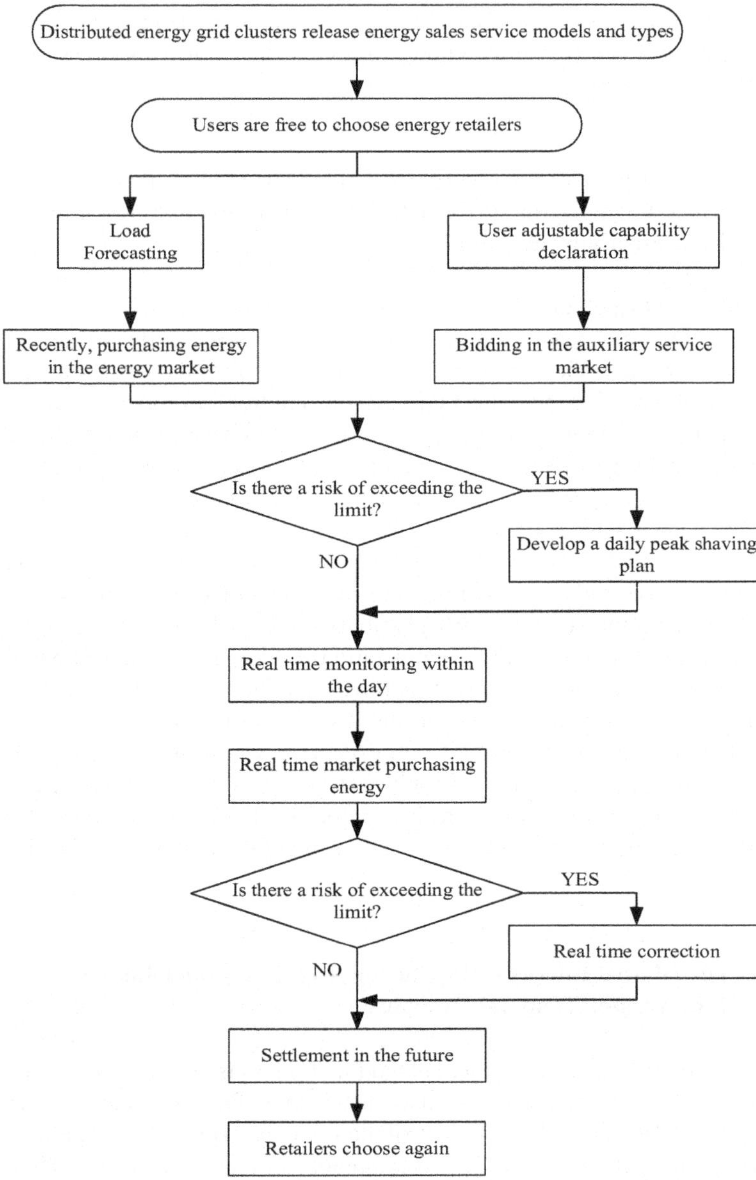

Fig. 7.4 Market participation process of distributed energy grid group including DR

The DR cost based on incentives includes the reduced income caused by load reduction and compensation for users who reduce their load. Its function expression is a quadratic function of the load reduction amount:

$$C_{IB,t} = k_1^* Q_{IB,t}^2 + (k_2 + r_{p,t})^* Q_{IB,t} \tag{7.7}$$

$$Q_{B,\min} \leq Q_{B,t} \leq Q_{B,\max} \tag{7.8}$$

where k_1 is the quadratic coefficient of the DR compensation amount based on incentives; k_2 is the first-order coefficient of the DR compensation amount based on incentives; $r_{p,t}$ is the retail electricity price in the electricity market during time period t; $Q_{IB,t}$ is the load reduction amount based on incentive DR during time period t; $Q_{IB,\min}$ is the lower limit of load reduction based on incentive DR; $Q_{IB,\max}$ is the upper limit value of load reduction based on incentive DR.

Price based DR refers to the behavior of reducing load by users in response to price changes after receiving price adjustment signals. The DR cost based on price is the change in energy sales revenue before and after price changes, and its function expression is a quadratic function of load reduction:

$$C_{PS,t} = r_{p,t}^* Q_{PS,B,t} - R^*(r_{p,t} + \Delta r_{p,t})^*(Q_{PS,B,t} - Q_{PS,t}) \tag{7.9}$$

$$Q_{PS,\min} \leq Q_{PS,t} \leq Q_{PS,\max} \tag{7.10}$$

where R is the proportion of preferential electricity prices provided by the seller to demand responsive users; $Q_{PS,B,t}$ is the baseline load of DR based on price for time period t; $Q_{PS,t}$ is the load reduction amount based on price DR during time period t; $Q_{PS,\min}$ is the lower limit of DR load reduction based on price; $Q_{PS,\max}$ is the upper limit of DR load reduction based on price.

(2) The Risk Cost of WP Generation

Due to the randomness of WP output, if the actual output of WP is higher than the planned output, the distributed energy grid may waste some resources due to the inability to absorb the excess electricity, and this part of the cost is included as WP abandonment cost; If the actual output of WP is lower than the planned output, it requires backup power sources such as ES to provide backup electricity, resulting in additional costs, which are calculated as penalty costs. Therefore, the risk cost of WP can be expressed as:

$$C_{w,t} = \begin{cases} \alpha_w(q_{wr,t} - q_{wp,t}), & q_{wr,t} \geq q_{wp,t} \\ \alpha_{pu,t}(q_{wp,t} - q_{wr,t}), & q_{wr,t} < q_{wp,t} \end{cases} \tag{7.11}$$

where a_w is the cost coefficient for WP curtailment; $a_{pu,t}$ is the penalty cost coefficient for time period t; $q_{wp,t}$ is the planned WP output for time period t; $q_{wr,t}$ is the actual output of WP during time period t.

(3) ES Device Model

As a backup within the distributed energy grid, ES devices purchase energy at low prices and sell energy at high prices based on the actual energy price situation, while meeting the energy balance constraints of ES devices and the internal backup needs of the distributed energy grid. ES needs to meet the constraints of power balance and output during operation:

$$E_t = E_{t-1} + \mu_t P_t \eta_c + \frac{(1-\mu_t)P_t}{\eta_d} \tag{7.12}$$

$$\max(-P, (E_{\min} - E_{t-1})\eta_d) \le P_t \le \min\left(P, \frac{E - E_{t-1}}{\eta_c}\right) \tag{7.13}$$

$$E_{\min} \le E_t \le E \tag{7.14}$$

where E_t represents the energy stored during the time period t; μ_t is the operating state variable of ES, with a charge of 1 and a discharge of 0; P_t is the ES charging and discharging power during time period t; η_c is the charging efficiency of ES; η_d is the energy release efficiency of ES; P is the rated power of ES; $E_{\min}$ is the lower limit of ES capacity; E is the rated value of ES capacity.

(4) Distributed Energy Grid Quotation Function

DEGs can choose to participate in energy market competition as energy generators or energy sellers based on the power generation situation of internal power generation units and the energy consumption situation of load users. The quotation of distributed energy grid adopts linear function form, and its specific expression is as follows:

$$\lambda_{vpp,t} = k^*_{vpp,t}(2a_{vpp,t}P_{vpp,t} + b_{vpp,t}) \tag{7.15}$$

$$P_{vpp,t,\min} \le P_{vpp,t} \le P_{vpp,t,\max} \tag{7.16}$$

where $k_{vpp,t}$ is the strategy pricing coefficient of the distributed energy grid during time period t; $a_{vpp,t}$ is the quadratic coefficient of the marginal cost of the distributed energy grid during time period t. It is positive when the distributed energy grid reports as an energy generator and negative when it reports as an energy seller; $b_{vpp,t}$ is the first-order coefficient of the marginal cost of the distributed energy grid during time period t; $P_{vpp,t}$ is the reported planned output of the distributed energy grid during time period t; $P_{vpp,t,\min}$ is the lower limit value of distributed energy grid output during time period t; $P_{VPP,t,max}$ is the upper limit value of the output of the distributed energy grid during the time period t.

7.2.3 Multi Agent Dual Layer Game Approach Considering Electric and Thermal Gas Collaboration

The distributed energy grid clusters system considering the coordination of electric and thermal energy consists of multiple distributed energy sources and electric, heating, and cooling loads. Firstly, analyze the operational characteristics of key unit components in the distributed energy grid system and establish a multi energy complementary lower level scheduling model based on cooperative game theory. Then, study the bidding evolution law of the upper level regional energy market. Considering the marginal cost quotation and the profit mechanism based on quotation settlement, the energy agent applies reinforcement learning algorithms to analyze limited historical information based on the scheduling results of lower level cooperative games, in order to select quotation strategies with different profit margins.

The game framework in the distributed energy grid clusters system is shown in Fig. 7.5. The upper layer is a non cooperative bidding game with incomplete information, and the game subjects are electricity/heat/gas companies, users, and energy suppliers. Users participate in the game based on their own DR characteristics, while energy suppliers calculate marginal operating costs and design corresponding pricing strategies based on their own unit operating characteristics. The market model of Pay as Bid Price (PAB) settlement can save the total energy purchase cost of the system compared to the marginal price mechanism under the same quotation. In order to ensure fairness and justice, the strategy reported by the stakeholders in each day ahead scheduling optimization is open to all stakeholders, and the true marginal cost curve of the unit is kept confidential.

The lower level is the cooperative game of day ahead dispatching of cooling and heating power. The main players of the game are energy suppliers and users, and the business department acts as the matchmaker. In combination with the quotation curves reported by multiple entities, the system cooperates with multiple entities to meet the needs of different loads with the goal of optimizing the system energy cost. In order to achieve a balance between energy production and sales and minimize the need for external energy purchases, the distributed energy grid clusters system first supplies distributed energy to users for consumption, and then sells excess power generation to the grid in cooperation between users and energy suppliers. Distributed energy obtains maximum energy sales revenue with minimal fuel costs by selling electricity and supplying cooling and heating, while considering interaction with electricity/heating/gas companies. On the basis of cooperation with energy suppliers, load users consider maximizing the on-site consumption of distributed energy by first purchasing the system's own production capacity, and then purchasing energy from electricity/heat/gas suppliers for the shortfall, minimizing the cost of purchasing energy without affecting production efficiency. Electricity/heat/gas companies can also interact with users through DR mechanisms, encouraging them to peak shaving and valley filling, reducing their own operational and scheduling costs, as shown in Fig. 7.6.

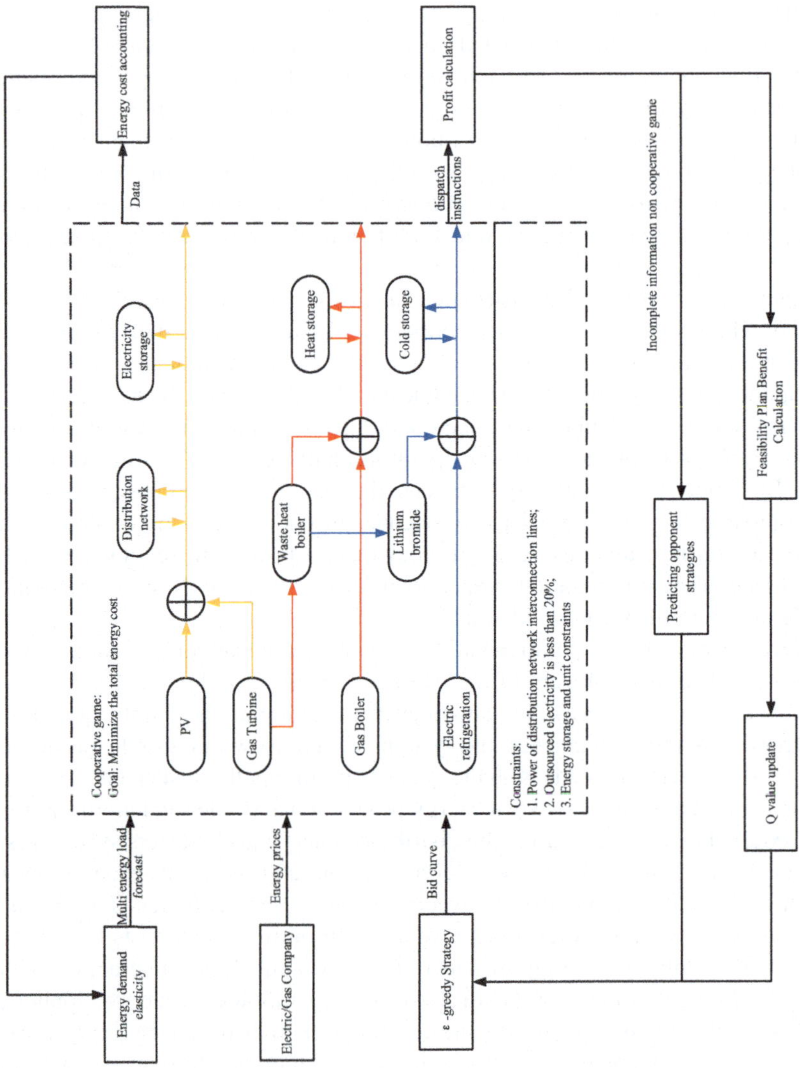

Fig. 7.5 Multi agent double layer game strategy

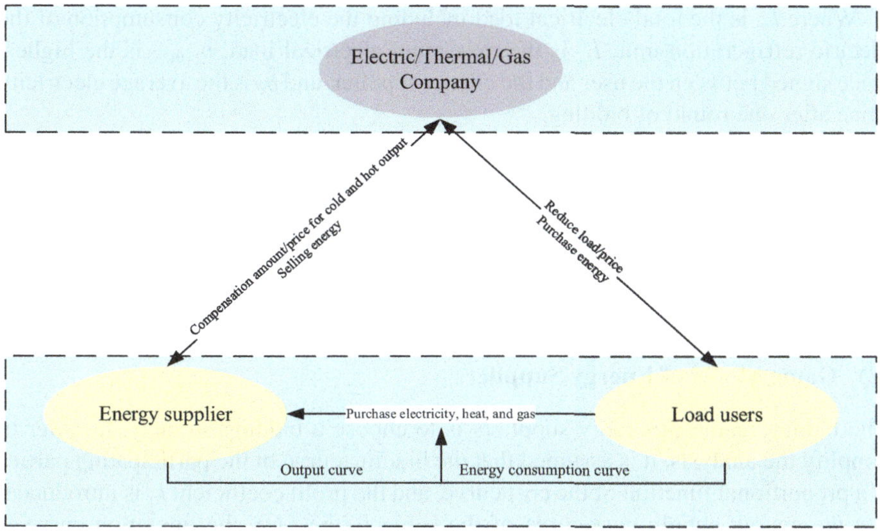

Fig. 7.6 Game framework of distributed energy grid group system

7.2.3.1 Upper Layer: Non Cooperative Bidding Game Under Incomplete Information

(1) **User Game Model**

To ensure that all Nash equilibrium points are within a reasonable range of energy consumption costs, the user representative signs the highest bid price with the energy supplier $b_{\max}$, b_{ref1} and b_{ref2} are the bidding price ranges for the elastic segment. And in order to prevent the occurrence of monopoly and unreasonable prices in the game, users participate in the game through their own DR characteristics. The energy price elasticity model for users of cooling, heating, and power is shown in Fig. 7.7.

Fig. 7.7 Multi energy flow user energy consumption price elasticity model

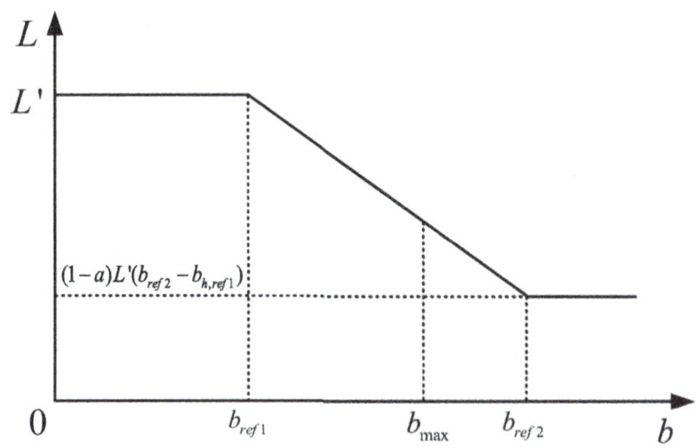

Where L_e is the total electrical load including the electricity consumption of the electric refrigeration unit, L_e' is the maximum electrical load, $b_{e,\max}$ is the highest price signed between the user and the energy supplier, and $\overline{b}_e$ is the average electricity price after one round of bidding.

$$L_e = \begin{cases} L_e' & \overline{b}_e < b_{e,ref} \\ L_e' - a_e\left(\overline{b}_e - b_{e,ref}\right) & b_{e,ref} < \overline{b}_e < r_e b_{e,ref} \\ L_e' - a_e(r_e - 1)b_{e,ref} & \overline{b}_e \geq r_e b_{e,ref} \end{cases} \tag{7.17}$$

The elastic models using cold and hot are as described above.

(2) **Game Model of Energy Suppliers**

The bidding game of energy suppliers is to choose a bidding strategy. In order to simplify the analysis, it is assumed that the bidding curve of the participating parties is a proportional function of the cost curve, and the profit coefficient k_j is introduced. Let the energy supply cost curve of the subject j be $f_j(t)$, the quotation curve is $b_j(t) = k_j f_j(t)$, when calculating costs, consider the current energy prices and working conditions unchanged.

In a distributed energy grid clusters system, there are multiple stakeholders involved, and a single topic often cannot obtain all the cost and capacity information of all stakeholders during bidding decisions. Therefore, the upper level bidding game adopts Q-Learning reinforcement learning method to simulate the bidding process in the energy market. The Q-Learning algorithm provides intelligent systems with the ability to select the optimal action based on the sequence of actions they have experienced in a Markov environment. Its advantage is that it does not require modeling of the dynamic environment in which it operates, and can treat the complex process of multi energy coupled bidding scheduling as a "black box", selecting the optimal action based on current information to change the impact on the dynamic environment. The algorithm only needs to input the state variables of the current round, which is faster than the Actor critic learning algorithm and can perform real-time online strategy game optimization.

The information set of multi-agent game process is (N, S, R, B, U), where $N = \{1, 2, \ldots, N\}$ is the set of agents participating in the game, and $S = \{s_i\}$ is the set of all discrete states of the game's bidding, including the bidding curves of each subject. The set of quotation curves with $B = \left\{b_{\min,}, \ldots, b_{\max}\right\}$ as the main body. $R = \left\{R_1, \ldots R_j, \ldots R_n\right\}$ is the set of payoff functions for the participants, which are related to the game state and the strategies of the participants $R_j : \left(s_i, b_j\right) \to \Re$. b_{-j} is the pricing strategy of the competitor, and the transition probability is defined as the transition probability $U\left(S, b_j, b_{-j}\right) \in [0, 1]$ in state s.

7.2.3.2 Lower Layer: Day Ahead Dispatching Cooperation Game of Cold and Heat Power

The system aims to minimize the energy consumption cost F of the park as the objective function, and coordinates the cooperation of various entities to optimize the system objectives.

$$\min \quad F = \sum_{j=1}^{m} f_j(t) \sum_{i=1}^{n} P_j^i(t) \tag{7.18}$$

Among them, $f_j(t)$ is the marginal cost of energy supply for participant j at time t, and $P_j^i(t)$ is the scheduling power of the i-th unit of the subject j at time t.

The cost of energy suppliers includes energy conversion costs $f_{e,j}(t)$ and operational management costs $f_{OM,j}$, and the marginal cost $f_j(t)$ can be expressed as:

$$f_j(t) = \frac{1}{\sum_{i=1}^{n} P_j^i(t)} \left[f_{e,j}(t) + f_{OM,j} \right] \tag{7.19}$$

$$f_{e,j}(t) = \sum_{i=1}^{n} C_j^i [P_j^i(t), t] \tag{7.20}$$

$C = C_j^i \left[P_j^i(t), t \right]$ is the energy conversion cost function of the i-th unit of the j entity, with the independent variables being the unit dispatch power $P_j^i(t)$ and time t.

$$f_{OM,j}(t) = C_{OM,j} P_j^N \tag{7.21}$$

where $f_{OM,j}$ is the operating and management cost of j, $C_{OM,j}$ is the operating and maintenance cost coefficient of j at time t, and P_j^N is the total capacity of the unit put into operation by the main body of j.

The energy conversion cost functions of each energy supplier are shown in Fig. 7.8.

7.3 Cooperative Game Bidding Model

From the perspective of organizational relationships within DEGs, the game relationships among DEGs can be divided into two types. The first is non-cooperative games, which primarily emphasize independent decision-making by individuals, without considering the influence of other entities in the strategic environment. Current research on the optimal operation of DEGs predominantly focuses on independent non-cooperative operation (Rahi et al. 2017), while often neglecting the potential for

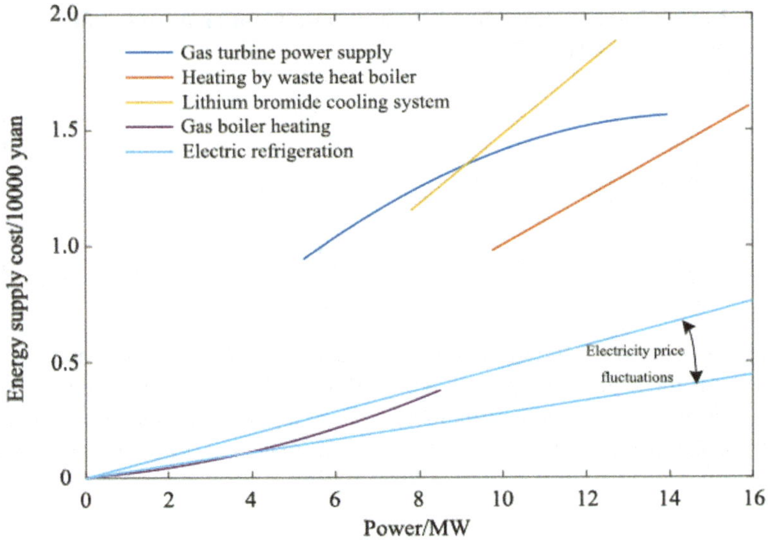

Fig. 7.8 Energy conversion cost chart of the unit

cooperative interactions among adjacent DEGs. In fact, considering the complementary characteristics of loads in different types of DEGs, cooperative operation among neighboring DEGs can not only promote multi-energy trading and mutual support in electricity, heat, and gas but also enhance the economic operation of controllable units within DEGs. This, in turn, helps individual DEGs achieve higher economic benefits.

Based on this, research on cooperative games has gradually gained attention in recent years due to the increasing complexity and diversity of organizational relationships within distribution systems. Cooperative games refer to scenarios where participants engage in alliances and cooperation, and the benefits obtained through cooperation exceed those gained by acting independently. However, the dynamic and complex formation of alliances makes it challenging to achieve a Nash equilibrium solution, which is one of the factors hindering the practical application of cooperative game research (Li et al. 2024). Current studies on cooperative games primarily focus on energy management between DEGs and electric vehicle charging stations (Parhizi et al. 2018).

7.3.1 Overview of the Cooperative Game Model

The outcome of a cooperative game is the formation of alliances. Similar to non-cooperative games, rationality and profitability are the foundational principles for the establishment of cooperative alliances. In cooperative games, all participants adhere to rational rules and pursue individual profit maximization, consistent with non-cooperative games. The key difference lies in the fact that cooperative games allow participants to achieve a "1 + 1 > 2" effect through agreements, thereby generating additional benefits.

The "rationality" in cooperative games encompasses both individual rationality and collective rationality. Individual rationality refers to the condition that participants will only choose to join a cooperative alliance if the benefits allocated within the alliance exceed the benefits obtained independently. Collective rationality indicates that an alliance will form if and only if the total benefit of the alliance exceeds the sum of the individual benefits each participant would achieve independently. Under the combined effect of individual rationality and collective rationality, cooperative alliances form spontaneously.

Non-cooperative games primarily focus on the competitive strategies and individual benefits of each participant, aiming to determine possible equilibrium states. In contrast, cooperative games additionally consider the cooperative relationships among participants, the overall benefits of alliances, and how these benefits or payments are distributed among alliance members.

Based on whether the total benefits of the alliance can be distributed among participants, cooperative alliances can be divided into transferable utility games (TU) and non-transferable utility games (NTU). According to different factors influencing the alliance's total benefits, TU can further be divided into partition function games (PFC) and characteristic function games. In characteristic function games, the total benefit of an alliance depends solely on its own actions. In PFC, however, the benefit of an alliance depends not only on its own actions but also on the behavior of other alliances.

It is critical to emphasize that the "rationality" in cooperative games includes both individual rationality and collective rationality. From the perspective of individual rationality, each member of the alliance aims to maximize its benefits or minimize its payments. If the alliance cannot allow all members to achieve optimal goals, those members failing to achieve their objectives will act in their best interests, causing the alliance to dissolve. Thus, in most cases, individual benefits are fundamental, while alliance benefits are not the essence. Therefore, a stable alliance not only requires increasing both the individual benefits of its members and the total benefits of the alliance but also ensuring that alliance members do not have the tendency to leave the alliance.

7.3.2 Transaction Model Based on Cooperative Game

This section aims to explore the multi-energy coordinated optimization problem for DEGs based on cooperative game theory, forming alliances among multiple DEG systems willing to cooperate. Considering the heterogeneity of different DEG operators, the Nash bargaining mechanism is employed to promote energy exchange and cooperative transactions among DEGs while ensuring the autonomy of their operations. To account for the decision-making independence and information privacy of different entities, the ADMM is introduced to achieve a distributed solution for the Nash bargaining cooperative model.

Assume there are multiple DEG systems willing to form a cooperative alliance, as illustrated in Fig. 7.9. Each region is managed by a DEG operator, and each DEG contains controllable units (e.g., CHP units and GBs), uncontrollable units (e.g., WT and PVs), storage systems (e.g., electric storage and thermal storage), and loads (e.g., electrical and thermal loads). With the inclusion of demand-side response, electrical loads are further divided into critical electrical loads and transferable electrical loads, while thermal loads include critical thermal loads and interruptible thermal loads.

Unlike the centralized scheduling model, this framework assumes that each DEG retains autonomous decision-making authority (Lu et al. 2017; Gao et al. 2014). Cooperation among DEGs is achieved solely through the exchange of public information. Consequently, each DEG is equipped with an independent EMS capable of optimizing the scheduling of energy conversion facilities within the region and making energy purchasing decisions from external electricity and natural gas companies.

In a cooperative alliance of DEGs, each DEG autonomously decides its operational scheme based on the internal energy demand of its users (Lee and Baldick 2003). The different DEGs can exchange excess energy, forming internal transactions within the virtual cooperative alliance, thereby reducing the amount of energy purchased from electric power or natural gas companies and minimizing the overall operational costs of the entire cooperative alliance. Therefore, the relationship between DEG operators is one of cooperative competition, where the operators need to negotiate the distribution of the surplus profits generated by the cooperation to ensure that they achieve increased benefits (or reduced costs) during the cooperation process.

Since a cooperative alliance of DEGs is typically composed of several rational players, the formation of such an alliance is based on the premise that the players reach a consensus on profit distribution. This is essentially a negotiation process. In other words, the cooperative operation among the internal members of the DEGs is based on transaction negotiations, which belongs to the typical "cooperation and competition problem," where cooperation means that members' gains are Pareto improved. However, different participants have different preferences for Pareto states, leading to changes in the equilibrium point of operation.

In non-cooperative games, each participant makes independent decisions, emphasizing individual rationality; whereas in cooperative games, participants make joint decisions, and agreements are binding for all participants, emphasizing collective rationality. While Nash equilibrium focuses on the equilibrium analysis of non-cooperative games, Nash bargaining is suited for the negotiation analysis of cooperative games. Therefore, it can be used for coordinating the operation of cooperative alliances formed by different entities.

(1) Nash Bargaining Model

Assume there are N rational individuals engaged in negotiations to form a cooperative alliance. The premise for the formation of this cooperative alliance is to determine

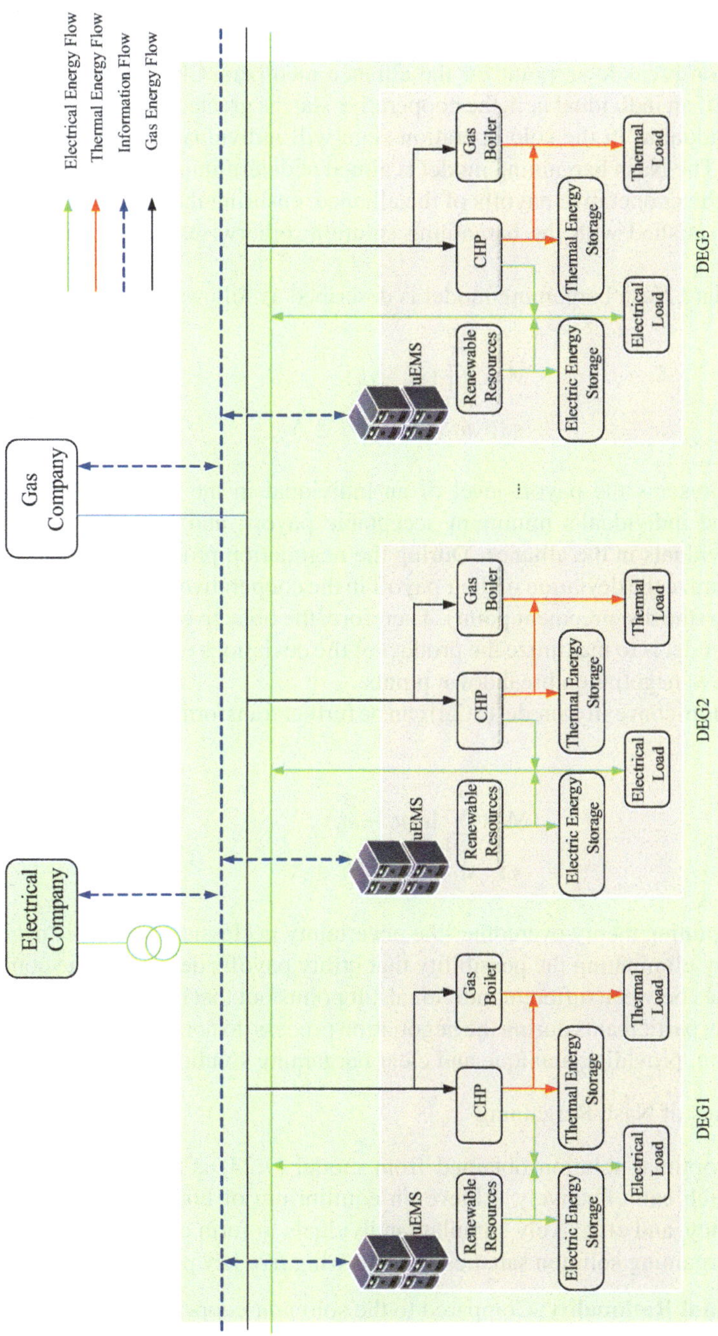

Fig. 7.9 Cooperative alliance framework structure

a globally recognized, clear bargaining solution. In the initial individual operating state, the payoff level for player i is d_i, and correspondingly, the starting payoff state for all rational individuals is $D = \{d_1, \ldots, d_N\}$. This point is also known as the negotiation breakdown point for the alliance members. Clearly, only when the payoff level of an individual u_i in the cooperative state is greater than the payoff level of that individual d_i in the solo operation state will individual i be willing to join the alliance. The Nash bargaining model is aimed at determining a clear distribution solution for the cooperative payoffs of the alliance, ensuring that all members of the alliance are satisfied with the bargaining solution; otherwise, the negotiation will break down.

The standard Nash bargaining model is described as follows:

$$\begin{aligned} &\text{Max} \prod_{i=1}^{N} (u_i - d_i) \\ &\text{s.t.} \quad u_i \geq d_i, \quad \forall i \in \mathcal{N} \end{aligned} \tag{7.22}$$

where u_i represents the payoff level of an individual in the cooperative state, d_i represents the individual's minimum acceptable payoff, and N is the number of rational individuals in the alliance. During the negotiation process, each individual aims to maximize the deviation of their payoff in the cooperative state from the worst-case scenario (the disagreement point). Therefore, the objective function of the Nash bargaining model is to maximize the product of the deviations of all individuals from their respective negotiation breakdown points.

To avoid non-convexity, model (7.22) can be further transformed into the following form:

$$\begin{aligned} &\text{Max} \sum_{i=1}^{N} \ln(u_i - d_i) \\ &\text{s.t.} \quad u_i \geq d_i, \quad \forall i \in \mathcal{N} \end{aligned} \tag{7.23}$$

Nash bargaining theory overcomes the uncertainty in classical economic negotiation theory by eliminating the possibility that utility payoffs depend on the comparison of utilities between different individuals. It points out that the voluntary concessions made by participants during the negotiation process do not affect the uniqueness of the solution, providing a unique and clear bargaining solution.

(2) Properties of Nash Bargaining

The unique optimal solution obtained from model (7.23) is the Nash bargaining solution, which can effectively achieve an equilibrium of both Nash fairness and Nash efficiency, and effectively stimulate individuals to form cooperative alliances. The Nash bargaining solution satisfies the following five key properties:

(a) **Individual Rationality**: Compared to the solo non-cooperative operation (i.e., the utility value at the negotiation breakdown point), all individuals' utility levels achieved through the negotiation and cooperation process are improved, i.e.,

$u_i^* \geq d_i, \quad \forall i \in \mathcal{N}$. Otherwise, the individual will not be willing to participate in the cooperative alliance.

(b) **Pareto Efficiency**: In the Nash bargaining model, all individuals reach the optimal state, meaning that no rational individual can find a solution better than the Nash bargaining solution. In other words, there is no solution $(u, v) \in S$ within the feasible region such that $u > u^*$ and $v > v^*$ (or $u \geq u^*$ and $u \geq u^*$)

(c) **Symmetry**: If the coalition members have identical negotiation breakdown points and utility functions, they will receive the same utility value regardless of their position within the coalition.

(d) **Independence of Irrelevant Alternatives**: The removal of certain irrelevant alternatives from the original feasible set of solutions does not affect the determination of the optimal solution in Nash bargaining.

(3) Sub-region Models

In the process of energy optimization allocation, the sub-region operator aims to minimize its own operational cost C_i, which specifically includes the transaction cost with the main grid, natural gas purchase cost, user utility function, internal operational cost of units within the sub-region, as well as the transaction cost with other sub-regions within the virtual energy system.

$$C_i = C_i^{\text{grid}} + C_i^{\text{gas}} - U_i^{\text{user}} + C_i^{\text{dg}} + C_i^{\text{vpp}} \tag{7.24}$$

$$C_i^{\text{grid}} = \sum_{t \in \mathcal{T}} \left(\lambda_t^{\text{bg}} P_{\text{e},i,t}^{\text{bg}} - \lambda_t^{\text{sg}} P_{\text{e},i,t}^{\text{sg}} \right) \tag{7.25}$$

$$C_i^{\text{gas}} = \sum_{t \in \mathcal{T}} \lambda_t^{\text{gas}} \left(P_{\text{g},i,t}^{\text{chp}} + P_{\text{g},i,t}^{\text{gb}} \right) \tag{7.26}$$

$$U_i^{\text{user}} = \sum_{t \in \mathcal{T}} \left[k_i \ln\left(1 + L_{i,t}^{\text{e}}\right) - a_i \Delta L_{i,t}^{\text{h}} \right] \tag{7.27}$$

$$C_i^{\text{dg}} = \sum_{t \in \mathcal{T}} \left[c_{\text{e},i}^{\text{chp}} P_{\text{e},i,t}^{\text{chp}} + c_{\text{h},i}^{\text{gb}} P_{\text{h},i,t}^{\text{gb}} + c_i^{\text{es}} \left(P_{\text{e},i,t}^{\text{ch}} + P_{\text{e},i,t}^{\text{dis}} \right) + c_i^{\text{hs}} \left(P_{\text{h},i,t}^{\text{ch}} + P_{\text{h},i,t}^{\text{dis}} \right) \right] \tag{7.28}$$

$$C_i^{\text{vpp}} = \sum_{t \in \mathcal{T}} \left[\left(\lambda_{\text{e},t}^{\text{bc}} P_{\text{e},i,t}^{\text{bc}} - \lambda_{\text{e},t}^{\text{sc}} P_{\text{e},i,t}^{\text{sc}} \right) + \left(\lambda_{\text{h},t}^{\text{bc}} P_{\text{h},i,t}^{\text{bc}} - \lambda_{\text{h},t}^{\text{sc}} P_{h,i,t}^{\text{sc}} \right) \right] \tag{7.29}$$

where, C_i^{grid} represents the transaction cost between DEG i and the grid, while $P_{\text{e},i,t}^{\text{bg}}$ and $P_{\text{e},i,t}^{\text{sg}}$ denote the electricity purchased from and sold to the main grid during the time period t, respectively. λ_t^{bg} and λ_t^{sg} are the corresponding unit electricity purchase and sale costs. C_i^{gas} represents the gas purchase cost for DEG i, with λ_t^{gas} being the natural gas price from the gas supplier. $P_{\text{g},i,t}^{\text{chp}}$ and $P_{\text{g},i,t}^{\text{gb}}$ represent the natural gas consumption of the CHP unit and the GBs within DEG i during the time period t, respectively; the user utility function U_i^{user} consists of two parts: the electricity consumption utility of the user $k_i \ln\left(1 + L_{i,t}^{\text{e}}\right)$ minus the discomfort cost $a_i \Delta L_{i,t}^{\text{h}}$.

$L_{i,t}^{\mathrm{e}}$ and $\Delta L_{i,t}^{\mathrm{h}}$ represent the electricity consumption and heat rejection of the user in DEG i during the time period t, while k_i and a_i are the corresponding preference coefficients. a_i, $c_{\mathrm{h},i}^{\mathrm{gb}}$, c_i^{es} and c_i^{hs} correspond to the unit operation costs of the CHP unit, GBs, electrical storage, and thermal storage devices within the DEG, respectively. C_i^{vpp} represents the transaction cost between DEG i and other DEGs within the virtual integrated energy system. $P_{\mathrm{e},i,t}^{\mathrm{bc}}$ and $P_{\mathrm{e},i,t}^{\mathrm{sc}}$ denote the electricity purchased from and sold to other DEGs during time period t with $\lambda_{\mathrm{e},t}^{\mathrm{bc}}$ and $\lambda_{\mathrm{e},t}^{\mathrm{sc}}$ being the corresponding community electricity purchase and sale prices. $P_{\mathrm{h},i,t}^{\mathrm{bc}}$ and $P_{h,i,t}^{\mathrm{sc}}$ represent the heat purchased from and sold to other DEGs during time period t, with $\lambda_{\mathrm{h},t}^{\mathrm{bc}}$ and $\lambda_{\mathrm{h},t}^{\mathrm{sc}}$ being the corresponding heat transaction prices.

The constraints include the energy balance constraint of the DEG, the power constraint of the internal energy transmission system, the operational constraints of the energy supply devices, the operational constraints of the ES devices, and the demand-side response constraints of the DEG. Since these are not the focus of this chapter, they will not be elaborated upon here.

(4) Nash Bargaining-based Joint Interaction Model for Multiple DEGs

Since the DEGs need to negotiate the distribution of the benefits gained from cooperation, this forms a bargaining relationship under cooperative game theory. It is assumed that the negotiation starting point for each DEG is the negative value of its non-cooperative independent operating cost C_i^{non}, i.e., $d_i = -C_i^{\mathrm{non}}$; while in the cooperative case, the operating cost of each DEG is C_i, and the corresponding utility level is $u = -C_i$. During the cooperative negotiation process, each DEG operator aims to maximize the benefits derived from cooperation as much as possible. Therefore, the cooperation alliance is constructed based on a Nash bargaining model with the joint interaction objective function as follows:

$$\operatorname{Max}\sum_{i=1}^{N}\ln\left\{\begin{array}{l} C_i^{\mathrm{non}} - \sum\limits_{t\in\mathcal{T}}\left(\lambda_t^{\mathrm{bg}}P_{\mathrm{e},i,t}^{\mathrm{bg}} - \lambda_t^{\mathrm{sg}}P_{\mathrm{e},i,t}^{\mathrm{sg}}\right) - \sum\limits_{t\in\mathcal{T}}\lambda_t^{\mathrm{gas}}\left(P_{\mathrm{g},i,t}^{\mathrm{chp}} + P_{\mathrm{g},i,t}^{\mathrm{gb}}\right) + \sum\limits_{t\in\mathcal{T}}\left[k_i\ln\left(1+L_{i,t}^{\mathrm{e}}\right) - a_i\Delta L_{i,t}^{\mathrm{h}}\right] \\ - \sum\limits_{t\in\mathcal{T}}\left[c_{\mathrm{e},i}^{\mathrm{chp}}P_{\mathrm{e},i,t}^{\mathrm{chp}} + c_{\mathrm{h},i}^{\mathrm{gb}}P_{\mathrm{h},i,t}^{\mathrm{gb}} + c_i^{\mathrm{es}}\left(P_{\mathrm{e},i,t}^{\mathrm{ch}} + P_{\mathrm{e},i,t}^{\mathrm{dis}}\right) + c_i^{\mathrm{hs}}\left(P_{\mathrm{h},i,t}^{\mathrm{ch}} + P_{\mathrm{h},i,t}^{\mathrm{dis}}\right)\right] \\ - \sum\limits_{t\in\mathcal{T}}\left[\left(\lambda_{\mathrm{e},t}^{\mathrm{bc}}P_{\mathrm{e},i,t}^{\mathrm{bc}} - \lambda_{\mathrm{e},t}^{\mathrm{sc}}P_{\mathrm{e},i,t}^{\mathrm{sc}}\right) + \left(\lambda_{\mathrm{h},t}^{\mathrm{bc}}P_{\mathrm{h},i,t}^{\mathrm{bc}} - \lambda_{\mathrm{h},t}^{\mathrm{sc}}P_{\mathrm{h},i,t}^{\mathrm{sc}}\right)\right] \end{array}\right\} \tag{7.30}$$

It should be noted that the non-cooperative operating cost C_i^{non} in the objective function (7.30) is a known input parameter, independently calculated by each DEG operator, and does not include the operational transaction cost with other DEGs C_i^{vpp}. In the cooperative interaction, the operating cost of each DEG includes the operational transaction cost with other DEGs, and the magnitude of these transaction costs will affect the internal operating plan of each DEG. Since there are no third-party profit-making entities, the transaction costs and volumes between DEGs are in an overall balanced state within the cooperation alliance: that is, at any given time $t \in \mathcal{T}$, the total alliance transaction energy for all buyer DEGs equals the total alliance transaction energy for all seller DEGs; and the total community transaction cost for all buyer sub-regions equals the total community transaction revenue for all seller sub-regions, as shown in Eqs. (7.31)–(7.34):

$$\sum_{i=1}^{N} P_{\mathrm{e},i,t}^{\mathrm{bc}} = \sum_{i=1}^{N} P_{\mathrm{e},i,t}^{\mathrm{sc}}, \quad \forall t \in \mathcal{T} \tag{7.31}$$

$$\sum_{i=1}^{N} P_{\mathrm{h},i,t}^{\mathrm{bc}} = \sum_{i=1}^{N} P_{\mathrm{h},i,t}^{\mathrm{sc}}, \quad \forall t \in \mathcal{T} \tag{7.32}$$

$$\sum_{i=1}^{N} \left(\lambda_{\mathrm{e},t}^{\mathrm{bc}} P_{\mathrm{e},i,t}^{\mathrm{bc}} - \lambda_{\mathrm{e},t}^{\mathrm{sc}} P_{\mathrm{e},i,t}^{\mathrm{sc}}\right) = 0, \quad \forall t \in \mathcal{T} \tag{7.33}$$

$$\sum_{i=1}^{N} \left(\lambda_{\mathrm{h},t}^{\mathrm{bc}} P_{\mathrm{h},i,t}^{\mathrm{bc}} - \lambda_{\mathrm{h},t}^{\mathrm{sc}} P_{\mathrm{h},i,t}^{\mathrm{sc}}\right) = 0, \quad \forall t \in \mathcal{T} \tag{7.34}$$

To simplify the variables, the electricity purchase quantity $P_{\mathrm{e},i,t}^{\mathrm{bc}}$ and the electricity sale quantity $P_{\mathrm{e},i,t}^{\mathrm{sc}}$ of each DEG at the same time t are transformed into a common variable—the alliance transaction electricity $P_{\mathrm{e},i,t}^{\mathrm{ex}} = P_{\mathrm{e},i,t}^{\mathrm{bc}} - P_{\mathrm{e},i,t}^{\mathrm{sc}}$, where a positive value of the alliance transaction electricity indicates that the DEG purchases electricity from the cooperation alliance, and a negative value indicates that the DEG sells electricity to the cooperation alliance. Similarly, the heat purchase and sale quantities of each DEG at the same time t are transformed into a common variable—the alliance transaction heat $P_{h,i,t}^{\mathrm{ex}} = P_{h,i,t}^{\mathrm{bc}} - P_{h,i,t}^{\mathrm{sc}}$. Thus, Eqs. (7.31) and (7.32) can be further transformed as:

$$P_{\mathrm{e},i,t}^{\mathrm{ex}} = P_{\mathrm{e},i,t}^{\mathrm{bc}} - P_{\mathrm{e},i,t}^{\mathrm{sc}} \tag{7.35}$$

$$\sum_{i=1}^{N} P_{\mathrm{h},i,t}^{\mathrm{ex}} = 0, \quad \forall t \in \mathcal{T} \tag{7.36}$$

Correspondingly, the electricity transaction cost between each DEG and other DEGs during the trading interaction is given by $C_{\mathrm{e},i,t}^{\mathrm{ex}} = \lambda_{\mathrm{e},t}^{\mathrm{bc}} P_{\mathrm{e},i,t}^{\mathrm{bc}} - \lambda_{\mathrm{e},t}^{\mathrm{sc}} P_{\mathrm{e},i,t}^{\mathrm{sc}} = \lambda_{\mathrm{e},i,t}^{\mathrm{ex}} P_{\mathrm{e},i,t}^{\mathrm{ex}}$. This equation is reasonable because the alliance transaction process is essentially a virtual clearing market, where the purchase price of energy faced by a DEG generally equals the sale price of energy at that moment. Similarly, the alliance heat transaction cost for each DEG is given by $C_{\mathrm{h},i,t}^{\mathrm{ex}} = \lambda_{\mathrm{h},t}^{\mathrm{bc}} P_{\mathrm{h},i,t}^{\mathrm{bc}} - \lambda_{\mathrm{h},t}^{\mathrm{sc}} P_{\mathrm{h},i,t}^{\mathrm{sc}} = \lambda_{\mathrm{h},i,t}^{\mathrm{se}} P_{\mathrm{h},i,t}^{\mathrm{ex}}$. Therefore, Eqs. (7.35) and (7.36) can be further transformed as:

$$\sum_{i=1}^{N} C_{\mathrm{e},i,t}^{\mathrm{ex}} = 0, \quad \forall t \in \mathcal{T} \tag{7.37}$$

$$\sum_{i=1}^{N} C_{\mathrm{h},i,t}^{\mathrm{ex}} = 0, \quad \forall t \in \mathcal{T} \tag{7.38}$$

Therefore, the objective function for the multi-region interactive model of the multi-energy DEG cooperation alliance is:

$$\text{Max}\sum_{i=1}^{N}\ln\left\{\begin{array}{l} C_i^{\text{non}}-\sum\limits_{t\in\mathcal{T}}\left(\lambda_t^{\text{bg}}P_{\text{e},i,t}^{\text{bg}}-\lambda_t^{\text{sg}}P_{\text{e},i,t}^{\text{sg}}\right)-\sum\limits_{t\in\mathcal{T}}\lambda_t^{\text{gas}}\left(P_{\text{g},i,t}^{\text{chp}}+P_{\text{g},i,t}^{\text{gb}}\right)+\sum\limits_{t\in\mathcal{T}}\left[k_i\ln\left(1+L_{i,t}^{\text{e}}\right)-a_i\Delta L_{i,t}^{\text{h}}\right] \\ -\sum\limits_{t\in\mathcal{T}}\left[c_{\text{e},i}^{\text{chp}}P_{\text{e},i,t}^{\text{chp}}+c_{\text{h},i}^{\text{gb}}P_{\text{h},i,t}^{\text{gb}}+c_i^{\text{es}}\left(P_{\text{e},i,t}^{\text{ch}}+P_{\text{e},i,t}^{\text{dis}}\right)+c_i^{\text{hs}}\left(P_{\text{h},i,t}^{\text{ch}}+P_{\text{h},i,t}^{\text{dis}}\right)\right]-\sum\limits_{t\in\mathcal{T}}\left(C_{\text{e},i,t}^{\text{ex}}+C_{\text{h},i,t}^{\text{ex}}\right)\end{array}\right\} \tag{7.39}$$

Due to the monotonic increasing nature of the ln function, Eq. (7.39) is equivalent to:

$$\text{Max}\sum_{i=1}^{N}\left\{\begin{array}{l} C_i^{\text{non}}-\sum\limits_{t\in\mathcal{T}}\left(\lambda_t^{\text{bg}}P_{\text{e},i,t}^{\text{bg}}-\lambda_t^{\text{sg}}P_{\text{e},i,t}^{\text{sg}}\right)-\sum\limits_{t\in\mathcal{T}}\lambda_t^{\text{gas}}\left(P_{\text{g},i,t}^{\text{chp}}+P_{\text{g},i,t}^{\text{gb}}\right)+\sum\limits_{t\in\mathcal{T}}\left[k_i\ln\left(1+L_{i,t}^{\text{e}}\right)-a_i\Delta L_{i,t}^{\text{h}}\right] \\ -\sum\limits_{t\in\mathcal{T}}\left[c_{\text{e},i,i}^{\text{chp}}P_{\text{e},i,t}^{\text{chp}}+c_{\text{h},i}^{\text{gb}}P_{\text{h},i,t}^{\text{gb}}+c_i^{\text{es}}\left(P_{\text{e},i,t}^{\text{ch}}+P_{\text{e},i,t}^{\text{dis}}\right)+c_i^{\text{hs}}\left(P_{\text{h},i,t}^{\text{ch}}+P_{\text{h},i,t}^{\text{dis}}\right)\right]-\sum\limits_{t\in\mathcal{T}}\left(C_{\text{e},i,t}^{\text{ex}}+C_{\text{h},i,t}^{\text{ex}}\right)\end{array}\right\} \tag{7.40}$$

That is to say,

$$\text{Max}-\sum_{i=1}^{N}\left\{\begin{array}{l} \sum\limits_{t\in\mathcal{T}}\left(\lambda_t^{\text{bg}}P_{\text{e},i,t}^{\text{bg}}-\lambda_t^{\text{sg}}P_{\text{e},i,t}^{\text{sg}}\right)-\sum\limits_{t\in\mathcal{T}}\lambda_t^{\text{gas}}\left(P_{\text{g},i,t}^{\text{chp}}+P_{\text{g},i,t}^{\text{gb}}\right)+\sum\limits_{t\in\mathcal{T}}\left[k_i\ln\left(1+L_{i,t}^{\text{e}}\right)-a_i\Delta L_{i,t}^{\text{h}}\right] \\ -\sum\limits_{t\in\mathcal{T}}\left[c_{\text{e},i}^{\text{chp}}P_{\text{e},i,t}^{\text{chp}}+c_{\text{h},i}^{\text{gb}}P_{\text{h},i,t}^{\text{gb}}+c_i^{\text{es}}\left(P_{\text{e},i,t}^{\text{ch}}+P_{\text{e},i,t}^{\text{dis}}\right)+c_i^{\text{hs}}\left(P_{\text{h},i,t}^{\text{ch}}+P_{\text{h},i,t}^{\text{dis}}\right)\right]\end{array}\right\} \\ +\sum_{i=1}^{N}C_i^{\text{non}}+\sum_{i=1}^{N}\sum_{t\in\mathcal{T}}\left(C_{\text{e},i,t}^{\text{ex}}+C_{\text{h},i,t}^{\text{ex}}\right) \tag{7.41}$$

Due to the transaction balance constraints between different DEGs at the cooperation alliance level, i.e., (7.37) and (7.38), and since C_i^{non} is a known constant, it can be further transformed into:

$$\text{Min}\sum_{i=1}^{N}\left\{\begin{array}{l} \sum\limits_{t\in\mathcal{T}}\left(\lambda_t^{\text{bg}}P_{\text{e},i,t}^{\text{bg}}-\lambda_t^{\text{sg}}P_{\text{e},i,t}^{\text{sg}}\right)-\sum\limits_{t\in\mathcal{T}}\lambda_t^{\text{gas}}\left(P_{\text{g},i,t}^{\text{chp}}+P_{\text{g},i,t}^{\text{gb}}\right)+\sum\limits_{t\in\mathcal{T}}\left[k_i\ln\left(1+L_{i,t}^{\text{e}}\right)-a_i\Delta L_{i,t}^{\text{h}}\right] \\ -\sum\limits_{t\in\mathcal{T}}\left[c_{\text{e},i}^{\text{chp}}P_{\text{e},i,t}^{\text{chp}}+c_{\text{h},i}^{\text{gb}}P_{\text{h},i,t}^{\text{gb}}+c_i^{\text{es}}\left(P_{\text{e},i,t}^{\text{ch}}+P_{\text{e},i,t}^{\text{dis}}\right)+c_i^{\text{hs}}\left(P_{\text{h},i,t}^{\text{ch}}+P_{\text{h},i,t}^{\text{dis}}\right)\right]\end{array}\right\} \tag{7.42}$$

From Eq. (7.42), it can be seen that the Nash bargaining-based multi-region joint interaction model essentially aims to minimize the overall operational cost of the integrated energy cooperation alliance. Since the social benefits of the entire cooperation alliance are the negative values of the operating costs of each DEG, this Nash bargaining model can maximize the social benefits of the entire alliance.

7.3.3 Case Analysis

An energy cooperation alliance consisting of four DEG operators is constructed. The economic parameters of the devices for each DEG are shown in Table. 7.1. The electricity and heat load data for each DEG are presented in Fig. 7.10, and the

Table 7.1 Economic and technical parameters of energy supply devices in each DEG

	CHP unit operating cost (yuan/kWh)	GB unit operating cost (yuan/kWh)	ES unit operating cost (yuan/kWh)	HS unit operating cost (yuan/kWh)
DEG 1	0.05	0.004	0.01	0.01
DEG 2	0.06	0.003	0.01	0.01
DEG 3	0.07	0.002	0.01	0.01
DEG 4	0.08	0.001	0.01	0.01

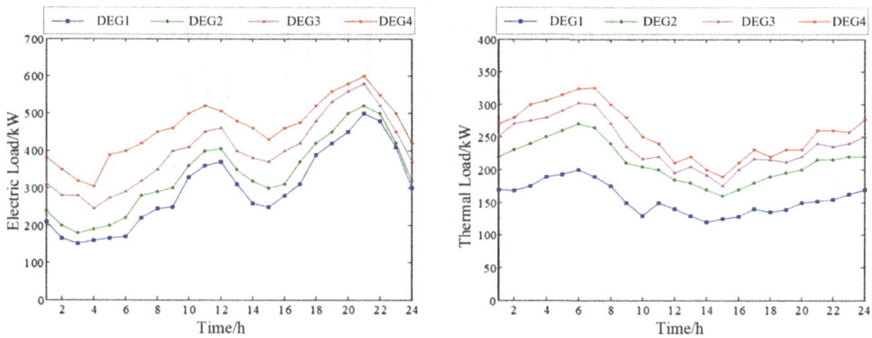

(a) Electric load curves of each sub-region (b) Thermal load curves of each sub-region

Fig. 7.10 Electric and thermal load curves

renewable energy output curves are shown in Fig. 7.11. The analysis is based on the AEMO (Australian Energy Market Operator) day-ahead energy price forecast data, with the electricity and natural gas prices shown in Fig. 7.12.

Figure 7.13 illustrates the electricity and thermal energy trading situations of each DEG under the cooperative alliance framework, where positive values indicate the

Fig. 7.11 Renewable energy output

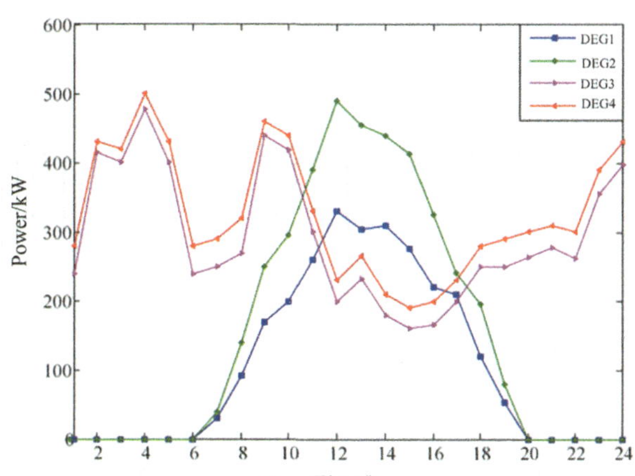

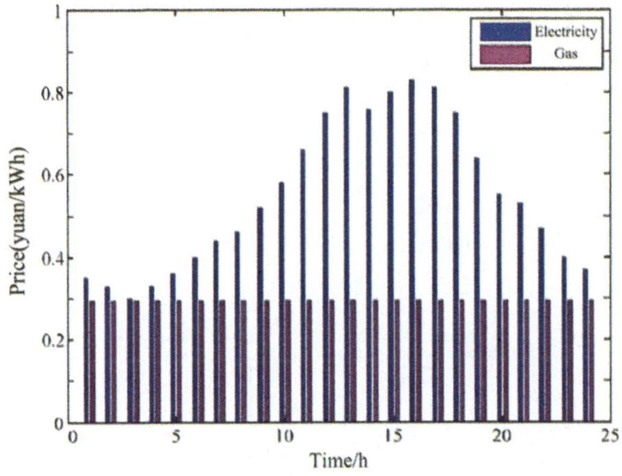

Fig. 7.12 Electricity and gas prices

purchase of energy, and negative values indicate the sale of energy. In the cooperative alliance framework, electricity trading between the DEGs primarily takes place between time periods 2 and 19, while thermal energy trading remains relatively stable across all 24 time periods.

In Fig. 7.13a, during time periods 2 to 7, DEG 1 and DEG 2 predominantly act as electricity buyers in the community exchange, while DEG 3 and DEG 4 mainly act as electricity sellers. This is because the renewable energy units in DEGs 1 and 2 are PV units, which cannot generate electricity at night. In contrast, the renewable energy in DEGs 3 and 4 is WP, which can sell excess electricity generated at night to DEGs 1 and 2.

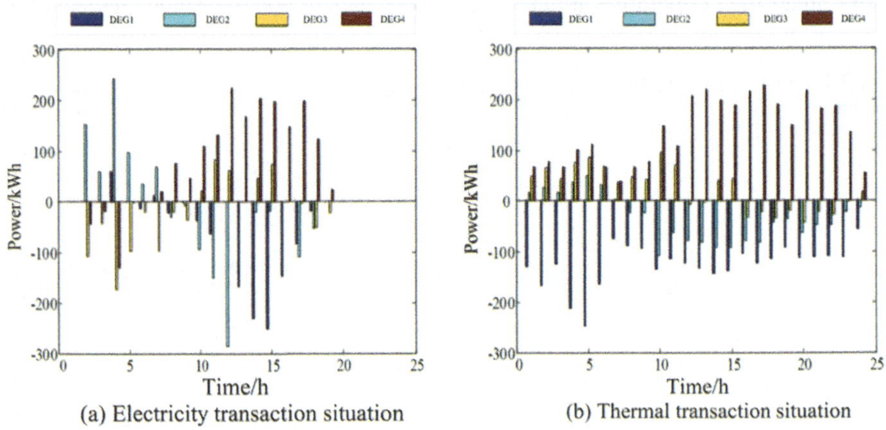

Fig. 7.13 The electricity and thermal energy trading situation between different DEGs

Table 7.2 Specific operational costs of each DEG

Cost (yuan)	DEG1		DEG 2		DEG 3		DEG 4	
	Sta.	Co.	Sta.	Co.	Sta.	Co.	Sta.	Co.
Electricity	1248.86	692.39	631.25	400.10	48.55	132.26	165.75	617.82
Gas	2036.48	3895.56	2736.48	3265.29	2971.48	2730.86	3319.60	1196.69
DG	110.44	217.38	168.82	206.79	202.32	190.04	263.84	59.29
Utility	2701.74	2701.21	2764.90	2764.90	2858.79	2858.82	2935.57	2935.69
Discomfort	11.19	14.50	12.44	12.69	12.81	11.28	10.98	10.15
Transaction	0	−1487.7	0	−410.66	0	96.4	0	1801.99
Total	705.24	630.89	783.69	709.32	376.38	302.02	824.61	750.25

During time periods 8 to 19, DEGs 1 and 2 begin selling electricity to DEGs 3 and 4. This change occurs because the WP output from DEGs 3 and 4 significantly decreases during the day, while the PV output from DEGs 1 and 2 significantly increases, causing a reversal in the roles of buyers and sellers.

Regarding the thermal energy trading situation shown in Fig. 7.13b, DEGs 1 and 2 primarily act as sellers in the thermal energy market (except for time periods 1 to 7, when DEG 2 also purchases a small amount of thermal energy from the community exchange). This is due to the relatively low operating cost of the CHP units in DEGs 1 and 2. Consequently, under reasonable trading price incentives, it is more economical for DEGs 3 and 4 to purchase thermal energy directly from the alliance rather than relying solely on their own units. Overall, under the cooperative alliance framework, DEGs 1, 2, 3, and 4 exhibit significant complementary advantages in both electricity and thermal energy trading. As a result, the amount of direct trading with the main grid by each DEG is likely to decrease.

Table 7.2 presents the specific operational costs of each DEG under standalone and cooperative operation conditions. It can be observed that although the purchase cost of electricity from the main grid for DEG 4 increases from 165.75 yuan under standalone operation to 617.82 yuan under cooperative operation, its total operational cost decreases from 824.61 yuan to 750.25 yuan, a reduction of 9.02%. Looking at the alliance transaction cost item in Table 7.2, positive values indicate that the DEG incurs transaction costs in cooperative operation, while negative values indicate that the DEG earns income from alliance transactions. It can be seen that DEGs 1 and 2 receive transaction income, whereas DEGs 3 and 4 only incur transaction costs. This suggests that DEGs 1 and 2 mainly act as sellers in the alliance transactions, while DEGs 3 and 4 primarily act as buyers. This is consistent with the electricity and thermal energy trading situations shown in Fig. 7.13. Meanwhile, when comparing the changes in the total operational costs of each DEG under standalone and cooperative modes, the total operational costs of DEGs 1, 2, 3, and 4 decrease by 10.54%, 9.49%, 19.76%, and 9.02%, respectively. This strongly demonstrates that the cooperative mode effectively reduces the operational costs of each DEG, encouraging the transition from standalone operation to the cooperative alliance mode.

References

Ai Q, Zhang J (2010) Optimization bidding strategies of microgrids based on multi-agent system (in Chinese). Power Syst Technol 34(12):46–51. https://doi.org/10.13335/j.1000-3673.pst.2010.02.008

Ai Q, Liu S, Wu R, Fang S (2016) Research and prospect of multi-agent system in energy internet system (in Chinese). High Volt Eng 42(9):2697–2706. https://doi.org/10.13336/j.1003-6520.hve.20160907002

Bai Y, Li A, Xia Q (2016) Electricity business marketing modes in the new environment and new electricity pricing systems (in Chinese). 44(5):10–16. https://doi.org/10.7667/PSPC201501

Chen P (2011) Micro-grid-based electricity market transactions and economic operation (in Chinese). Beijing Jiaotong University

Cui H, Liu J, Guo K, Kang K (2016) Ruminations on electricity marketing strategies and the law of supply and demand (in Chinese). Mod Market 1:57. https://doi.org/10.3969/j.issn.1009-2994.2016.01.041

Duan J, Sun Q, Wu C, Yang Y, Tu S (2024) Non-cooperative game based optimal dispatch model for integrated energy systems. CSEE J Power Energy Syst 10(3):1195–1207. https://doi.org/10.17775/CSEEJPES.2020.06460

El Rahi G, Etesami SR, Saad W et al (2017) Managing price uncertainty in prosumer-centric energy trading: a prospect-theoretic Stackelberg game approach. IEEE Trans Smart Grid 10(1):702–713. https://doi.org/10.1109/TSG.2017.2750706

Gao S, Chau KT, Liu C, Wu D, Chan CC (2014) Integrated energy management of plug-in electric vehicles in power grid with renewables. IEEE Trans Veh Technol 63(7):3019–3027. https://doi.org/10.1109/TVT.2014.2316153

Gkatzikis L, Koutsopoulos I, Salonidis T (2013) The role of aggregators in smart grid demand response markets. IEEE J Sel Areas Commun 31(7):1247–1257. https://doi.org/10.1109/JSAC.2013.130708

Jenkins S, Annaswamy A, Hansen J, Knudsen J (2015) A dynamic model of the combined electricity and natural gas markets. In: 2015 IEEE power & energy society innovative smart grid technologies conference (ISGT), Washington, DC, pp 1–5. https://doi.org/10.1109/ISGT.2015.7131879

Lee K-H, Baldick R (2003) Solving three-player games by the matrix approach with application to an electric power market. IEEE Trans Power Syst 18(4):1573–1580. https://doi.org/10.1109/TPWRS.2003.818744

Li C (2016) A study of pricing strategies in electricity marketing (in Chinese). J Shandong Ind Technol 24:185+211. https://doi.org/10.16640/j.cnki.37-1222/t.2016.24.169

Li Y et al (2024) Mixed game-based two-stage planning of PV and energy storage in distribution system with multiple integrated energy microgrids. IEEE Trans Ind Appl 60(6):8054–8066. https://doi.org/10.1109/TIA.2024.3428442

Lu T, Wang Z, Ai Q, Lee W-J (2017) Interactive model for energy management of clustered microgrids. IEEE Trans Ind Appl 53(3):1739–1750. https://doi.org/10.1109/TIA.2017.2657628

Ma L, Fang Z, Guo L, Xue S, Li K (2014) Latest development trends of international electricity markets and their enlightenment (in Chinese). Autom Electr Power Syst 38(13):1–9. https://doi.org/10.7500/AEPS20140520007

Parhizi S, Khodaei A, Shahidehpour M (2018) Market-based versus price-based microgrid optimal scheduling. IEEE Trans Smart Grid 9(2):615–623. https://doi.org/10.1109/TSG.2016.2558517

Peng H (2014) Analysis of electricity prices in an electricity market economy (in Chinese). Mod Econ Inf 20:383–384. https://doi.org/10.3969/j.issn.1001-828X.2014.20.301

Syothert A, MacLeod I (2000) Competitive bidding as a control problem. IEEE Trans Power Syst 15(1):88–94. https://doi.org/10.1109/59.852105

Wu L, Sun K, Chen Y (2015) A comparison of renewable energy policies in imperfect competition electricity market (in Chinese). China Popul Resour Environ 25(10):53–60. https://doi.org/10.3969/j.issn.1002-2104.2015.10.008

Zakariazadeh A, Jadid S (2014) Smart microgrid energy and reserve scheduling with demand response using stochastic optimization. Int J Electr Power Energy Syst 63:523–533. https://doi.org/10.1016/j.ijepes.2014.06.037

Chapter 8
Outlook on Modeling and Collaborative Control Technology for Future Distributed Energy Grid Cluster Operations

This section focuses on the modeling and collaborative control technology prospects of distributed energy grid clusters and explores the future technological development trends in this field. With the comprehensive development of the power big data platform, the amount of data obtained from power marketing, production scheduling, and other systems is growing exponentially. These data not only include telemetry and remote signals but also encompass various forms of data such as operation tickets and videos. Faced with such a large and heterogeneous dataset, how to transform these "data raw materials" into effective driving forces has become the core challenge.

Current research often adopts model-based methods, but traditional models are often deterministic and low dimensional, making it difficult to fully absorb massive data and handle heterogeneous data, which can easily lead to erroneous conclusions. With the development of artificial intelligence and machine learning, new ideas have been provided to solve this problem. This chapter will delve into the application of machine learning in distributed energy grid clusters, particularly by constructing Random Matrix Models (RMMs) to handle big data systems, utilizing RMT to analyze certain random features in large-scale data systems, and thereby achieving state perception, anomaly detection, and construction of operational status indicator systems for power systems.

In addition, the specific applications of artificial intelligence technologies such as expert systems, artificial neural networks, evolutionary algorithms, fuzzy logic theory, and MASs in distributed energy grid clusters will be introduced, aiming to establish a robust and event-processing capable artificial MAS to meet the intelligent control needs of distributed energy grid clusters.

T. Lu et al., *Operation Modeling and Collaborative Regulation of Distributed Energy Grid Clusters*, Smart Energy Systems, https://doi.org/10.1007/978-981-96-9942-1_8

8.1 Technological Development Trends

8.1.1 *Artificial Intelligence and Machine Learning Optimization Modeling and Control Technology*

Currently, the construction of big data platforms for electricity is in full swing. Big data platforms for power marketing, production, scheduling, and other systems will truly record telemetry, remote signaling, and even operation tickets, videos, and other data generated by various ports (Yang et al. 2013). Some platforms can even collect data from different ports (heterogeneous data). The core challenge in implementing machine learning is to convert the collected 'data raw materials' into effective driving forces.

The classic solution for current research is model-based, and models are often deterministic and low dimensional. This means that specific models often cannot absorb more data and heterogeneous data; If abnormal data is chosen to be brought into the model, it often leads to incorrect conclusions. The current mainstream data processing methods are still model-driven: for example, PCA aims to select "high-quality" data from massive amounts of data and bring it into the model, while intelligent algorithms such as neural networks identify or correct model parameters through training based on a large number of samples.

On the other hand, data will become increasingly accessible, and the amount of data generated by power grid operation and equipment monitoring will grow exponentially; It is urgent to explore a driving tool that uses data as the "raw material" to transform data resources into effective information, to model, analyze, and classify tasks in distributed energy grid groups, and to expand the capabilities of MASs through artificial intelligence learning and memory, to better adapt to intelligent control of distributed energy grid groups. Establish an artificial MAS for coordinated control of distributed energy grid groups, determining the roles and responsibilities of each distributed power source in the distributed energy grid group. The system has strong robustness and event processing capabilities, and has strong stability through network interconnection and conflict resolution mechanisms (Zakariazadeh et al. 2014). The autonomy of each agent enables it to process different events in parallel, while its learning and memory abilities enable it to respond quickly to repeated events.

(1) **The Application of Machine Learning in Distributed Energy Grid Clusters**

Model-based and data-driven are two typical SA techniques applied in machine learning. Each driving method has its application scenarios, and cross-analyzing and mutually verifying the two modes can ultimately achieve an understanding of the power grid.

In terms of a model-driven approach, based on multi-source data analysis and time series neural network algorithm, a distributed power generation long/short/ultra short-term prediction model is established; By using correlation analysis, regression analysis, neural network data mining, and analysis methods, sensitive factors affecting rigid and flexible loads are identified, and the potential for flexible loads to participate in system regulation is explored (Li et al. 2017). Long/short/ultra short-term load prediction models are established to achieve multi-time scale and multi-agent comprehensive prediction.

In terms of a data-driven approach, big data analysis methods are adopted to obtain high-dimensional indicators reflecting changes in the state of the power system; Combining dynamic equivalence and other mechanism modeling methods with energy methods, calculate low dimensional indicators that characterize changes in the power system; The two indicators are compared, verified and integrated to form a new indicator that combines the advantages of the two methods and is used to characterize the evolution of the energy Internet. Figure 8.1 shows the energy Internet awareness method based on the data-driven and mechanism model.

Model-driven is a method based on optimization theory and intelligent algorithms, which determines the objective function of a problem and sets corresponding constraints to ultimately obtain the optimal solution or solution set of the problem. In the field of power grid state perception, it can be roughly divided into two parts: static stability and dynamic stability analysis. The currently mature methods include the calculation of static and transient stability domains, as well as the proposal of relevant safety margin indicators (Gkatzikis et al. 2013). For example, research on the static voltage stability of power systems has been very in-depth. The traditional calculation of transient feasible region mainly includes a direct method based on Lyapunov stability theory, transient energy function method, mixed method, trajectory analysis method, and analysis method based on trajectory sensitivity. On the other hand, in

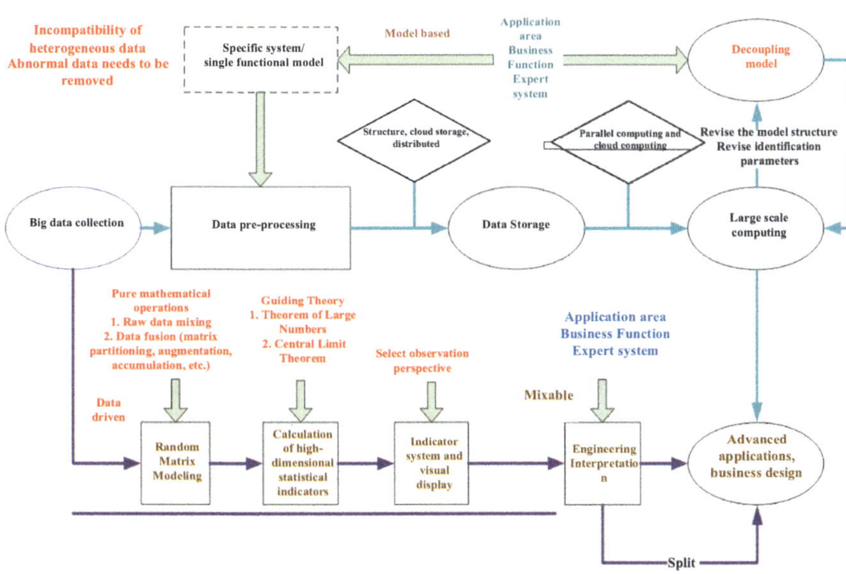

Fig. 8.1 Energy Internet awareness method based on data-driven and mechanism model

terms of big data systems, as big data has just emerged, there is currently no unified and clear definition in mathematics. Therefore, based on the current work being carried out, the following definition is made.

At a certain point in time, the sampled data in the system can be arranged into an n-dimensional vector, that is $x_i \in R^n$; The system can generate a database $\Omega \in R^{n \times t}$ by observing a series of data over a while (such as t). Big data is such a system:

(1) The data sampling time t of the system is sufficiently long;
(2) Functions $f(x_1, x_2, \ldots, x_T)$ can be defined for sampled data at any given time in the system.

The machine learning scheme in this chapter will establish a random matrix model (RMM) for big data systems using RMT. RMT processes data in units of matrices and can handle independently identically distributed (IID) data without requiring distribution or features (such as Gaussian distribution, Hermitian matrix, etc.). It only requires data to be large enough rather than infinite, making it highly adaptable and suitable for handling most engineering problems, especially for analyzing large-scale data systems with certain randomness. The theory of random matrices holds that when there is only white noise, small perturbations, and measurement errors in the system, the data will exhibit statistical randomness; When there are event/signal sources in the system, the operating mechanism and internal mechanism of the system will change under their influence, and its random statistical characteristics will be broken (Nikmehr and Ravadanegh 2015). The single loop law and Marchenko Pastur (MP) law are both significant breakthroughs in the theoretical system of random matrices. Based on these theories, further research can be conducted on the LES of random matrices, and MSR is a specific LES.

(1) Characteristics of Model Driven/Data Driven

With the rise of intelligent algorithms, model-driven methods will be suitable for modeling and computing large and complex system problems, and their detailed and comprehensive system design applied to various scenario analyses. On the one hand, the energy internet has changed the operation mode of the traditional power grid. The DG with the characteristics of production and consumption integration makes the system more flexible. The model-driven method combined with some new technologies, such as dynamic mechanism modeling, can meet the needs of static and dynamic analysis of the system. On the other hand, model-driven methods require long-term preliminary research and extensive offline calculations for different scenarios. At the same time, the power system in the context of the energy Internet is essentially a volume system, which will generate different types of data every moment. These data need to be processed in time, and the data obtained will inevitably have errors and missing due to sampling, communication, human, and other factors, and the results require authenticity and reliability. Therefore, the power system has the property of 4Vs, which is a typical open information energy fusion system. In terms of processing and analyzing a large number of real-time data, the model-driven method is not flexible enough. Based on the relevant theories of big data and RMT, an RRM mapping power grid system is established to describe the overall system situation through the correlation between data, forming a new high-dimensional spatial observation perspective of the power grid and providing an advanced means for power system cognition. It is the core technology for the future development and construction of smart grids. Therefore, data-driven methods can complement and validate model-driven approaches. The advantages of a data-driven approach are as follows:

(1) Directly describing the situation by analyzing the correlation of data rather than establishing mathematical models avoids the problems of difficult modeling or inaccurate models caused by the complexity of power grid topology, diversification of components, adjustability and uncertainty of renewable energy and flexible loads, greatly reducing hardware resource requirements and improving the credibility of analysis.
(2) Describing problems through high-dimensional correlations between data rather than causal relationships, quantitatively defining the correlations between events, directly identifying the source of faults or events, and avoiding the problem of difficulty in describing causal relationships due to system uncertainty, contingency, and multiple repeated and complex recursive relationships. The causal chain of summer light intensity on the power system can be analyzed using "light intensity → solar power generation → power system". On the other hand, the light intensity can also affect the power system by influencing human behavior, such as "light intensity → temperature → air conditioning load → power system". Therefore, it is difficult to describe the impact of a physical parameter on a system based on assumptions or simplified models, and even if it can be described, the errors implied by this description will continue to accumulate as the causal chain passes.
(3) Data-driven analysis primarily relies on the theory of random matrices. RMT is a universal theoretical system for studying big data. This theoretical system is based on data analysis, and its application scope is not limited to specific network structures or mechanisms, with high universality.
(4) The data analysis process based on random matrices is an independent mathematical process that can be directly combined with parallel computing or distributed computing at the algorithm level to solve the problem of the curse of dimensionality. It can perform real-time overall situation analysis on systems with tens of thousands of nodes, greatly reducing the time and resource requirements.
(5) Due to the large amount of data, the system will have very high redundancy, and data-driven analysis, especially unsupervised analysis, will have excellent robustness. Unsupervised system analysis does not rely on manual training processes or individual sample selection. By extracting high-dimensional parameters from all parameters to represent the system's operating situation, it can effectively solve the problems that traditional statistical methods find difficult to solve and avoid, such as noise accumulation, false correlation, and accidental correlation, which occur with the increase of data in big data systems.

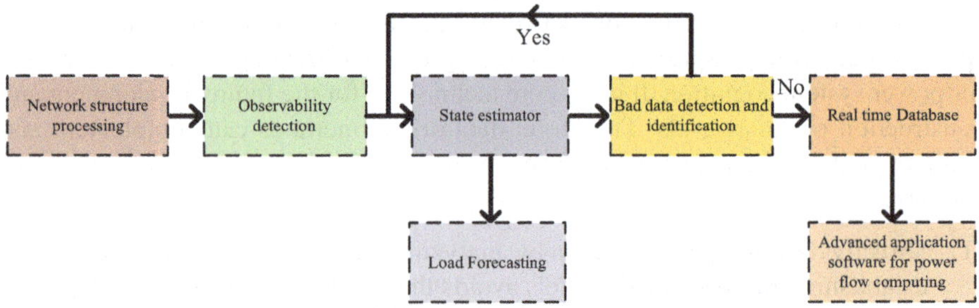

Fig. 8.2 Traditional power system state estimation function flowchart

Especially in cases where the system has bad data or even core area data loss, information can still be accurately mined.

(2) **Model-driven/data-driven SA**

1. State estimation of the power system

Power system state estimation refers to the use of various measurement devices such as SCADA, PMU, etc. in the power system to collect system information and estimate the current operating state of the power system. Accurate and reliable state estimation is the foundation for advanced applications such as economic optimization and the operation of power systems. Figure 8.2 shows a typical state estimation process (Nikmehr and Ravadanegh 2016). Model-driven power system state estimation generally includes the following steps: model assumptions, state estimation, detection, and identification. Currently, mature state estimation algorithms such as Weighted Least Squares (WLS) are available.

In the context of the energy Internet, distributed energy producers and consumers are freely connected to the grid, which makes the system operation more complex. In this section, static and dynamic feasible domain models are constructed through offline operations, and real-time data collected is inputted into the model to achieve rapid estimation of the system state based on the relative position of the current operating point in the feasible domain. Figure 8.3 shows the variation of the L1 norm distance from the steady-state operating point of the system to the boundaries of various static voltage safety domains over time. The following are three specific scenarios for implementing state perception based on model-driven methods. The research content and general steps are as follows:

- Static feasible domain modeling based on the topology structure of large-scale interconnected systems
  a. Identification of weak nodes;
  b. Static feasible domain modeling based on weak node power injection or phase space;
  c. Establishment of safety margin indicators;
  d. Realize system security operation status awareness.

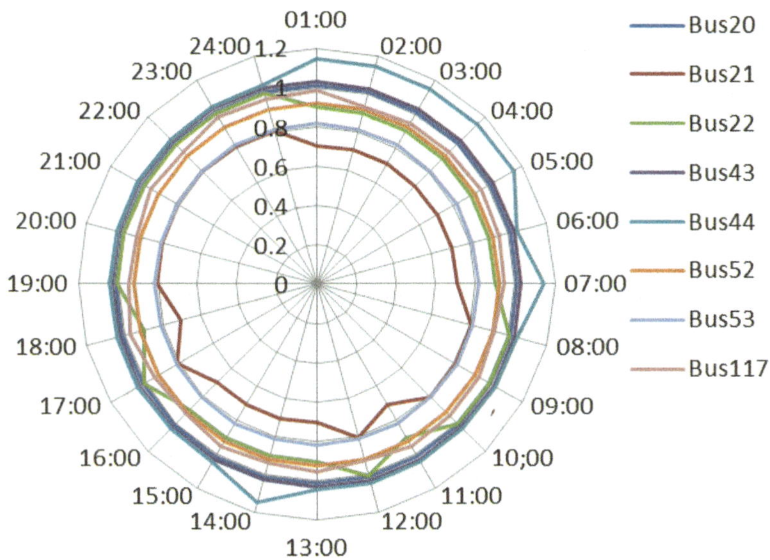

Fig. 8.3 L1 norm distance from the operating point to the SVSR boundary from 01:00 to 24:00

- Modeling of feasible area for power exchange of energy Internet with multiple DEGs
  a. Distributed energy grid group operation SA based on uncertain factor prediction;
  b. Perceive the internal device information of the distributed energy grid group and determine the feasible range of power exchange in the distributed energy grid group;
  c. Measure the power absorption capacity of large interconnected systems, consider the feasible area of distributed microgrid power exchange, and formulate the economic security optimization scheduling strategy of energy Internet;
  d. System security status perception based on security reserve capacity.
- Transient feasible region modeling based on dynamic mechanism model parameter identification
  a. Dynamic mechanism modeling for research objects;
  b. Establish a dynamic feasible region model based on energy function method and mechanism parameter identification;
  c. Perception of system transient security status based on current data.

Compared to model-driven state estimation methods, data-driven methods have better robustness in handling erroneous data (anomalies, missing). By monitoring various operational indicators, especially statistical indicators such as LES, in the power grid, the status and trends of power grid operation can be obtained, and the online operation status of the system can be evaluated to understand the power grid and provide decision-making support for power grid dispatch personnel, improve power supply quality and stability, and prevent potential crises. Regarding data-driven state perception methods, this project proposes two research objectives:

a. For a specific system, set a threshold value to determine the safe/stable feasible operating range of the system. Enable dispatchers to obtain system safety/stability margin information based on the relative position of the current operating point and boundary of the system;
b. Combining and verifying the big data cognitive methods of the power system with traditional classical cognitive methods (such as trajectory method, energy method, etc.) to achieve advanced applications such as automatic system warning; At the same time, provide auxiliary information for dispatchers to enhance their perception and understanding of the information.

2. Anomaly detection of distributed energy grid group

Distributed energy grid clusters are event-triggered systems, but in the early stages of events, their characteristics are very weak and uncertain. If its features can be amplified and extracted, disturbance scenes can be obtained promptly, and event types, locations, triggers, etc. can be identified, it will further enable precise SA of the power system, and elevate the security and stability control of distributed energy grid groups to a new level (Tian et al. 2016). The remaining part of this section demonstrates how to implement anomaly detection from both model-driven and data-driven perspectives.

a. Determine if there are any abnormalities in the system

According to the model-driven method, the implementation of system state perception can be roughly divided into the following two steps: (1) Based on the relevant principles of transient analysis and steady-state analysis of distributed energy grid systems, establish feasible regions; (2) Determine the relative position information of the system operating point within the feasible domain, achieve state perception and abnormal state recognition.

The implementation of system state perception based on data-driven methods includes the following steps: analyzing the data streams collected from distributed energy grid measurement devices to obtain their LES, and further mining indicators that characterize the system's operating status. When an event (abnormality or malfunction) occurs, the abnormal parts in the collected data can trigger significant changes in these signal indicators, thereby serving as a warning. LES, on the other hand, is directly driven by the data collected through sampling rather than based on evaluation models (often requiring assumptions, dimensionality reduction, etc.).

b. Abnormal identification and analysis, including the type, location, and causes of anomalies

In terms of model driving, different types of feasible domains can be established for system anomaly types, such as line thermal stability domain, static voltage safety domain, voltage stability domain, dynamic safety domain, etc. Determine the abnormal situation of the system by judging the type of domain boundary crossed by the operating point of the system. The data-driven approach uses correlation analysis to identify the relevant information of events, including location, type, and cause, and mutually confirms it from multiple sources to achieve power system cognition.

3. Constructing an indicator system for the operational status of microgrid systems

A complete indicator system modeling is beneficial for comprehensively perceiving the operating status and trends of the system. The indicator sets constructed based on model-driven methods can be roughly divided into two categories: static indicator sets and dynamic indicator sets. The static indicator set is mainly used to measure safety margins such as voltage safety, voltage stability, and line thermal stability. This part can be determined based on the norm distance between the system operating point and the corresponding feasible domain boundary. The dynamic indicator set is mainly used to measure the transient stability level of the system. According to traditional transient research methods such as the equal area method and Lyapunov energy function method, the analysis results of these methods are very reliable when dealing with single-machine infinite bus systems. Therefore, the dynamic equivalence mechanism modeling method is considered to equivalently convert the system to be studied into a single-machine or multi-machine system, obtain the system parameters required by the energy method, and achieve transient stability research of the system.

The indicator set obtained based on model-driven methods often has clear physical meanings but requires high data accuracy, and system dimensionality reduction and simplification are necessary when necessary. On the contrary, most indicator sets based on data-driven methods have no physical meaning and are only statistics, but they have good robustness in handling erroneous data (anomalies, missing data). With the development of sensors and communication technology, for multi-temporal and spatial span data, data-driven methods that fully utilize data resources, use pure mathematical steps, and do not rely on sampling processes and mechanistic models have better representation power for large-scale interconnected power grids. So model model-driven indicators and data-driven indicators can complement each other, constructing a clear and comprehensive indicator system with clear meaning.

4. System state perception based on the indicator system

In terms of system perception based on model-driven methods, conventional low dimensional indicators are usually used to measure the steady state of microgrid systems, including voltage safety margin, voltage stability margin, and line thermal stability margin. The indicators for measuring power system transients include transient energy function values, etc.

Compared to low-dimensional source data (such as V), data-driven high-dimensional linear eigenvalue statistical indicators LES (such as K_{MSR}) have better performance and robustness in the system. By combining the geographical and zoning information of the power grid, it can be visualized by zone to form a Power Map for observing the system situation. The geographic location of information is used as the basic two-dimensional information, while the third dimension (or more high-dimensional) can be used to observe statistical measures. Especially when combining LES with visualization, even if the data in the core area of the event is completely lost, analyzing only the data in the surrounding area and visualizing the obtained LES can still obtain real results, that is, discovering the existence of events in the system and locking the event area, while low dimensional data based on interpolation appears weak.

(3) Model-driven/data-driven distributed energy grid group state perception and anomaly detection method

1. Model-driven approach

Firstly, by studying the impact of node power injection on the system's security indicator set, identify the weak nodes in the system. Secondly, combined with the traditional static feasible region model, a static feasible region under low dimensional power injection space is constructed. Then, using the hyperplane approximation of the static feasible domain boundary, based on continuous power flow, the operating point iteratively calculates the approximate feasible domain boundary of the system and characterizes the complete static safety domain from the points and surfaces. Finally, based on the relative position of the operating point from the feasible boundary of the large-scale interconnected system, the security situation of the system is determined and whether the system is abnormal is judged. Regarding visualization, system dimensionality reduction can be achieved through weak node power injection. Figure 8.4 shows the feasible domain model in three-dimensional space, which represents the distance from the system operating point to the nearest safe domain boundary hyperplane. The calculation formula for the safety margin is as follows:

$$\xi_{d,i} = \frac{\left|\eta_{d,i1} d_1^0 + \cdots + \eta_{d,i,n_d} d_{n_d}^0 - L_{d,i}\right|}{N_{d,i}} \tag{8.1}$$

where $D_{0_0} = \left[d_1^0, \ldots d_{n_d}^0\right]$ represents the operating point of the system based on multidimensional space.

2. Data-driven approach

The first step in implementing event discovery is to design the basic architecture of distributed energy grid group application big data, mapping the distributed energy grid to a random matrix, which can be preprocessed to some extent, such as augmentation, partitioning, and other operations as shown in Fig. 8.5. The power system is a distributed interconnected real-time system that prioritizes safety and stability and has very high requirements for diversity. Some functions, such as transient analysis, also have high requirements for velocity. By using the moving split window (MSW) method and proposing LES such as mean spectrum radius index (MSR), the power

Fig. 8.4 Schematic diagram of the system running point to the nearest security domain interface

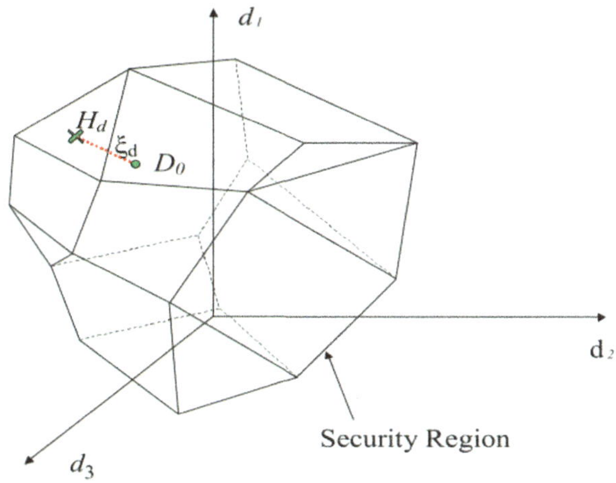

grid is connected to big data theory and algorithms through real-time data, forming a power system from the perspective of big data. A RMM of the power system is established. Research on big data algorithms, especially universal algorithms based on RMT that can handle large-scale general matrices (only requiring matrix rows and columns to be sufficiently large), such as the single loop law, and their role and performance in power RMMs. Compare distributed big data with traditional methods such as distributed PCA, clustering, and regression, analyze their feasibility, advantages, and applicability, and form an algorithm library for the power system. When the system runs into specific problems, call corresponding conventional/big data methods from the database to solve them. Further develop the integration of big data and power systems, and develop a visual power map for the power system; Develop event localization function using partitioning technology; Develop event cause locking function using augmented matrix technology and conduct correlation analysis; By observing LES, the function of quantifying the delay between two systems (caused by matrix synchronization, system inertia constant, communication time, etc.) can be achieved.

The basic implementation scheme for event discovery is to use the moving window method to process real-time and historical sampling data, forming a natural matrix with data type as the vertical axis and time as the horizontal axis. The eigenvalues and LES of the random matrix corresponding to different spatiotemporal data can be calculated, and the results can be compared with the single loop law/M-P law in RMT to discover whether there is a signal (not completely random) in the selected cross-sectional window, that is, whether there is an event in the corresponding system.

(3) **The Application of Artificial Intelligence in Distributed Energy Grid Clusters**

Artificial intelligence is an interdisciplinary field. At present, artificial intelligence technologies widely used in distributed energy grid clusters include expert systems,

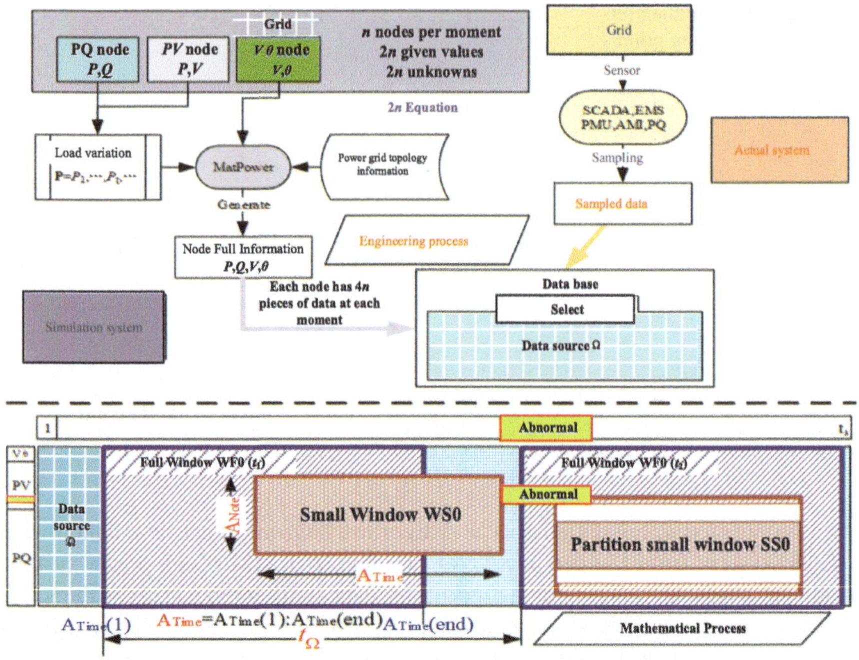

Fig. 8.5 Implementing anomaly detection using data-driven methods

artificial neural networks, evolutionary algorithms (genetic algorithms, particle swarm optimization algorithms, simulated annealing algorithms, etc.), fuzzy logic theory, and MASs. These technologies play an irreplaceable role in DG forecasting, energy management, monitoring, and early warning in distributed energy grid clusters. The following will briefly introduce these 5 artificial intelligence technologies:

(1) Expert system. The expert system is the most active and widely used method in artificial intelligence, and it is also one of the earliest branches of artificial intelligence. It is an intelligent computer system that contains a wealth of knowledge and experience from experts in the field. This knowledge and experience can be used to solve some complex problems in this field.
(2) Artificial neural network. Artificial neural network is an important branch of artificial intelligence, which abstracts human brain neurons, establishes simple models, and forms networks through different connection methods. It is a very effective predictive model and a research hotspot in the field of artificial intelligence.
(3) Evolutionary algorithm. Evolutionary algorithm, also known as evolutionary algorithm, is a simulation of the biological evolution process in nature. These evolutionary operations generally include gene coding, cross-compilation, population initialization, and so on. Compared with traditional optimization

algorithms, evolutionary algorithms are global optimization algorithms with high robustness and wide practicality. Due to its advantages of self-organization, adaptability, and self-learning, evolutionary algorithms can effectively handle complex problems that traditional algorithms find difficult to optimize, such as parameter optimization, scheduling, resource allocation, etc. Currently, the most commonly used evolutionary algorithms include the genetic algorithm, particle swarm optimization algorithm, and simulated annealing algorithm.

(4) Fuzzy logic theory. Fuzzy logic theory can enable computers to consider and reason about some real-life problems that are expressed in language according to human thinking patterns. If there is no clear standard for distinguishing between concepts such as fast and slow, the essence of fuzzy logic theory is to use defuzzification rules to enable computers to handle these fuzzy language expression problems.

(5) MAS. MAS technology integrates multiple disciplines such as artificial intelligence, distributed computing, and system theory. It goes beyond the ability of a single agent to handle problems and can be used to solve larger and more complex problems in real life. Multi-agent, in short, is the combination of multiple intelligent agents, to simplify large and complex systems into small ones that can quickly communicate and collaborate, and are easy to manage. This is beneficial for solving real-world decision-making problems with uncertainty and environmental complexity.

The proposal of distributed energy grid clusters can solve the problems of grid connection difficulties encountered by large power grids and is an important development direction for the intelligent future power system of distributed energy grid clusters. In the development process of intelligent distributed energy grid clusters, artificial intelligence is indispensable. Below are several main applications of artificial intelligence.

(1) Identification of risk factors. There are numerous risk factors in the operation of distributed energy grid clusters, and current research focuses on studying the risk factors of new energy power systems from a single factor, with little systematic analysis of these factors.

(2) Prediction. Both distributed power generation equipment generation forecasting and short-term load forecasting play a crucial role in energy management and optimization of distributed energy grid operations. In artificial intelligence technology, both neural networks and evolutionary algorithms have high efficiency and good usability in prediction. Therefore, how to improve and combining these algorithms to propose more accurate prediction algorithms is one of the important research directions for future distributed energy grid group regulation technology.

(3) Intelligent scheduling optimization. Whether it is a distributed energy grid group or a large power grid, it must meet the constraints of supply–demand balance, but at the same time, it also needs to take into account the goals of economic and environmental benefits, which form a single or multi-objective optimization

problem. Since it is an optimization, it inevitably relies on artificial intelligence technology.

(4) Fault diagnosis. More than 50% of faults in the power grid are caused by electrical equipment, so fault diagnosis of electrical equipment in distributed energy grid clusters is crucial for their safe, stable, and efficient operation. The IoT technology can monitor various operational data of power equipment, extract useful information from these data, and accurately determine the status of power equipment. This requires the use of big data technology and artificial intelligence technology, which complement each other to promote decision-makers to make reasonable and accurate judgments and reduce the failure rate of distributed energy grid group equipment.

(5) Risk dynamic assessment. The application of artificial intelligence technology can enable robots to replace humans in performing high-risk tasks such as high-altitude work and live work. Even through deep learning algorithms, distributed energy grid groups can achieve real-time identification and diagnosis of risks within the network, and achieve dynamic risk assessment.

Among many AI applications, intelligent sensors, control, and software applications are used to connect hundreds of millions of equipment and systems at the energy production end, transmission end, and consumer end, forming the "perception layer" of the ubiquitous power Internet, which is the key to the ubiquitous power IoT. There are huge challenges in China's distribution energy grid and distributed energy grid clusters. For example, an ordinary distribution community has over 200 distribution devices, but these devices are all invisible and intangible sub-terminals. From a national perspective, there are 4.8 million charging stations in China that are in an unconnected and unmanaged state, with a low degree of digitization. Any problems that arise require individual replacement and manual repair, which has a significant impact on the operation of the power grid. Especially when equipment or technology needs to be upgraded, the cost is high and the efficiency is low. The collaborative solution of cloud, distributed energy grid, edge, and end can significantly improve the cost of manual maintenance, failure rate, average power outage time, and so on.

Cloud: Building a distributed energy grid group AI brain with artificial intelligence as the core provides a solution to the above problems. It is a comprehensive platform that integrates various AI technologies such as machine learning, image recognition, and natural language processing. Based on meeting the requirements of traditional distributed energy grid automation systems, equipment asset management systems, data connectivity, and information fusion, it also has the characteristics of ubiquitous interconnection, open applications, collaborative autonomy, and intelligent decision-making.

Network: complete the access and transmission of data from the perception layer through the Internet, mobile communication network, satellite communication network, and other infrastructure network facilities. Connect the perception layer and platform layer through the network to form the network layer of the ubiquitous power IoT. It is the communication hub between DEGs and between DEGs and distribution networks.

Edge: establish basic AI middleware, study knowledge maps in scheduling, operation inspection, marketing, and other fields, and solve problems locally with edge computing. Deploy intelligent control terminals for distributed energy grid clusters to enhance on-site data processing capabilities.

End: Build a comprehensive data perception with a panoramic view, enabling data-driven artificial intelligence to truly perceive the actual state. It is reported that State Grid has 540 million smart meters for end customers nationwide, which is a huge quantity. As the terminal of the power grid, the number of distributed energy grid groups will also increase sharply in the future. State perception and data collection for distributed energy grid groups will be important tasks in the future.

8.1.2 The Deepening Application of Digital Twin Technology

The concept of digital twins can be traced back to Professor Grieves' "Mirror Space Model" proposed in the Product Lifecycle Management (PLM) course at the University of Michigan in 2003. It is defined as a three-dimensional model that includes physical products, virtual products, and their connections. Digital twin: It is a simulation process that fully utilizes physical models, sensor updates, operational history, and other data, integrates multiple disciplines, physical quantities, scales, and probabilities, and completes mapping in virtual space to reflect the full lifecycle process of corresponding physical equipment. It is a concept that transcends reality and can be seen as a digital mapping system of one or more important, interdependent equipment systems (Bui et al. 2016).

Since its proposal, the concept of digital twins has attracted widespread attention from various sectors in China, including government, industry, academia, research, and application, and has sparked a wave of research and construction. Digital twins have become the mainstream mode of choice for local governments to promote the construction of smart cities, and the industry also sees it as a barometer of technological innovation and a new opportunity for development. The application of digital twins has taken the lead in some fields. As an important component of smart cities, distributed energy grid clusters, combined with ubiquitous IoT construction, will inevitably become one of the important development directions for future distributed energy grid clusters through digital twins of the distributed energy grid cluster system.

Ubiquitous perception and intelligent facility management platforms are the link between terminal devices and intelligent applications. It is the fundamental support platform for the digital twin distributed energy grid group, based on global IoT perception and intelligent facility access, with overall construction and operation services as the core, and the concept of open and shared business empowerment. It serves various participants such as device developers, application developers, business managers, and operation and maintenance service providers, connects devices downwards, is compatible with various protocol interfaces, provides access and aggregation of perception data, supports multi-level distributed deployment, promotes intensive construction of information infrastructure, and realizes device overall management and collaborative linkage; Open up and share data upwards, empower various IoT applications, and support the cultivation of innovative IoT data applications. The IoT perception platform is the gateway between the digital twin distributed energy grid group and the real world, and the ubiquitous perception granularity determines the level of refinement of the digital twin microgrid.

In digital twins, all anomalies that may affect the working status of equipment will be examined, evaluated, and monitored. Data twin is the integration of sensor data, historical maintenance data, and related derived data generated through mining from an embedded integrated health management system. By integrating the above data, data twins can continuously predict the health status, remaining service life, and probability of successful task execution of equipment or systems. They can also anticipate the system response to critical safety events and reveal unknown issues in system operation by comparing it with the system response of entities. Data twins can mitigate damage or degrade the system by activating self-healing mechanisms or suggesting changes to task parameters, thereby increasing the lifespan and probability of successful task execution. Specifically, the application of digital twins in distributed energy grid clusters includes the following aspects:

a. System and equipment digital modeling: Based on data-driven design, a virtual model of the system that meets technical specifications is created, accurately recording various physical parameters of the equipment, visually displaying them, and verifying the accuracy of the design through a series of verification methods.
b. Simulated operation simulation: Through a series of repeatable, variable parameter, and accelerated system operation simulations, the performance and performance of distributed energy grid clusters in different external environments are verified, and the adaptability of products is continuously verified during real-time operation. A virtual operation can also be used to simulate the operating conditions under different devices, parameters, and external conditions, enabling early prediction of system stability, efficiency, and potential production bottlenecks.
c. Digital control: Integrating various instructions and states during the scheduling operation phase, such as power flow, equipment operating conditions, and electricity price signals, into a closely coordinated operation process through digital means, and automatically completing scheduling and control under different combinations of conditions according to established rules, achieving an automated production process; Simultaneously record various data during the production process to provide a basis for subsequent analysis and optimization.
d. Key indicator monitoring and operation evaluation: By collecting real-time operation data of various devices under real scheduling and control instructions, visual monitoring of the entire operation process is achieved. Monitoring strategies for key equipment parameters and inspection indicators are established through experience or machine learning, and abnormal situations that violate

operational constraints are promptly handled and adjusted to achieve safe, stable, and continuously optimized distributed energy grid group operation.

e. Remote monitoring and predictive maintenance: By reading various real-time parameters of sensors or control systems in intelligent distributed energy grid groups, constructing visual remote monitoring, and providing collected historical data, constructing a hierarchical health indicator system for components, subsystems, and even the entire equipment, and using artificial intelligence to achieve trend prediction; Based on the predicted results, optimize the construction strategy and equipment health management strategy to reduce and avoid losses caused by user failures.

f. Optimizing customer energy consumption packages: For energy users with different energy consumption characteristics, the rationality of energy consumption packages and parameter settings, as well as their adaptability under different production conditions, often determine the user's energy consumption cost and DR effect. The distributed energy grid group control system can build empirical models for different application energy scenarios and operating processes through massive data collection, helping its customers optimize the parameter configuration of energy packages, reduce energy costs for energy users, and improve the efficiency of DR for distributed energy grid groups.

8.1.3 Low Carbon and Zero Carbon Distributed Energy Grid

With the increasingly clear global carbon reduction targets, how to build and optimize low-carbon and zero-carbon DEGs has become one of the core issues in achieving sustainable energy transformation. In this context, low-carbon and zero-carbon DEGs are not only key technological means to address climate change, but also infrastructure to promote energy structure upgrading, improve energy utilization efficiency, and achieve energy autonomy and intelligence. In the distributed energy grid, the application of low-carbon and zero-carbon energy can not only significantly reduce carbon emissions, but also improve the reliability, flexibility, and intelligence level of the distributed energy grid by optimizing the allocation of energy resources.

The core concept of low-carbon and zero-carbon DEGs usually refers to the application of renewable energy (such as solar energy, wind energy, hydro energy, etc.), clean energy (such as nuclear energy, hydrogen energy, etc.), and efficient energy conversion technologies (such as hydrogen FCs, battery ES technology, etc.) to achieve the full process from energy production, transmission to consumption, and minimize or eliminate carbon emissions (Che et al. 2015). The characteristics of these low-carbon and zero-carbon energy sources are that they have low environmental impact, can significantly reduce the proportion of traditional fossil fuels (such as coal, oil, and natural gas) in the energy supply, reduce environmental damage, and thus reduce greenhouse gas emissions. With the advancement of technology and policy support, the cost of low-carbon and zero-carbon energy continues to decrease and efficiency continues to improve, gradually becoming an indispensable component of various DEGs.

In the context of DEGs, low-carbon and zero-carbon DEGs have unique advantages. The distributed energy grid itself emphasizes user-centered energy supply and demand regulation, achieving localized energy supply and intelligent scheduling through multiple small-scale and flexible energy production and consumption units. In this process, low-carbon and zero-carbon DEGs can optimize distributed energy systems through reasonable technological means, enabling efficient allocation and utilization of various energy resources. Especially when dealing with issues such as energy supply volatility, timeliness, and the impact of climate change on energy demand, low-carbon and zero-carbon DEGs can achieve energy balance, stability, and efficient operation through intelligent scheduling and collaborative control. Specifically, the application of low-carbon and zero-carbon energy grids in DEGs is mainly reflected in the following aspects.

Localization of energy production and supply: An important feature of DEGs is the localization of energy production and consumption. By installing renewable energy equipment such as solar and WP on the user side, energy self-sufficiency can be achieved, reducing dependence on traditional power grids. Especially for remote areas and regions with unstable energy supplies, low-carbon and zero-carbon energy girds can solve energy shortages through distributed renewable energy systems, reduce carbon emissions, and improve energy reliability and autonomy.

(1) Intelligent dispatching and optimized management: With the development of energy Internet and smart grid technology, low-carbon and zero-carbon DEGs can be efficiently managed through intelligent dispatching systems. By using big data and artificial intelligence algorithms, real-time monitoring and data collection of various devices in the distributed energy grid can be carried out. Combined with load forecasting, energy production forecasting, and other information, the optimal scheduling plan for energy can be accurately calculated. The intelligent scheduling system can not only cope with fluctuations in energy supply and demand but also adjust energy scheduling strategies in real-time based on external conditions such as weather changes and user demand, ensuring the reliability and stability of energy supply.
(2) Efficiency of ES and regulation: ESSs play a crucial role in DEGs. Through technologies such as battery ES and hydrogen ES, excess energy can be stored when renewable energy is overproduced, and released during peak energy demand or insufficient renewable energy generation to balance grid load fluctuations. The implementation of low-carbon and zero-carbon DEGs relies on efficient ES technologies and intelligent ES scheduling systems. ESSs can not only improve the utilization rate of renewable energy but also reduce carbon emissions while ensuring grid stability.
(3) Multi-energy complementarity and collaborative optimization: The construction of low-carbon and zero-carbon DEGs not only relies on a single form of energy but also achieves optimal energy allocation through the complementarity and collaborative scheduling of multiple clean energy sources. For example, the

volatility and instability of wind and solar energy can be supplemented and regulated through various ES devices such as batteries and hydrogen energy. In addition, the application of smart grid technology enables real-time information sharing between distributed energy systems, achieving coordinated operation among multiple energy systems, further improving energy utilization efficiency, and reducing carbon emissions.

In current research, scheduling optimization methods for low-carbon and zero-carbon DEGs have involved the application of advanced technologies such as machine learning, artificial intelligence, and deep learning. By collecting and analyzing real-time data from multiple links such as distributed energy production, storage, consumption, and transmission, and using big data technology to accurately model energy consumption trends, load forecasting, etc., it is possible to effectively achieve refined scheduling of low-carbon energy, maximize energy utilization, reduce energy waste, and reduce overall carbon emissions through intelligent decision-making processes. Especially when facing the uncertainty brought by distributed energy access, multi-factor collaborative optimization algorithms can significantly improve the flexibility and scheduling capabilities of DEGs, ensuring stable and reliable energy supply under various load and energy fluctuation conditions (Wang et al. 2016).

Although low-carbon and zero-carbon DEGs have shown great potential and broad prospects in DEGs, there are still many challenges and problems to achieving their comprehensive application and promotion. Therefore, future research work will mainly focus on the following directions:

(1) Innovation and breakthroughs in low-carbon and zero-carbon energy technologies: With the development of technology, low-carbon and zero-carbon energy technologies are constantly advancing, but there are still many key technologies that urgently need to be broken through. For example, the production, storage, and transportation technology of hydrogen energy still faces technological barriers, and the capacity, lifespan, and cost of battery ES also need further improvement. In the future, new low-carbon and zero-carbon energy technologies such as solid-state batteries, hydrogen ES, and artificial photosynthesis may provide new solutions for DEGs.
(2) Improvement of intelligent scheduling and optimization technology: In the future, the intelligent scheduling technology of low-carbon and zero-carbon DEGs will rely more on the application of advanced technologies such as big data, artificial intelligence, and deep learning. Through precise data analysis and prediction, combined with complex energy flow and scheduling models, future scheduling systems will be able to achieve optimal scheduling under more complex environmental conditions, ensuring efficient energy utilization and low-carbon emissions.
(3) Wide application and optimization of ES technology: ES technology is an indispensable component of future low-carbon and zero-carbon DEGs, but existing ES technologies still have room for improvement in terms of cost, capacity, lifespan, and efficiency. Future research needs to focus on the development

of new ES materials, such as solid-state batteries, sodium-ion batteries, and superconducting magnetic ES. In addition, the integration and scheduling optimization of ESSs are also key areas. Through intelligent EMS, the charging and discharging strategies of ES devices can be dynamically adjusted to maximize the efficiency of ESSs and adapt to the diverse needs of DEGs.

(4) Multi-energy synergy and optimization of cross-regional DEGs: The development of low-carbon and zero-carbon DEGs in the future requires better coordination and complementarity among various forms of energy. The joint application of various clean energy sources such as solar energy, wind energy, hydro energy, and hydrogen energy can achieve optimal allocation and comprehensive utilization of energy through advanced optimization scheduling algorithms. At the same time, the construction of DEGs is not limited to local areas, but should also explore the interconnection and intercommunication of cross-regional DEGs. With the support of the smart grid and energy Internet technology, different regions can realize real-time energy sharing and dynamic scheduling, alleviate the imbalance between energy supply and demand, and further improve the flexibility and reliability of DEGs.

(5) Construction of digital twins and virtualized DEGs: Digital twins technology, as a technology that deeply integrates the physical world and digital space, provides a new solution for optimizing and scheduling low-carbon and zero-carbon DEGs. By constructing a digital twin of the energy system, precise modeling and simulation of the actual system can be carried out, real-time monitoring of the system's operating status, prediction of potential faults, and optimization of energy flow paths. Future research can further explore the application of digital twin technology in energy scheduling, equipment management, and carbon emission monitoring, while combining virtual reality and augmented reality technology to provide more intuitive and efficient tools for energy management.

(6) Carbon footprint assessment and optimization of DEGs: The core goal of low-carbon and zero-carbon DEGs is to minimize carbon emissions; therefore, further improvement of carbon footprint assessment and optimization methods is needed in the future. By establishing a high-precision carbon emission measurement model and combining multidimensional data such as energy flow, equipment operation, and user behavior, the carbon emissions of the entire DEGs can be dynamically evaluated. On this basis, optimize energy dispatch strategies, prioritize low-carbon energy and efficient equipment, and further reduce the carbon emission intensity of the energy system.

(7) Policy support and market mechanism innovation: The development of technology cannot be separated from the support of policies and markets. In the future, the government needs to establish clearer goals for low-carbon and zero-carbon energy development, and provide sustained financial support and incentive policies, such as carbon taxes, green electricity subsidies, carbon trading markets, etc. At the same time, it is necessary to promote innovation in energy market mechanisms, such as achieving transparency and efficiency in energy transactions through blockchain technology, encouraging users to participate in

distributed energy production and scheduling, and jointly building low-carbon and zero-carbon DEGs.

(8) Public participation and social awareness enhancement: The construction of low-carbon and zero-carbon DEGs cannot be separated from the support and participation of the public. Future research needs to focus on how to enhance the public's low-carbon awareness and promote the popularization and application of low-carbon energy technologies. For example, by promoting the environmental and economic benefits of low-carbon and zero-carbon energy, encouraging residents to install distributed PV systems, or adopting low-carbon energy consumption methods such as electric vehicles. At the same time, we will promote the development of energy management platforms, provide personalized energy optimization solutions for users, and enhance their sense of participation and responsibility in the DEGs.

(9) Global cooperation and experience sharing: The construction of low-carbon and zero-carbon DEGs is a key measure for the world to jointly address climate change, therefore it is necessary to strengthen international cooperation and exchange. In the future, research institutions, governments, and enterprises can collaborate across borders to jointly develop advanced low-carbon and zero-carbon energy technologies, share successful practical experiences, and jointly formulate technical standards and development roadmaps for global energy transition.

Low-carbon and zero-carbon DEGs, as important components of future distributed energy grid operation and scheduling, not only have significant environmental benefits but also bring economic and social benefits. Through the widespread application of renewable energy, the support of efficient ES technology, the optimization of intelligent scheduling, and the drive of policies and markets, low-carbon and zero-carbon DEGs will play an increasingly important role in DEGs. Future research requires comprehensive exploration in various aspects such as technology, management, policy, and public participation to comprehensively promote the development of low-carbon and zero-carbon DEGs, and provide solid support for achieving global carbon neutrality goals and building sustainable energy systems.

8.2 Key Challenges Faced

With the widespread application of DEGs and multi-energy grid clusters, real-time monitoring and processing of massive amounts of data have become core elements in improving the intelligence, sustainability, and efficiency of energy systems. However, the existing traditional data processing technologies and architectures struggle to handle the growing volume and complexity of data in the energy sector, especially under the constraints of ensuring data security, privacy protection, and efficient processing. The challenges are particularly severe in this context.

In 2020, the "Opinions on Building a More Perfect Mechanism for the Market-Based Allocation of Production Factors" clearly stated that "data as a new type of production factor" is critical to the development of the digital economy. The efficient allocation of data resources has been identified as a key issue in the development of the digital economy. With the gradual implementation of national policies, the circulation and sharing of data have become central issues driving the digital economy. Against this backdrop, real-time data processing for DEGs and multi-energy grid clusters faces the following challenges.

(1) Security and Privacy Protection of Multi-node Data Sharing

In distributed energy systems, vast amounts of data from multiple nodes need to be shared and circulated among different participants. This not only requires efficient data processing and analysis but also ensures data security and privacy protection across nodes. Currently, data sharing faces a range of challenges, particularly in how to balance the security, privacy protection, and compliant sharing of data. The lack of effective technical pathways and collaborative mechanisms creates trust barriers among participants, especially when it comes to issues such as benefit distribution and sensitive data leakage. This trust gap further complicates the data-sharing process.

In response, emerging technologies such as blockchain and privacy computing have been proposed as solutions for ensuring the security of data sharing. Blockchain technology, with its decentralized and immutable characteristics, holds promise for securing data flow in DEGs. Privacy computing, on the other hand, enables cross-domain data sharing while protecting user privacy. While these technologies have made some progress, the key challenge remains how to achieve efficient data circulation while ensuring data security and privacy protection.

(2) Real-Time Monitoring and Processing of Massive Data

With the application of multi-energy grid clusters and DEGs, real-time monitoring and processing of energy data have become key to improving grid reliability, optimizing energy dispatch, and implementing DR. However, the rapid generation and real-time processing of massive data present unprecedented challenges to traditional IT architectures and data storage. Especially in complex scenarios such as multi-energy flows, DG, and load forecasting, efficiently processing vast amounts of data from various energy systems has become a bottleneck for improving system intelligence.

Currently, research on real-time monitoring of massive data focuses on the following areas:

(a) Data Redundancy Removal and Noise Reduction: The vast amounts of data generated by energy systems often contain redundant information and noise. In particular, during the fusion of multi-source data, efficiently removing redundant data and improving data quality has become a key research direction.
(b) Data Storage and Fast Retrieval: With the rapid growth of data volume, traditional storage methods can no longer meet the demands. Researchers have

proposed layered storage architectures and high-speed indexing and storage methods based on stream data to address the challenges of real-time data access.

(c) Distributed Computing and Parallel Processing: When dealing with massive data, distributed computing and parallel processing become effective solutions. Distributed computing frameworks, such as MapReduce and Hadoop, have achieved significant results in many big data fields. However, their real-time application in multi-energy grids still faces performance bottlenecks, especially regarding data real-time processing and scalability.

These challenges have driven the rapid development of emerging technologies, such as edge computing and artificial intelligence, in real-time data monitoring. Edge computing moves data processing closer to the network edge, reducing data transmission delays and improving real-time responsiveness. Artificial intelligence, particularly deep learning, has shown tremendous potential in energy data analysis, helping predict energy demand, optimize power generation schedules, and enhance fault prediction accuracy.

In summary, research into the security and privacy protection of multi-node data sharing and the real-time monitoring and processing of massive data will be crucial not only for improving the operational efficiency of DEGs and multi-energy grid clusters but also for playing an increasingly important role in the development of future smart grids and energy internet systems.

8.2.1 Security and Privacy Protection Issues in Multi-node Data Sharing

With the deep integration of DEGs and superior energy grids, the complexity of data flow and collaborative applications has been continuously increasing. Against this backdrop, data security sharing technologies such as privacy computing, blockchain, and cloud environments are gradually becoming key solutions for ensuring the security of energy data. However, each of these technologies still has specific shortcomings.

(1) Challenges and Risks of Privacy Computing in Energy and Power Data Security Sharing

Privacy computing technology provides strong technical support for data security sharing. However, in the application of DEGs and superior energy grids, the following risks still exist:

(a) Privacy leakage risks from attackers or semi-honest servers: In DEGs, model parameters uploaded by devices may be accessed by attackers or semi-honest servers. Using the latest privacy attack techniques (Fredrikson et al. 2015; Tramèr et al. 2016), such as Generative Adversarial Networks (GANs), attackers may infer the device's private data. In open-edge networks, there is a risk that different devices may collaborate to steal the confidential data of a target device. This is especially problematic in scenarios like collaborative optimization and DR, where attackers may exploit weak connections and protocol vulnerabilities between devices to steal data.

(b) Insufficient privacy protection in the federated learning framework: Recent privacy attacks have demonstrated that the current federated learning framework cannot fully protect user data privacy. For example, Bonawitz et al. (2017) used Generative Adversarial Networks to infer user data from the model gradients uploaded by users. This issue is particularly prominent in the energy and power sector, especially when sensitive data such as user consumption behavior and device operating status are involved.

In practical applications, the challenges faced by privacy computing-based data security sharing technology can be summarized in the following two areas:

(a) Efficiency issues: In federated learning, it is assumed that local data from each client follows an independent and identically distributed (IID) pattern, which ensures the unbiasedness of gradient updates. However, in reality, data in DEGs often exhibits a non-independent and identically distributed (Non-IID) characteristic, especially energy data from different geographic locations and devices. The Non-IID property causes local model weights to diverge, particularly during the training process of horizontal federated learning. This divergence increases the gap between the final global model and the ideal model, affecting model training efficiency, slowing down convergence, and impairing optimization results.

(b) Potential security issues: Traditional federated learning systems adopt a centralized architecture, where data flows are asymmetrical, easily triggering the curiosity or malicious behavior of device nodes, leading to single points of failure and privacy breaches. In distributed energy systems, devices have high autonomy and freedom, especially in edge computing environments, where malicious nodes may falsify data or interfere with model training, compromising system security.

(2) Risks and Challenges of Blockchain-based Data Security Sharing

In DEGs, blockchain-based data security sharing technology also faces certain risks, mainly in the following aspects:

(a) Blockchain public ledger risks: Blockchain, as a public ledger, makes all transaction data visible to all nodes, allowing malicious nodes to access the entire ledger and conduct deep analysis to obtain user identity information or other sensitive data. Although data in the blockchain can be transmitted in encrypted form, there is still a risk of decryption through analysis. Especially in large-scale DEGs, the openness of blockchain data may lead to additional privacy breaches.

(b) Key leakage risks: Blockchain can to some extent address the issues of single points of failure and untrusted authorization bodies. However, in a multi-authority framework, where each authorization body holds a separate private key, the risk of key leakage still exists, potentially affecting the entire data-sharing system's security. This is especially true in multi-energy complementary energy systems, where any key leakage can lead to severe security problems.

The challenges in blockchain technology applications can be summarized in the following four aspects:

(a) Blockchain efficiency issues: Blockchain-based data security sharing schemes need to balance consistency, availability, and latency issues in both theory and practice (Gilbert and Lynch 2012; Abadi 2012). Due to the real-time data flow requirements in the energy and power sector, blockchain's efficiency issue is particularly prominent, especially in large data and distributed environments. How to optimize consensus protocols and network topologies to improve efficiency is a major challenge.
(b) Data control security closure issues: Blockchain technology ensures that data cannot be tampered with, but it cannot verify the authenticity of the data. In the energy and power sector, blockchain platforms cannot judge the authenticity of uploaded data, which may result in inaccurate or malicious data entering the shared system, further affecting the optimization and scheduling of the power grid.
(c) Privacy leakage issues: Blockchain, as a public ledger, allows participants to analyze ledger data to mine private information. In energy and power systems, sensitive data such as user consumption behavior and load data may be analyzed by malicious nodes, leading to privacy leaks.
(d) Blockchain infrastructure security issues: The security of blockchain infrastructure is critical. If the platform is attacked or has vulnerabilities, the entire data-sharing system faces significant security threats. Especially in distributed energy systems, where blockchain serves as the core data validation layer, any insecurity in its infrastructure can lead to system collapse or data leakage.

(3) Risks and Challenges of Cloud-Based Data Security Sharing

With the development of cloud computing technologies, cloud-based data security sharing has been widely applied in distributed energy systems. However, this technology also faces numerous challenges:

(a) Privacy protection issues in cloud environments: In the energy and power sector, user consumption data, device status data, etc., often need to be uploaded to cloud servers for storage and analysis. However, due to the separation of management rights and ownership, data owners lose full control over their data. Cloud service providers may be subject to internal threats, leaking or losing user data. Additionally, there are risks of unauthorized access or malicious attacks in cloud environments, which may lead to the leakage of user privacy information.
(b) Security risks of terminal devices: Cloud computing relies on the computational and transmission capabilities of terminal devices, and the security of these devices directly affects the overall system's security. Especially browsers and client applications are prone to vulnerabilities, increasing the risk of user attacks, and thereby affecting data security in the cloud environment.

The challenges of cloud-based data security sharing technology can be summarized in the following three aspects:

(a) Privacy protection issues: In cloud environments, users' identity information and data are easily exposed to cloud service providers and even malicious attackers. Moreover, excessive exposure to user privacy could lead to privacy abuse. This is particularly concerning in electricity data scenarios, where malicious users could use false information to leak the privacy of other users.
(b) User management issues: There are many users in the data-sharing system, and effective control over different users' access to cloud servers is required. How to prevent unauthorized users from damaging the cloud system and ensure data security is an urgent issue.
(c) Data access issues: The massive amount of data stored in cloud environments needs to be managed through encryption. Authorized users face challenges in extracting the required information from encrypted data quickly and accurately. Furthermore, balancing storage space, computing power, and data access efficiency is a critical factor affecting the performance of data-sharing systems.

With the construction of the new power system, future energy and power data circulation and collaborative applications will face more complex security challenges. Therefore, further exploration of the following directions in data security sharing technology within the energy and power sector is needed:

(1) Zero-trust power system security defense: With the increasing frequency of cyberattacks, zero-trust security architectures will be widely applied in integrated energy systems. In the future, how to achieve secure sharing and access control of energy and power data under a dynamic defense system is a pressing technical challenge.
(2) Self-controlled data security sharing technology based on trusted computing: Strengthening the application of domestic cryptographic algorithms can help improve the security of power data and avoid excessive dependence on foreign technologies. How to achieve secure sharing of power data while protecting privacy will be a key issue in the development of future power systems.
(3) Intelligent data security sharing in multi-cloud scenarios: With the popularization of cloud computing, future power systems will face more complex multi-cloud scenarios, such as cloud-network collaboration and cloud-edge collaboration. How to adapt to different data security sharing needs in these complex scenarios and build an intelligent adaptive data security management system will be an important direction for future technological development.

8.2.2 Real-Time Monitoring and Processing of Massive Data

Existing data center technologies face numerous challenges when handling massive data, particularly in complex systems such as DEGs and multi-energy grid clusters, where traditional IT architectures are increasingly inadequate. To address these challenges, a revolutionary restructuring of existing IT infrastructure must be considered. Additionally, the growth of storage capacity often cannot keep pace with the explosive increase in data, making the design of efficient hierarchical storage architectures crucial for improving system performance. Data transmission and movement, particularly in scenarios involving extensive data flow across upper-level energy grids, have become the largest cost in IT systems. Currently, the most efficient method for transmitting massive data is still through physical transportation of disks by aircraft or ground vehicles, rather than relying on network communication (Agrawal et al. 2012; Manyika et al. 2011; Lohr 2012; Genovese and Prentice 2011).

In the era of massive data, the architecture of IT systems must shift from the traditional model of "data revolving around processors" to "processing power revolving around data," i.e., pushing computation to the data sources instead of pushing data to computing centers. This shift is especially critical in the real-time data analysis of multi-energy grid clusters and DEGs, where system scalability becomes a core demand. The scale of concurrent task processing needs to expand from the current level of tens of millions to billions (Dealing with data 2011).

In addressing the technological challenges of massive data processing, the following issues are particularly important.

(1) Redundancy Removal and Noise Reduction in Massive Data

Massive data typically comes from multiple sources and is often generated in dynamic data streams. In the application scenarios of DEGs and multi-energy grids, this data often includes noise and redundant information. Noise sources may include equipment faults, sensor biases, etc., while redundancy arises from overlapping information across multiple data sources. To improve data quality and reduce storage costs, advanced data deduplication and noise reduction technologies must be employed. This is essential not only for optimizing the operation of DEGs but also for improving the data processing efficiency of EMS in multi-energy grid clusters.

(2) Novel Data Representation Methods for Massive Data

Current traditional methods for representing massive data may not adequately reflect the actual significance of the data, especially when dealing with the complex data structures of multi-energy grid clusters and DEGs. To effectively utilize the data and extract knowledge from it, more suitable data representation methods need to be developed. Inappropriate representation methods may lead to missing patterns, causal relationships, and associations within massive data. Therefore, researching more intuitive and efficient data representation techniques, especially those tailored to the specific needs of energy systems, has become a significant technical challenge.

(3) Efficient and Cost-Effective Storage of Massive Data

In energy internet architectures, the way massive data is stored directly impacts the efficiency of subsequent data analysis and processing, as well as storage costs. Especially for complex systems like DEGs and multi-energy grid clusters, it is essential to explore efficient storage solutions that accommodate their multi-source, multi-modal data. Research into high-speed indexing for streaming data, approximate computing for low-quality data, and other related technologies is fundamental to improving data storage and analysis efficiency.

(4) Effective Integration of Massive Data

Massive data often comes from multiple heterogeneous platforms and different data formats. How to effectively integrate these data becomes the key to enhancing intelligent energy management. Data integration and information exchange are especially important in the integration of DEGs and upper-level energy grids. To address this, the promotion of platform-independent data formats and the standardization and unification of data are necessary. This approach will enable the creation of intelligent systems that fuse human, machine, and object interactions, improving the overall efficiency of systems within multi-energy grid clusters.

(5) Efficient Processing of Unstructured and Semi-structured Data

Statistics show that over 85% of the energy data collected today is unstructured or semi-structured, and traditional relational database technologies are ill-equipped to handle the processing demands of such massive data. In the environments of DEGs and multi-energy grid clusters, non-relational database technologies such as MapReduce and Hadoop must be relied upon, especially for parallel processing and large-scale data set analysis, to meet the system's demands for high scalability and efficiency.

(6) Massive Data Mining Tools and Development Environments for Multi-energy Grids

For massive data analysis tools and development environments tailored to different energy types (such as electricity, heat, hydrogen, etc.), further research and innovation are needed. Cross-domain data sharing and analysis, especially the data integration between power grids, gas networks, and heat networks, form the foundation of intelligent operations for multi-energy grid clusters. Researching efficient data mining tools designed to meet these specific needs will enable smarter integration across domains.

(7) New Technologies to Reduce Data Processing, Storage, and Communication Energy Consumption

The acquisition, transmission, storage, and processing of massive data typically consume large amounts of energy. In the context of global energy transformation and low-carbon development, developing energy-efficient data processing and transmission technologies is critical. Through innovations in data compression, distributed computing, edge computing, and other technologies, energy consumption during data processing can be reduced, particularly in the efficient operation of DEGs, which can significantly lower the overall energy consumption of the system.

While massive data presents significant opportunities, it also introduces major challenges in engineering technology, management policies, and talent cultivation. Only by solving these foundational issues can the full potential of massive data in multi-energy grid clusters be realized. Therefore, national policies must be formulated to promote the deep integration of massive data with energy systems, enhancing its application value in future smart grids and energy internet scenarios.

Our proposals for mass data research programmers and initiatives are as follows.

(1) Prioritize Support for Network-based Massive Data Research

Massive data is widely used in fields such as physics, biology, neuroscience, medicine, environmental protection, economics, culture, and security. In the context of energy internet and multi-energy grid clusters, network data is particularly important, especially due to its close relationship with various data exchanges within DEGs and human activities. As energy systems become more intelligent and diversified, network data science and engineering have emerged as a new interdisciplinary field that integrates information science, technology, and social sciences, with profound impacts on national economic development and social stability. Therefore, support for research on network-based massive data should be increased, particularly in applications related to energy management and environmental protection. Currently, research on massive data should focus on areas closely related to national life, such as scientific decision-making, environmental and social management, financial engineering, emergency management (e.g., energy emergency dispatch, disaster prediction, and control), and the knowledge economy.

(2) Basic Research on Massive Data Science

Whether it is government-led massive data research programs abroad or R&D investments by large companies, the current focus of research is on data analysis algorithms and system efficiency. However, while the industry concentrates on solving the engineering technical problems of massive data, the scientific community should strengthen basic theoretical research. As an emerging interdisciplinary field, the theoretical foundations of massive data science will involve multiple domains, including computer science, statistics, artificial intelligence, and social sciences. For the energy sector, especially in DEGs and multi-energy grid clusters, researching the inherent mechanisms of massive data is essential. This includes the data lifecycle, evolution and propagation patterns, and the interaction mechanisms between energy data and economic/social activities. At the same time, attention should be given to computational models, representation methods, and complexities related to massive data, exploring ways to improve computational efficiency in multi-energy flows and distributed energy environments. Additionally, relationships between massive data and knowledge discovery (e.g., intelligent scheduling and optimization, semantic analysis), as well as issues of data security and privacy protection, also require further investigation.

(3) Organizational Approaches to Massive Data Research

Since October 2012, the China Computer Federation and the China Communications Society have established expert committees on massive data, promoting academic, industrial, and application activities. However, the efforts of a single organization are still insufficient. It is therefore recommended that the Chinese Academy of Sciences, the Ministry of Science and Technology, and the National Natural Science Foundation of China jointly establish a national-level research platform for massive data science, further promoting collaborative innovation in massive data applications within the energy sector (especially in DEGs and multi-energy grid clusters). Additionally, a national-level massive data-sharing alliance should be established, uniting industry, academia, and government sectors to address cross-domain data sharing, processing, and storage challenges, providing data resources and technical support for the intelligent development of energy systems. Furthermore, a national-level open-source community should be established, particularly to provide open-source tools and platforms for massive data research and applications in the energy sector, and recommend domestic experts to the international open-source community, ensuring that China's voice is heard in the global massive data research field.

(4) Resource Support for Massive Data Research

To promote the scientific research and application of massive data, it is recommended to launch the "China Massive Data Science and Engineering Research Plan," systematically planning short-term and long-term goals for the production, academia, and application of massive data in China. This includes establishing special funding programs, such as the "973" or "863" major science and technology projects, focusing on innovative research related to massive data in areas like multi-energy grids and DEGs. Additionally, the state should provide strong financial, material, and human resource support for the construction of massive data platforms, the development of typical energy application scenarios, and professional talent cultivation, promoting the deep integration of energy systems and massive data to achieve sustainable energy management and scheduling optimization.

References

Abadi D (2012) Consistency trade-offs in modern distributed database system design: CAP is only part of the story. Computer 45(2):37–42

Agrawal D, Bernstein P, Bertino E et al (2012) Challenges and opportunities with big data. Cyber Center Technical Reports. http://docs.lib.purdue.edu/cctech/1

Bonawitz K, Ivanov V, Kreuter B et al (2017) Practical secure aggregation for privacy-preserving machine learning. In: Proceedings of the 2017 ACM SIGSAC conference on computer and communications security. ACM, pp 1175–1191

Bui VH, Hussain A, Kim HM (2016) A multiagent-based hierarchical energy management strategy for multi-microgrids considering adjustable power and demand response. IEEE Trans Smart Grid (99):1-1

Che L, Shahidehpour M, Alabdulwahab A, Al-Turki Y (2015) Hierarchical coordination of a community microgrid with AC and DC microgrids. IEEE Trans Smart Grid 6(6):3042–3051
Dealing with data (2011) Science 331(6 018):639–806
Fredrikson M, Jha S, Ristenpart T (2015) Model inversion attacks that exploit confidence information and basic countermeasures. In: Proceedings of the 22nd ACM SIGSAC conference on computer and communications security. ACM, pp 1322–1333
Genovese Y, Prentice S (2011) Pattern-based strategy: getting value from big data. Gartner Group Press Release
Gilbert S, Lynch N (2012) Perspectives on the CAP theorem. Computer 45(2):30–36
Gkatzikis L, Koutsopoulos I, Salonidis T (2013) The role of aggregators in smart grid demand response markets. IEEE J Sel Areas Commun 31(7):1247–1257
Li, Yu X, Yu W, Chen G, Wang J (2017) Efficient computation for sparse load shifting in demand side management. IEEE Trans Smart Grid 8(1):250–261
Lohr S (2012) The age of big data. New York Times
Manyika J, Chui M, Brown B et al (2011) Big data: the next frontier for innovation, competition, and productivity. McKinsey Global Institute
Nikmehr N, Ravadanegh SN (2015) Optimal power dispatch of multi-microgrids at future smart distribution grids. IEEE Trans Smart Grid 6(4):1648–1657
Nikmehr N, Ravadanegh SN (2016) Reliability evaluation of multi-microgrids considering optimal operation of small scale energy zones under load-generation uncertainties. Int J Electr Power Energy Syst 78:80–87
Tian P, Xiao X, Wang K, Ding R (2016) A hierarchical energy management system based on hierarchical optimization for microgrid community economic operation. IEEE Trans Smart Grid 7(5):2230–2241
Tramèr F, Zhang F, Juels A et al (2016) Stealing machine learning models via prediction APIs. In: Proceedings of the 25th USENIX security symposium. USENIX, pp 1–18
Wang D, Guan X, Wu J, Li P, Zan P, Xu H (2016) Integrated energy exchange scheduling for multi-microgrid system with electric vehicles. IEEE Trans Smart Grid 7(4):1762–1774
Yang P, Tang G, Nehorai A (2013) A game-theoretic approach for optimal time-of-use electricity pricing. IEEE Trans Power Syst 28(2):884–892
Zakariazadeh, Jadid S, Siano P (2014) Smart microgrid energy and reserve scheduling with demand response using stochastic optimization. Int J Electr Power Energy Syst 63:523–533

The manufacturer's authorised representative in the EU is Springer Nature Customer Service Centre GmbH, Europaplatz 3, 69115 Heidelberg, Germany. If you have any concerns regarding our products, please contact ProductSafety@springernature.com

Printed and bound by CPI Group (UK) Ltd, Croydon, CR0 4YY
30/09/2026
02225303-0011